# The Food Delusion

## How to Avoid the Evolutionary Clash of Food and Genes

Farid I. Zapparov

*In loving memory of my parents, Ismagil and Magzuna*

# Table of contents

Preface......................................................................................................... vii
1.  Decoding the Future ........................................................................... ... 1
    1.1   The anthropological principle of nutrition ................................... 1
    1.2   The trouble ahead ........................................................................ 4
    1.3   What to do?.................................................................................. 7
    1.4   The energy challenges ................................................................ 17
2.  The Nonlinear universe of Food and Body Interaction...................... 27
    2.1   Evolutionary disanthropy............................................................ 27
    2.2   Thermodynamics of anthropy..................................................... 31
    2.3   Strange attractor of metabolism ................................................ 33
    2.4   Evolution of the strange attractor............................................... 39
    2.5   Genesis of human metabolism.................................................... 41
    2.6   Flicker-noise from our food ....................................................... 43
    2.7   The cumulative interactions....................................................... 49
3.  The Nonlinear Biochemistry of Eating and Living ........................... 53
    3.1   Biochemical aspects of the strange attractor
          of the metabolism ...................................................................... 53
    3.2   The fabric of life.......................................................................... 60
    3.3   Cybernetics of living molecules.................................................. 65
    3.4   Life of viscoelastic beings ........................................................... 86
    3.5   Aging of viscoelastic beings ..................................................... 110
    3.6   Extracellular food effects .......................................................... 124
    3.7   Food and genes interaction in viscoelastic beings.................... 142
    3.8   Food, cancer, diabetes............................................................... 194
4.  The Demon of Toxicity ..................................................................... 209
    4.1   Allergy and toxins...................................................................... 209
    4.2.  The evil of small doses .............................................................. 216
    4.3.  Radioactive life supplements..................................................... 222
5.  The Nonlinear Code of the Gut ....................................................... 235
    5.1   Bacteria files .............................................................................. 235
    5.2.  The food digestion .................................................................... 237
    5.3   The Anthropy of the Gut .......................................................... 240

6. Decoding the Truth ................................................................. 259
    6.1   The apocalypse of modern food ........................................ 259
    6.2   Superfoods' glitter isn't gold? ........................................ 273
    6.3   The hidden charm of cacao ............................................ 283
    6.4   DS and vitamins—a conceptual disaster .......................... 297
    6.5   Are the commercial superfoods really that super? ............. 318
7. The Light at the End of the Tunnel ........................................ 323
    7.1   The theory and practice of anthropic diets ...................... 323
    7.2   Smart technology for smart food ................................... 355
Conclusion. ............................................................................ 359
Literature. ............................................................................. 361

# Preface

A few years ago by the will of the Fates, I was thrown into the research and promotion of superfoods in Russia. I devoted my time to observing the current market and the products that were available. However, this left me buzzing with questions to which there were no conclusive, but more importantly, persuasive answers:

How should we eat, and how did our evolutionary ancestors eat back when they couldn't pop over to the supermarket or grab a McDonald's Big Mac?

Scientists tell us we can live up to 150 years, so why are our lives much shorter? Where do degenerative diseases, which kill so many young people, come from, and how can we prevent them?

- How did "civilization" change our food habits? Why does it not allow us to enjoy (hopefully) 150 years on this Earth?
- Why don't multivitamins, "bio active" supplements, and commercially available superfoods live up to their allegedly life-enhancing claims?
- How does food interact with the human organism?
- Why is it that the undisputed progress in research of evolution and bio-chemistry (including genetics) has not led to the creation of magical new foods that give a considerable boost to our health?
- So what are the key differences that scientists are ignoring in life systems? How did life appear on earth, and how is this question related to the nutritional processes of modern man?
- And the most important question of all: how can we live longer and healthier lives?

You'll find my incomplete, fragmented, and completely debatable answers to all of these questions nestled in the pages of this book.

# I

# Decoding the Future

## 1.1 The anthropological principle of nutrition

To improve our understanding of the general basis of mankind's existence in nature and nutrition, let's start *ab ovo*—i.e., from the beginning. What is the meaning of the mysterious "anthropological principle of modern natural sciences"?

In the middle of the last century, scientists began pondering ideas about the links between our universe and the existence of intelligent life—the observer. This was the birth of the anthropic principle: the parameters of the universe must be defined by conditions needed for the observer to exist. The scientific community is still concerned with the question of the experiential significance of the anthropic principle.

Some researchers do not assign any insightful value to this principle, but only see it as a proof of our existence. Others seek to understand the existence of physical constants and the laws of nature based on the theory that humans are the final aim, so to speak, of the universe; this theory includes the idea of God. This is known as anthropocentrism. Our knowledge in this sphere of science prevents us from analyzing all of the facts to choose between the concepts because our comprehension of nature is more limited than you may believe.

The role of the observer is fundamental for investigation based on what we know about the evolution of our universe and the theoretical extrapolation of existing knowledge to build a theory of trajectories for its possible evolution. As far as the quantum states are concerned, the observer determines the outcome of the experiment. The fact that the observer exists, leads to a change in the state of quantum particles (trajectories). Thus by applying the same ideas, we can conclude that the existence of the observer changes the condition of the universe. The evolution of intelligent life in our part of the universe has a multitude of effects on the evolutionary processes of matter. For example, widespread use of radio in the last 100 years has caused changes in the background radio waves within a radius of 100 light years from Earth. Yes, this change is indeed very small! But if you compare the history of the universe to how long humanity has been in existence on Earth (100,000 to 1 million years), then this change is highly significant.

If we think about it, then, in the vicinity of our planet (about 100 light years), a small change occurred for the first time; an observer's action has caused a change in the course of development of the universe. On a smaller planetary scale, in the last 10,000–15,000 years, humanity has had a remarkable effect on the evolution of eco-systems. These changes can be described as anthropogenic. The rate of this influence is progressing at a tremendous speed. To imagine the scale of anthropogenic influ-ence on the interplanetary and interstellar environment in the coming millennia is no mean feat. However, human intervention will be able to take place in the future only on one condition—if it continues to exist. The tiny bubble of the universe that surrounds the Earth has changed from natural processes of evolution to a stage that is affected by anthropogenic factors. I shall call this process "nooevolution" (from the Greek "-noos," meaning "mind"). We shouldn't look at it as something unnatural for the universe. For the universe which that obeys the anthropological principal, nooevolution is a natural path for its development.

According to the ingenious definition given by Erwin Schrödinger: "life—the ability of a system to lower its entropy at the expense of the environment" [9]. The definition of intelligent life is the additional ability to lower entropy by using the intellect. On the basis of the above hypothesis, intelligent life, as was predicted by the distinguished Russian scientist Vladimir Vernadsky [1], will in the end be the leading force driving the evolution of the planet and the surrounding universe. If intelligent life is a necessary stage or a result of this process, then it can be considered essential for the existence of the universe. Figuratively speaking, the universe decided to observe itself by using a part of itself—intelligent life, which has the quality of reflecting itself in conscious-ness, as its own eyes.

One important difference in the process of nooevolution from that of biologi-cal evolution is the fact that in biological evolution the acquired factors are not inherited (in the long term), but in the process of nooevolution acquired factors may be passed down the generations.It is true that all or almost all acquired factors in nooevolution, such as knowledge and skills, are inherited by and are the achieve-ment of humanity. An obvious example of such an inheritance is the invention of the wheel. Unfortunately, this does not always hold true in such important aspects of nooevolution as politics. Politicians, as is historically witnessed, like to make the same mistakes over and over and arrogantly lead our civilisation with the blind confidence of true barbarians.

Relative to biological evolution, nooevolution can be fast. In the last centuries, many changes have taken place very rapidly—even on the scale of the length of hu-man life—which significantly affect our lives. Global warming is one such example.

In the processes of nooevolution of nature, as in the processes of natural evolution, there is a possibility for different tangents and consequences to exist. Especially in the transitory stage of changes of evolutionary models, failures are possible. From the point of view of the existing intelligent species, these failures are in the process of transition to another form of intelligent life. Because of its lack of knowledge about the nature of things, humanity is developing on the basis of the known empirical method of trial and error in the process of nooevolution. It is possible that there are many such errors: the anthropogenic changes of the environment and the climate; erroneous and uncontrolled use of pharmaceuticals; possible epigenetic (relatively weakly inherited) changes in the latest generations due to the successes of birth assistance and lowered infant mortality; and artificial fertilization—i.e., complete disregard for natural selection. In the same list are changes in food habits due to changes in food production and cultivation; widespread and uncontrolled methods of genetic engineering; bacterial resistance to antibiotics; chemical pollution of the environment by modern agriculture and manufacture; and the use of barbaric technologies in food production.

The development of the visible universe is determined by several things: the physical laws of the quantum behaviour of particles, gravity, and the emergence and existence of thermodynamic non-equilibrium structures with locally lowered entropy, which provide the structure of the flow of energy and matter. By approaching the problems of stability in terms of non-equilibrium thermodynamics, Ilya Prigogine, together with Paul Glansdorff [2], developed a theory of local potential. In the first approximation it describes, the conditions of stability of such structures for a wide range of thermodynamic objects. All morphological structures in cosmology and material structures on a smaller scale can be viewed as dissipative structures in a state of local equilibrium with minimal entropy. The evolution of nature complicates these structures to a point further away from thermodynamic equilibrium.

Structures of movement of energy and matter that had evolved to a certain stage caused the appearance of living organisms and further intelligent life in the universe. If before intelligent life, evolution of the universe was unfolding according to the laws of physics (in the wider understanding of this science), then with the appearance of intelligent life a new stage of evolution of nature has begun, nooevolution. This stage is characterized by the appearance of structures whose appearance was almost impossible within the parameters of physical evolution. It is truly fascinating that the evolutionary processes in nature have led to the creation of the alphabet and other similar artifacts. Moreover, the alphabet or the times tables are effectively virtual objects, which can only exist in a universe with intelligent life. Consider also the modulation of electromagnetic emission, including radio waves and a vast array of other factors, which are the result of intelligent activity.

From the physical point of view, it is astonishing that the human brain activity, which transfers miniscule amounts of energy and material, can lead to flows of energy and matter that have a global impact. And this happens in a very small period of time relative to the natural processes of the universe. In this way, during nooevolution of the universe, absolutely new artifacts appear. These are the virtual, material, and energetic objects (cars, mechanisms, and physical and chemical processes).

The creation of such objects and the systems in which they function based on possibilities of intelligent life to know and shape the world will lead to nooevolution determining the evolution of the universe by a certain type of management of processes. The most important of these processes is the self-evolution of intelligent life. That is, the appearance and existence of "free will" (with certain possible boundaries to this concept [12]) as a manifestation of intelligent life, will change everything, including the form of existence of intelligent life itself.

However we must still consider that as of yet we don't know how stable nooevolution itself is. If this is a fluctuation in the parameters of the universe that is unstable and may fade out, then possibly the existence of civilization itself is a temporary phenomenon.

We can view the emergence of intelligence at a particular stage of its existence as being a fluctuation in the state of the universe as a whole. I.e. as fluctuation, that leads to the emergence of irreversible division along the trajectory of the existence of the universe. Then the evolution after the point of this bifurcation will continue according to the laws of the 'new' thermodynamics, i.e. thermodynamics in which the factor of intelligent life occurs (free will) on the state of the universe, at least on the macro levels. It may seem incredible, but today humanity is successfully experimenting with small-distance teleportation of photons and atoms. More than that, biochemists have come close to creating artificial (synthetic) life. K. Venter's research (at Maryland State University) allows the creation of new bacteria by transferring DNA. Work is also being done on organs by inserting microchips and attaching them to the nervous system, something along the lines of creating 'bionic' people. Such methods are already used to return vision and hearing, and the control of synthetic heart and limbs is carried out by computers and other instruments.

## 1.2   The trouble ahead

The above discussion may seems irrelevant in a book about nutrition. This is not so. The main idea behind this book is the fact that we as a species are undergoing a non-biological transformation in the process of nooevolution. How can this transformation affect our health? That is the main question that puzzles millions of people worldwide. The leading force behind this change is mankind itself, with its uncontrollable drive to "better" everything around it. In the short term, for a lot of people living today, a worsening of living conditions is possible in the broad sense. Humanity doesn't have solutions to global challenges that are born out of its own actions. The basis of our evolution is the trial and error method. As humanity develops, the cost of these errors will grow, as is happening now with social and economic globalization as the population of the planet increases [10].

From the start of nooevolution, humanity has had a significant influence on itself. Aesthetic preferences of our civilization in the last million years, in sexual relationships for example, have already changed the physical appearance of people. It is a known fact that the average breasts of contemporary female primates weigh much more than those of our ancestors (0.5–1 million years ago). That is, the superficial preferences of the males lead to an enlargement of female breast size as the result of nooevolutionary selection. At the same time, many experts agree that such a large breast size is unnecessary for feeding the offspring. The average mammary gland size in relation to the weight of modern women is close to that of modern primate females. In this case social preferences have changed women's appearance in a relatively short time. By the way, it is not known whether or not this is good for women or their offspring. It is possible that many such changes have taken place in relation to structures of internal organs, hair color, and skin color that are directly related to population migration in the last tens of thousands of years.

The sex life of humans has also drastically changed in the last 10,000 years. The widespread matriarchy in early humanity is explained by the original polygamous nature of man. In the early hunter-gatherer societies, the division of responsibilities within the existing social structures led to women being responsible for rearing the young, and men for acquiring food. It may not have been certain who the father was, but the mother was always known. That is why the bloodlines were determined by the females.

The transition to a settled lifestyle of an agricultural society—with all the consequences of the economic and technological order—led to the nooevolutionary transition to a new social structure in which monogamous relationships dominated. Many scientists agree that humanity is paying a price for this in the form of neurotic conditions, especially early impotency among men and the like. In this sphere of life, there are contradictions between the biological and social nature of man. Under natural conditions of evolution in nature, like many other mammals, our species is guaranteed to survive the next 3–5 million years. However, in the presence of nooevolution, the time span of existence for the part Homo in Homo Sapiens may be much more limited. In the transitory stage, which has already begun, even in the event of success in nooevolution as a whole (i.e., in the development of our offspring into a new and possibly better species of Sapiens), at separate stages of development many generations of the current biological type of human can suffer from the accepted erroneous concepts.

Today in the United States and Europe, for example, nationwide vaccinations against some types of cervical cancers are given to teenage girls. But nobody is sure if this leads to a fall in cancer rates. We will start finding that out, if we are lucky, in about 25–30 years' time, and we will know for sure in 50–70 years' time when the vaccinated subjects will be at the end of their lifespan. In this case it is clearly an experiment to validate an existing theory on humans that cannot be predicted on the basis of reliable research data. Who knows how many such experiments are carried out by governments of different countries? From the recent past we know about mass experiments on eugenics and an attempt to wipe out whole races by the Nazis. Another social experiment which had heavy consequences for a large number of people is communism, which was started in 1917 in Russia by the followers of Karl Marx's economic theories (which were designed to improve life for the masses) and ultimately led to the deaths of tens of millions of people in the Soviet Union within a few decades.

Two more recent examples are genetically modified food and pharmaceuticals. We are still in the dark about the long term consequences of their use. Here we have an added risk of "horizontal" transferal of genes between plants and animals. Scientists suppose that at the dawn of life, the horizontal transferal of genes between the bacteria from the "original soup" gave a powerful shock to evolution. However, the possible consequences of such transferal between the cells of genetically modified plants and the cells of a human organism (including through the cells of animals in the food chain) are unpredictable and potentially dangerous for the existing species. But it is already known from animal experiments that there is a possibility of drastic change in the metabolic processes of the organism in the process of classic or epigenetic (weakly inherited) mutation of the genome due to ingestion of genetically modified products.

Also, a relatively frequent horizontal transferal of genes between the mitochondria (cell organell) of similar species is a known fact [17]. What is stopping these genes being transferred from species to species in the food chain until they reach humans? As I show in the following chapters, this possibility is real. But even without this, genetic system of plants and possibly animals used as our food, may control our metabolism.

The media reported that in one of the former Soviet republics, Belarus, human genes were injected into the goat genome, which can be passed down genetically. A herd of such goats exists with separate human genes. Belarus is a relatively poor country and cannot afford comprehensive research projects. Yet technologies, knowledge, and the people who have that knowledge left over from the powerful scientific development of the Soviet era allow for such experiments to take place but without needing to control possible consequences. It is absolutely possible for this herd of goats with human genes to become a breeding ground for pathogens which are potentially hazardous to human health. I am not stating that this particular herd is an immediate danger to humans, but it is important to see the main point—the potential hazard is much bigger than Monsanto's activity. Experiments like this that can be taking place in countries with socially and politically irresponsible governments and totalitarian regimes represent a huge hazard. The real-life examples of this are endless.

So humanity started nooevolution from itself. About 3 million years ago, with the appearance of stone weapons, our ancestors relatively quickly transformed themselves from being prey in the food chain into the top predator on Earth. These changes stimulated the development of language and complex (compared to animals) elements of cognitive social interaction amongst the first people. Further on, in approximately the last 100,000 years (especially last 40,000 years), humanity, in the process of changing from natural evolution of primates to the nooevolution of Homo Sapiens, has transformed the environment, and has destroyed plants, animals, and maybe even another related species of primates, the Neanderthals. But most importantly it has in the last 10,000 years radically changed its living conditions and food habits.

At the same time, the biological evolution that is happening at a natural rate is not keeping up with the environmental changes caused by human activity, specifically the activities of the human intelligence. The anthropogenic load on the biosphere has increased tenfold and is also ten times over the limit. Humans are currently using up 40–45 percent of renewable resources, while the maximum allowed norm is 6–8 percent. In the United States, only 5–6 percent of unspoiled natural territories remain and the country is consuming more than a billion tonnes of oxygen, which is produced by the flora in the neighbouring countries' oceans. American industry is using up to five times more oxygen in manufacturing than the citizens use for breathing. All of the economically developed regions are also regions of environmental malaise; the environment is suffering as a consequence of industry. In the BRIC countries (Brazil, Russia, India, and China), these zones take up 15 percent of the territory where 70 percent of their citizens live. A toxic environment is the reason for degraded health and the decrease of natural life expectancy by 10-15 percent. In Russia, Norilsk is a good example of one of the most environmentally unfavourable areas on the planet. Because of high sulphur and heavy metal concentration in the atmosphere in the winter months, the region has black snow, and the factory workers are expected to live 12-16 years shorter than Europeans.

## 1.3 What to do?

A question arises: As ordinary people who only live once, what should we do? Should we pass the time from generation to generation until human evolution stops and a new form of intelligent life is born? Or, having understood that the uninterrupted existence of intelligence in the universe does not actually guarantee the comfortable existence for individual carriers of this intelligence, should we undertake measures to increase our life expectancy, to increase quality of life by eliminating degenerative diseases which make our life worse in old age and lead to premature death?

It is widely thought that the main technology of the modern age is social. We generally assume that technology organizes society in such a way that it acquires a necessary flexibility, creates a surplus of alternative decisions, and quickly adapts to changed circumstances. However, it is also obvious that the social technologies of our times do not solve and will not in the foreseeable future solve the problems that we humans face. To achieve change in our lives, we should improve the quality of our environment, stop the uncontrolled medical and genetic experiments by the biotech industry, and make other crucial systemic changes—all of which are only possible through political change in global human society. According to the experience of many societies [10], this is a near-impossible task. However, there is an element of our interaction with nature that we as individuals can control. It is the process of mass energy exchange with the environment—that is, eating. To understand how to carry out this process in the best possible way, let's have a quick look at the natural evolution of man and his nooevolution in the last 100,000 years. Man, like all the other living beings on this planet, went through a long period (more than 3 billion years) of evolution, starting from the simplest organic molecules to the existing forms of life. All this time, life on Earth was developing in the deep symbiotic interactions of all her forms. Nature provided the predecessors of modern man everything necessary for optimal function of the metabolism, which ensured survival and prosperity from food. In a relatively short period of time that intelligent man (as a biological type) has existed, there has been no significant change in cellular and intercellular metabolism as far as nutrition is concerned. But nutrition itself has changed.

Let's go back to Schrödinger's definition of life as a system that lowers entropy at a constant interaction with the environment, and look at it in relation to nutrition. A human being in the process of eating literally devours part of his environment. Evolution has put man on top of one of the food chains, which means he uses other beings—bacteria, plants, and animals—for food. And from their flesh, he receives all the bioorganic and mineral substances which are necessary for life. In order to maintain minimal entropy, the organism has to ingest foods that have elemental and phase states which are the most efficient in metabolism and are transformed into structures within the human organism to provide the necessary material and energetic balance. Also, other flows of elements and energies which are connected to other processes, such as breathing and thermal exchange with the environment, take place, but they are not relevant to this book. I consciously skip these important aspects of the livelihood of the human organism in order to focus the discussion exclusively on nutrition.

Deviation of entropy of the human organism from the appropriate lowest value manifests in the form of disease. If the growth of entropy exceeds a certain value, then it is classed as the death of the organism. When we eat, we eat food which is alive or has been alive; therefore, the entropy value of the food is important. In other words, how useful and metabolically efficient are the organic molecules, bioorganic substances, and minerals for the human organism? Evolution has made the human body more suited to the types of foods that provide minimum entropy of the body and expend a minimum amount of energy during digestion. Intuitively it is clear that the most suitable foods are those which are close to minimal entropy—i.e., live organisms—or in states of some fermentation. This ensures maximization of created negative entropy—a unit offered by Schrödinger in the Boltzmann-Gibbs equation for describing the states of live systems.

Schrödinger's approach implicitly assumes that the organism receives nutrition that thermodynamically provides minimal entropy, and the situation boils down to interactive processes with the environment. But in reality the most useful foods for a human being are the ones that can be efficiently assimilated and that contain the right nutritional elements. So the entropic state of the food is important but is not a satisfactory factor if we are talking about optimizing nutrition. For example, poisonous plants or animals are in a state of minimal local entropy, but when they get into the organism, they cause more entropy leading to death as they go through the metabolic processes. This means that the thermodynamic description of the state of the food must be enriched with such a characteristic as food entropy. That is, the process of achievement of genetic or evolutionary minimum entropy of a human body only happens during the consumption of anthropic food, appropriate for human consumption. If entropy is a measure of some abstract order, then anthropy is a measure of an order that is determined by the minimum entropy of the organism during digestion and assimilation into the human metabolism.

Anthropic characteristics of food appeared in the process of the biological evolutionary processes of the formation of nutritional networks, on top of one of which is man. As will be discussed further on, the concept of anthropy turns into entropy when we look at the metabolism of the human being in combination with the metabolism of the environment. However because in practical life we separate ourselves from the environment, in the broadest sense of the word the concept of anthropy, I hope, has a certain heuristic meaning as far as questions of nutrition are concerned.

Currently, within the processes of inanimate nature, man's anthropic influence is not big. For example, humanity can have quite a substantial influence on the gaseous content of the atmosphere, but cannot change the basic characteristics of the planet: the volume of the atmosphere, speed of flight in orbit, mass, etc. And that which we call substantial influence on the gaseous content in absolute terms is quite low. But because the biosphere in the whole traditional state turned out to be exclusively semi-stable to the external agitations then the process of our livelihood is significantly affected. However, if we look at it from cosmological scopes, then the changes are very small. For example, nothing will change in the solar system if the temperature on Earth changes by 10–20 degrees, but for us it's a question of life and death.

As already stated, nooevolution has had a cardinal effect on the living conditions of humanity and the environment. The main reason for this has been the increase in population, which still continues. With that increase the characteristics of food have changed substantially. Modern methods of farming and animal husbandry create plants and animals that have different values of minimal entropy. The change in minimal levels of entropy in these plants and animals means that these are slightly different organisms not quite appropriate for humans. This means that that the inner structures of all levels of these organisms one way or another are altered. In absolute majority this change is manifested as states of lower levels of resiliency of cultivated organisms of plants and animals compared to their wild relatives. In some cases it is practically a transformation of one species into another. As this is happening, anthropic qualities are lost because humans have never faced such changes during the process of evolution and are not well adapted to them. We cannot adapt to them in the process of biological evolution because of such different time scales.

On the basis of this, as far as daily nourishment is concerned, an anthropic or anthropological principal may be formulated—what I will refer to as APN (anthropological principal of nutrition). For a human being, optimal (anthropic) nutrition involves foods that are in a state which provides minimal entropy of the organism on the genetically (evolutionary) determined level. Evolutionarily, it turns out that these are live products which man consumed at the dawn of his history. Theoretically, we cannot exclude the possibility that as science and biochemical research develops, humanity will be able to control the anthropic characteristics of food. But due to how complicated metabolic processes are, it will take many more generations to achieve success in this sphere. Another theoretical possibility is the change of human structure and organization of nutrition through some other sources of energy. Firstly, this is the least likely scenario (I hope); secondly, the resulting creature will not be a human being in the conventional sense (it would be another type of Sapiens), and that's why this possibility is beyond the scope of this work.

One of the main purposes of this work is the study of the experimental value of applying the APN to the analysis and correction of modern man's dietary habits. We shouldn't think that the lives of higher primates and ancient peoples were trouble-free even when they were eating according to the APN. Due to climate change in those days, there were possibilities of food sources disappearing or predatory populations increasing. However, as the equilibrium was restored, the environment would provide sources of food that would provide the greatest possible survival rates for the individual and the population in the given conditions.

As a result of evolution of live matter, about 3 million years ago, stone weapons were invented. Approximately 100,000 years ago was born intelligent life on a relatively well-developed social level, which could also have a significant impact on its environment and active control of it. Of course in the beginning, this control was manifested in quite simple forms—increased meat consumption and other natural products, for example—because of highly effective hunting and gathering techniques. At this stage, human society had for the first time secured itself on top of the food chain. Further along in time, humanity, due to its high efficiency of consuming the surrounding food sources, started increasing in numbers much faster than the local external resources allowed for in the long term. In the end, population growth has led to a situation where there is not enough food for everyone. Humanity has conquered these negative changes with migration and in the end has populated the whole planet. From this moment on, humans were mastering new territories due to an ever-increasing power of the collective social consciousness and the ability to accumulate knowledge. However, the processes of population growth and migration, on the whole, weakened the specific product base for the individual and the population because it was necessary to migrate to areas less adapted for the livelihoods of hunters during the times of the great ice age and frequent climate changes. That is, even then a slow, invisible process started taking place, which included a narrowing of the variety of natural food products that was, in one way or another, compensated for by the intellectual power to use available resources with acceptable efficiency.

Further on (approximately 10,000–13,000 years ago) as the society was becoming more complicated with the establishment of hierarchy, it gave rise to agriculture and animal farming. The necessity to feed an ever-growing number of people (population growth) and the emergence of social add-ons who did not take part in creative labor directed the vector of development in the direction of monoculture farming of plants and animals. For the survival of a society where a considerable number of its members took part in intellectual activity (religion and government), which needed ever-increasing amounts of food, food quality and diversity decreased as did the mineral content compared to food gained from wild sources. Farmers prefer carbohydrate-rich cultures. But the combination of wild plants and animals that was the food of the remaining hunter-gatherers gives more protein and provides a better balance of other nutritional elements. Modern hunter-gatherers, the African Bushmen, consume regularly more than 150 different types of plants, animals, and insects. Quantity and quality of nutritional elements in the food of these "undeveloped" representatives of humanity exceeds that of the food of the rest of the modern people. Their diet has provided them with survival in the last 100,000 years in the conditions of the semi desert, in the absence of clothes and medicine. And it is important to note that a large part of their hunting grounds is taken up by the representatives of the farming culture. Anthropologist Peter Farm writes that the real primitive people have the best nutrition on Earth, and are the healthiest.

It is obvious that those who live in the wild eat organic food that is unpolluted by chemical preservatives, pesticides, and other additives. The description of the diet of primitive people is evidence that they all comply with the standards of the National Research Council of the United States as far as minerals, proteins, and vitamins is concerned and the worsening of the quality of nutrition only happens after external interference, which introduces technological agriculture, bovine cattle, and mining. Also, because of their healthy diet, peaceful life, and an unpolluted environment, the primitive people cannot fall victim to modern diseases such as cancer, heart disease, hypertension, and diabetes. High cholesterol levels are unheard of. Research of different indigenous populations of South America had shown that infectious diseases such as influenza, epidemic parotitis, polio, and smallpox can happen, but they do not spread. Blood pressure is normally low, so such intestinal disorders such as appendicitis, tuberculosis, and colon cancer do not occur until "civilised" nutrition is introduced.

In the real conditions of the evolution of humanity, it is difficult to assume that ancient people lived in bad climates with bad plants. It is more likely that they lived on the borders of the tropical forests and the savannahs, where hunting and gathering provided them with perfect nutrition. This is what paleontological data gathered from the tests of skeletons of hunter-gatherers of the pre-Neolithic agricultural revolution is telling us. One study [11] has data of the height of hunter-gatherers of the modern Mediterranean. Towards the end of the ice age, the average height of men/women hunter-gatherers was 175/165 cm. With the transition to farming, the average height in the same territories has decreased to 160/152 cm. Research shows that ancient farmers paid dearly for the new way of acquiring food. Compared to hunter-gatherers, the farmers had a 50-percent decrease in quality of tooth enamel, which tells us of malnutrition. They had a fourfold increase in anaemia (the evidence can be seen from the state of bones, a condition called parotid hyperostosis). There was a threefold increase in cases of bone damage as a result of infectious diseases and a deepening of degeneration of the spine, which was probably the result of hard physical labour. Paleontological studies of ancient bones of people who lived approximately at the same time in the same region show that unlike hunter-gatherers, the farmers were susceptible to diseases such as rickets, which means they lacked in vitamin D; night blindness because of lack of vitamin A; pellagra, which is caused by a lack of vitamin PP –nicotinic acid. The existence of these diseases in itself shows malnutrition and poor variety of available food, which is carbohydrate-based and less balanced compared to the nutrition of hunter-gatherers.

"The average life expectancy in the period preceding agricultural societies was around twenty-six years, but in post-agricultural society it was nineteen years," writes Jared Diamond [11]. So we can see that all these problems of nutrition and disease had a significant influence on their survival potential." A seemingly small difference of seven years practically doubled the working life expectancy of a hunter-gatherer compared to that of the farmer. Also, it is important to note that farmers died mainly of diseases that had to do with their lifestyle and carbohydrate-based nutrition, and that the hunters probably died in a healthy state, which probably had more to do with the occupational hazards that went with their lifestyle. The hunter-gatherers had a large variety of food available to them, but farmers consumed mainly one or several starchy crops that they were growing. Due to inferior nutrition, the farmers were not getting an adequate amount of vitamins, amino acids, animal protein, and an endless number of other nutrients that a human being needs. Apart from that, agriculture helped spell another curse for humanity: hunter-gatherers had very little or no stored food—they lived on wild plants and animals that they hunted for daily. That is, they consumed their food fresh or in a state of relatively low fermentation in natural conditions. Farmers had a necessity to store food reserves. When food is stored for a long time, it is subjected to damage by pests and acquires pathogenic qualities. The level of anthropy of fresh food, the measure of how good it is for the organism, exceeds the level of anthropy of food that has been damaged during storage.

The farmers got their calories from grain, which do not contain all the before-mentioned substances in adequate amounts. Also, when the crops grow the density at which they are sown is much higher than in natural conditions. This resulted in a lack of nutritional elements in the plants and a lack of minerals in the grains. After frequent use and intensive cultivation of the same fields, the mineral content in the soil was impoverished. The nutritional ration of household animals was poorer, which in turn caused the bioactive qualities of meat to suffer, resulting in an insufficient concentration of necessary nutrients for the humans. As Diamond writes [10, 11], "archaeologists which study the problem of emergence of agriculture, reconstructed a decisive moment in history, when humanity committed the biggest mistake. Being faced with a difficult decision to choose between population reduction and increasing food production, we chose the latter and doomed ourselves to hunger, war and tyranny."

Having started with the Neolithic revolution of domestication of wild plants and animals, these processes are continuing with an ever-greater intensity. Currently, the number of people living on Earth exceeds the collective number of people who have ever lived on the planet. If, for example, in ancient Egypt the mineral content of the soil was enriched yearly by the floods of the Nile, which brought sediments from the upper Nile during the rainy season, then in modern agriculture there is a huge erosion of the soil that is compensated for mainly by chemical fertilizers. According to one source [11], the simple facts that agriculture helped to increase population density and separate societies started participating in goods exchange with other heavily populated societies led to the spreading of parasites and infectious diseases. Some archaeologists think that it is the increase in population density, and not agriculture, that lead to the spreading of diseases, but this question reminds us of the problem of the chicken and the egg, because population increase develops agriculture and vice versa. It was difficult for diseases to reach epidemic proportions when people lived in separate groups, constantly changing dwelling sites. Tuberculosis and diarrhoea had to wait until the appearance of agriculture; measles and bubonic plague had to wait for the appearance of big cities. What shall we expect from the constant destruction of the biocoenosis of the planet? What kinds of diseases are waiting for us when the population exceeds 8–9 billion? As the result of "war" of humanity with the biocoenosis that gave birth to it, there is an ongoing destruction of the planet's biodiversity. Planetary ecosystems are being obliterated (global warming, holes in the ozone layer), and huge amounts of toxic waste of modern civilization are drawn into the atmospheric cycle, including millions of tons of chemical products used for fertilization and pest control. The use of chemical hormonal additives, antibiotics, other medicines in animal farming leads to the production of poisoned meat and assists in the mutation of flu viruses and other illnesses. All these processes lead to the disanthropy of our existence—i.e., our habitat becomes degraded and food is destructive to our health.

The cells of the body and the human organism on the whole are mortal because of genetically programmed cell death (apoptosis of the cells). However, scientists agree that according to this genetic programming, our life expectancy is much higher than what is seen in reality. Also, as is seen from scientific observation of cell division, a cell's state does not really depend on how many times the cell divided from the moment of its birth. Genetically the cell can divide, for example, N times. Most of the cells of vital organs function perfectly until division N-1, and then they die straight after N. This suggests we could all live in good health until a very old age. It is known that most centenarians die mainly of muscular dystrophy, and not from degenerative diseases, like most people in earlier age. In the course of a lifetime for such people, aging happens because gene expression, which is in charge of age-related changes, is altered in the organism. Later, I will present data which shows that gene expression (activation) in the cells depends on the food we eat. Obviously, it is very difficult, in such a complex organism as that of a mammal, especially primates, to single out an element in food which affects longevity. Most likely, such a component doesn't exist. The expression of thousands of genes happens gradually as the organism ages. But when correct (anthropic) food that has a large number (wide spectrum) of active elements is ingested, the process is carried out according to genetic programming, without the detrimental effect on the metabolism because of poisonous substances or malnutrition.

However, in reality we see a different picture—the human organism towards the end of its life cycle (which is much shorter than the genetically programmed life cycle of the cells) starts deteriorating. Then diseases appear and life ends with a slow process of dying from numerous maladies or from a severe dysfunction of one of the vital organs (the heart, the arteries, the liver, etc.) of a relatively young organism. Furthermore, the risk of diseases such as cancer, diabetes, and other ailments that are hard to cure increases with age.

All the substances that enter the organism through eating and breathing must either take part in building cells and energy production or leave the organism. This is how nature programmed all the existing beings on earth in the process of evolution. However, this can only be seen in the organisms in wild, unpolluted natural habitats. In environments where there is naturally occurring or anthropogenic pollution, a breach of ANP, people live much shorter lives and the early process of decrepitating is related to the relatively fast processes of the accumulation of changes in their cells and tissues.

A disanthropic habitat of the modern man does not stand up to any criticism from the point of view of living and eating conditions that are suitable for the life of higher primates. At the beginning of twentieth century, a clever Russian scientist, Ilya Mechnikov, was the first to suppose that the unnatural early aging happens due to accumulation of non-active or active mildly poisonous toxins in the intercellular space of human beings. Toxins are chemical substances that are not suitable for the organism, the source of which is the polluted environment and nutrition in the broadest sense of this word. To avoid the formation of these toxins, according to the APN, it is necessary to consume anthropic food that comes from nature. When such foods are consumed, there is no formation of toxins in the organism and often, already existing toxins in the intracellular space are drawn into biochemical reactions which facilitate their excretion from the organism.

The human genome contains genetic information that determines the flow of a plethora of metabolic reactions of the organism. The functioning of the genome in the metabolic pathways can be considerably altered epigenetically by receiving distorted signals in the form of disanthropic chemical compounds. In this case the metabolism malfunctions. To put it simply, the metabolism has to reform in order to be able to react with the incoming unusual substances. However, the process of this reform knocks out cascades of trajectories of reactions of transforming substances in the organism. As a result, poisonous disanthropic compounds are formed in the organism. The type, the combination, and the quality of food that we consume completely determines our physical and mental states. The makeup of our organism on the atomic level is determined by the atomic makeup of the food and water consumed plus breathing. However, on the molecular level each type of living being has its own chemical composition that is determined by the genetic programming of inner chemical reactions of the organism. Also, in the process of evolutionary interaction of different organisms in the natural food web, their minerals and polymolecular chemical compounds began to mutually complement each other and create conditions for symbiosis necessary for survival. There are many types of these chemical compounds and their preferential interactions. The spectrum of these compounds is exceptionally wide. Many of them appear and disappear during the intermediate stages of intra-cellular chemical reactions.

Modern food, because of the methods and the conditions in which it is grown and the methods used in culinary processing, does not have natural qualities that are suitable for man. Culinary processing destroys and/or alters the structure of organic molecules and bioorganic substances of food, the character of mineral interaction and their compounds. All of this results in the loss of anthropic qualities of food. It is evident that the unnatural animal feed that is rich in genetically modified products, monstrous doses of "vitamins," growth hormones, antibiotics, other medicines, together with the unnatural processes of animal breeding and chemical processing of carcasses, leads to the production of meat and meat products that are not only low in quality, but also have disanthropic qualities that degrade people's health. The appearance of the bird flu and the swine flu virus results from the widespread use of antibiotics, remantadin, immunomodulators, and other medicines in huge dosages in order to reduce mortality rates of birds and swine which are being bred in the worst conditions of the modern intensive animal farming. In this way, the powerful anthropogenic influences on the growth processes of animals has a negative, or disanthropic, effect on the eating processes on the level of the individual and humanity as a whole through the pandemic spread of dangerous diseases. It is a known fact that the antibiotics, growth hormones, and other chemical additives in beef production lead to an increase in cases of leukemia and other type of cancers in cows. These sick animals do not die of cancer, but instead are sent to the slaughterhouses. In some regions of developed countries, according to private scientific reports, up to 30 percent of animals slaughtered have leukemia. Nobody knows the possible consequences of mass consumption of animals with cancer. It is possible that the increase in cancer numbers among people who eat red meat could be attributed to this fact. The meat lobby is actively resisting any research in this area.

It cannot be said that even milk is safe. Apart from the presence of hormone traces, antibiotics and other "trash," the conditions of crowded farming of cows lead to genetic mutations and the appearance of cows of type A1 and A2. In the milk of the majority of European and American cows, beta-casein of type A1 prevails, which, according to the research of New Zealand scientists [300], who collected material from the twenty richest countries in the world, has a significant influence on the rate of increase of heart disease and diabetes in children. The nutritional value of cow's milk for a human organism today is under a big question mark according to many researchers. Cow's milk is intended for calves, which have completely different metabolic needs and requirements to humans. Another significant argument is that there is not an animal in nature that consumes milk outside the lactating period, especially from a source belonging to another type of animal. Milk, undoubtedly, is the best source of calcium. However, even though the United States is one of the leading countries in dairy consumption, this country is also the world leader in osteoporosis incidence, which is caused by calcium deficiency. This means that calcium from the milk is not assimilated by the human body, but is quickly expelled by it.

Many scientists believe that many carbohydrates have a high glycaemic index—i.e., when they are consumed, the level of glucose in the blood increases rapidly. Produce made from wheat flour, rice, sugar, and milk are considered to have a high glycaemic index. These foods induce the production of a high level of insulin. An increase in insulin content in the blood causes a drop in the level of glucose, which increases the subjective feeling of hunger. In order to keep the insulin levels low, specialists recommend consuming products with a low glycaemic index such as fruits and vegetables, seafood, and meat with a low fat content. When we talk about the adverse effect of milk consumption, we refer to adults who consume large quantities of milk. There is some data that shows that the workers of dairy factories, who consume a lot of milk at work, as a result often suffer from high to medium levels of insulin resistance—diabetes, obesity, arterial hypertension, and coronary heart disease. Also, a high level of dairy consumption is linked to some types of cancer. For example, it is known that high dairy consumption is connected to prostate, ovarian, and breast cancers. It is known that a high level of calcium in the blood is linked to prostate cancer metastasizing. This may have to do with the high content of saturated fats in dairy products.

## 1.4   The energy challenges

Humanity is faced with challenges on a global level that are endangering the existing model of its development. Among these challenges are the unwise policies of most governments, the depletion of natural resources used in the modern economy, population growth, destruction of natural habitats, a lack of good-quality food manufacturing, a relative lack of scientific development, and many others. All of these challenges demand a change in the model of humanity's development. Such these changes are happening, but slowly, and they are not keeping up with the rate at which problems emerge. In particular, the energy crisis, which has already permeated the global economy, and is sure to increase in the near future, will demand a considerable intensification of the agricultural sector. Unfortunately, experience shows that this intensification can only be achieved at the expense of the quality of food products.

There is also a large number of challenges that most scientists believe can be overcome by relying on new technologies, discoveries, and the sheer ingenuity of humanity. It is true that until now humanity has been dealing with such problems quite successfully. The existence of the "golden billion," fast growth of developed countries, and scientific and technological progress in general are all evidence of this. However, by studying collapses of previous, non-global civilizations [10], we are not given any reason to think that our civilization is guaranteed a trouble-free existence in this era of globalization. Moreover, we see the same carelessness or political motivation, the inability to foresee the unfolding of events, commitment to out-of-date values, and foolishness of those in power that destroyed previous societies.

To evaluate the effect of possible challenges to our existence, as an example let us look at the problems of the energy sector in world manufacture, which can potentially increase the disanthropy of our existence with consequences on our health and life expectancy. I want to note that this is only one of the problems; others include climate change, destruction of the environment, lack of clean water, inequalities in development, and many more.

Here I will touch on the geo-economics of oil and gas. We will not run out of oil and gas in the observable future. However, global oil production, from 1900 until the expected level in 2030 has a peak characteristic, which according to different forecasts will reach its maximum in 2013-2017. The problem is not that we will run out of oil, but that there will not be enough oil to supply the global economy at the rate that it is growing. A 5-percent drop in manufacture and supply of oil in the 1970s led to a quadruple rise in prices. A sharp 10–15-percent decrease in supply will lead to a destruction of many developed economies and thus to poverty. According to authoritative geologists, when the oil production is going to start falling, it will do so at the annual rate of 1.5–3 percent. Many economists believe the current energy intensive economy to be a relatively short-lived fluctuation in the history of humanity. It is widely known that the peaking character of oil production in the United States was forecast by a geologist called M. Hubbert in the mid 50s. The peak itself happened exactly as predicted by Hubbert in 1970. Hubbert's peak for the global oil production will lead to consequences that are hard to predict. These consequences will not only affect motorists. According to the data [631], approximately 10–13 calories from oil are used up in manufacturing, storage, and delivery of 1 calorie of food to the consumer. Not only is food manufacture using up raw materials for energy. Manufacturing a car uses up the energy equivalent of double the car's weight in oil. Manufacturing a computer also uses up double its weight in oil. To produce 1 gram of microchips, it is necessary to burn 630 grams of raw fuel in the oil equivalent [8,631]. For example, the internet in the United States uses up to 10 percent of the country's electricity.

If we carefully consider alternative sources of energy—sun batteries, wind mills, nuclear power stations, biodiesel, etc.—we find that all these sources and the technologies which are the base of their production depend on base metals such as lead, copper, aluminium, lithium, nickel, and others. The manufacture of these base metals utilizes huge amounts of fossil fuels. For example to manufacture 1kilogram of copper, 280 litres of oil are used [631]. Aluminium—twenty times more. Even though it is mainly hydro energy that is used in the manufacture of aluminium, its use for this purpose creates a deficit elsewhere. On the whole, according to the experts on electric cars, the production and usage of the most modern electric car for seven years uses up more fossil fuels than a normal gasoline car. This is also true for the modern hybrid models.

Let's consider biodiesel. Full or considerable replacement of oil with the production of biodiesel is an even less realistic alternative for many reasons. First, the production of biodiesel or bioethanol, without the use of other energy sources, is in fact unprofitable. This means that if we used biofuel throughout the whole cycle of production—i.e., the agricultural process, including fermentation, refining, and distribution—we would get less fuel than was used up. According to data from research on the production of ethanol from corn, 70 percent more energy is needed for the production of ethanol than is contained in the produced ethanol. This means that the production of such ethanol has economic limitations. Also, a more or less significant output volume needs a huge amount of agricultural land, and this will cause considerable ecological and political problems. This kind of enterprise could be more or less profitable in Brazil because of the need to use huge amounts of waste from sugar cane production. Another very important issue is the fact the production of biodiesel and ethanol causes considerable and irreversible depletion of the soil within 5–6 years of growing crops for this purpose [37, 38]. The reason is that biofuel is produced from all parts of the plant and there is nothing left on the fields to replenish the soil. This may have grave consequences on food production in the agricultural sector.

Another favourite of the "energy optimists" is shale gas. The production of shale gas in the United States has increased considerably from 2006-2009 thanks to the great efforts of the oil and gas corporations and the federal government. However, it is impossible to deny the fact that the extraction of shale gas requires constant investment, uses up too much water, and is considerably more expensive than the extraction of natural gas. The ecological problems are also immense. It is almost impossible to undertake such projects in Europe due to high costs, high population density, and very reasonable protests from the ecologists. Today there is only one tried and tested technology of shale gas extraction—multistep hydro fracturing. This method is extremely cost inefficient and causes serious damage to the environment. It requires huge amounts of water, plus a lot of capital to prepare the production fields, which involves building roads and a network of pipelines to collect relatively small amounts of gas from each well. The process also uses up large land areas. And a large proportion of the extracted gas is methane. Normally the gas that is extracted from natural gas fields is propane-butane, which when burned releases three times as much energy as methane. The final cost of shale gas extraction including the comparative efficiency is US$120-140 per thousand cubic meters at the well. This exceeds the Gazprom cost indices tenfold, if we overlook the theft and corruption which is usial in Russia.

Food production is not the only sector dependant on the availability of cheap energy sources. The modern global financial system is completely reliant on the supply of oil and other energy sources. Banks are a good example—they make money by providing much more money in credit than they have as deposits. All the world banks do business this way: by using complicated instruments of the stock market and of derivatives, they assume that economic growth is indefinite (with some interruptions during recessions). In fact, they use the expected future production and monetary base expansion as security for today's credit. However, if their predictions are wrong, this leads to the lowering of the value of the collateral. If the collateral's value drops below a critical level as a result of the lowered liquidity of derivatives, then we have a collapse of the banking system followed by a collapse of the manufacturing sector, the real estate market, and the stock market. Understanding this cycle of interdependency makes clear that the trigger for one of the biggest economic recessions of the last eighty years was the sharp increase in oil prices in the spring-summer of 2008. Stabilization was reached only through the unprecedented government financial intervention in most of the largest countries. But what does this stabilization mean? Only one thing: in the very near future (1–3 years) we can expect a sharp rise in inflation and an increase in prices for all products, including food. Inflation that is over the current level means that the non-liquid assets of the world financial system will be paid for by the ordinary consumer in the form of lowered living standards. Let us congratulate ourselves with the "successful" acquisition of out-of-date derivatives.

Accumulated disproportions and mistakes in financial management caused a huge global crisis in 2008. It is possible that humanity is in a relatively unstable state right now. In such systems the significance of small fluctuations increases, which then evolve into sustained finite fluctuations. When an economy nears the stability boundaries, there is a phenomenon of the smaller fluctuations energetically feeding the larger ones and the consequent acceleration of growth of the latter. In this sense if we take humanity as a whole, the existence of many disproportions on the "micro" level—international trade, finance, allocation of resources and production facilities, capital replacement, living standards, ecology, etc.—lead the global economy into an economic crisis. However, even this jolt on the basis of the world financial system is still a relatively small deviation from equilibrium. If we add to this the possible disruptions in the next decade—local wars, social unrest or aggressive international relationships, sharp increases in prices of raw materials, dwindling energy resources, sizeable technological changes, etc.—the initial disturbances can be amplified considerably.

It is not even that important that the current crisis will be more or less overcome. Its virtual existence in people's minds along with the fluctuations in the life of society can and will lead to revolutionary change in the model of humanity's existence in the next few decades. The ghost of the Great Depression still instils fear in many economists. The memory of the current recession is also going to cause fear, at least in the next 20–30 years. The superposition of these fears to any current unfavourable event is quite an explosive combination in the current and especially future conditions of globalization. Brilliant models of the collapse of "closed" societies are given in Jared Diamond's [10] book studying the examples of Polynesian island societies.

Economic activity is a process with a fading memory. That is, the more distant the process is in the past, the less effect it has on the current conditions. But the fear that is caused by some current events together with the recollections of the past existing in the mass consciousness can lead to an overreaction (panic) to events, thus amplifying the fluctuation in the conditions of low stability of the system of social interaction, and the system will jump to the new equilibrium. It is difficult to predict the details of this transition of civilization to another state, but we can definitely say that it will be painful for the individual, and for the large social groups.

The internet magazine of the American Democratic Party published a very interesting article about the definite relationship between oil and money in the economy [7], which analyses the wealth of modern America and its connection to the supply of oil. The article states that the development of modern technology in the twentieth century led to an increase in the efficiency of the use of fossil fuels and does not have anything to do with developing alternative sources of energy, excluding nuclear. To expect some breakthrough solutions in alternative energy without breakthrough scientific discoveries analogous to quantum mechanics and the theory of relativity in the beginning of the twentieth century is unrealistic. There are no scientific breakthroughs on such a level, which would have to involve a pivotal change in our understanding of the structure of matter and the space-time continuum. Even if by God's will they will appear in the near future, the time lag between the discovery itself and the full-scale engineered application may extend for decades.

The theoretically possible growth of nuclear energy to equate to 50–60 percent of energy production will need huge investments and a long time. Even that will not cover the increase in demand for energy in the next twenty to thirty years if the average rate of growth of the world economy is 2.5–3 percent annually and if even the most optimistic programs of energy production are successful. There is no point even talking about the rise in price of this extra energy. Even though nuclear energy is considered to be quite cheap, this is happens during a normal economic cycle of gradual building and quite inexpensive financing, because the cycle of investment recoupment is quite long. If global simultaneous construction of nuclear stations occurs, this will increase the capital intensity, which will in turn raise the price of borrowing, raise the price of building materials and after launching, lead to the increase in the price of uranium and the problems of waste storage and disposal. Basically, the above is true for nuclear power stations with fast neutron reactors, as far as the length of time and capital intensity are concerned. The length of time needed will increase because there is still no large commercial fast neutron reactor. The development and economic testing will take at least ten years before it will be possible to build one, and another ten years to explore large-scale exploitation. We should not even mention the capital costs—numbers on such a scale are used only in astronomy and in the international calling card codes of IP telephony.

Of course this is not a law of physics, but the current existing law in the world of economics is that, with a certain coefficient, money is equivalent to energy. The real liquid wealth of the modern world is its energy resources. It can be exchanged for everything that can be bought for money: any labour, goods, and services. Almost all the real expenditure of any economic activity is the expenditure of energy on this activity. Some may argue that there is an intellectual investment in the production activity. However, the value of this activity is dependent on the current state of the materials and energy. The intellectual part of the cost is spent on supporting the livelihoods of the individual rights owners and creators, which is also mainly spent on raw materials and energy. This point can be demonstrated with the example of civilian aircraft production and exploitation—a good with a high intellectual component. If we take a Boeing aircraft from the moment it is produced and throughout its time in service and spread the cost across the world aviation industry, then we see that profitability is only 1–2 percent, and the main expense is fuel and servicing. The profit margin, even of the manufacturers does not exceed 10 percent—i.e., the cost of the intellectual investment is not very large.

But the wealth may suddenly decrease or practically disappear if, say, the cost of energy (oil, gas, nuclear fuel, etc.) increases. In the aviation industry, encompassing manufacture and transport, Boeing must operate for some years at a loss, and more often to an almost chronic negative profitability of air carriers. A partial restoration of balance will happen through hyperinflation and decrease in manufacturing output, and with such an intensity of the process that the famous stagnation of the Western economy in the 1970s (which is seen today by the majority of the economists as a grave illness in comparison to the comatose state of the Great Depression) will seem like sniffles. It is often thought that the economy creates additional value [12]. This is possible and true in theory. In reality, my personal observations and analysis of the Western economy in the last several decades create an impression that, through derivative instruments and badly secured loans, it only generates high-risk liabilities. The liquidity of these liabilities is secured by the expectations of considerable economic growth with high leverage to the value of the current product. The evil wizard Margin Call appears as soon as there is a sharp rise in oil prices, and wealth disappears at the blink of an eye. The latest crisis is the perfect illustration of this pattern.

The cause of existence of such a system in the modern world is quite obvious: the greed and dishonesty of the managers of financial institutions. The system which values their activity (which, by the way, they made up themselves) allows them to receive huge salaries and annual bonuses in cash for fictitious economic achievements of the companies, banks, and funds which they manage. In reality it is fraud. The magnitude of this theft is such that the financial institutions spend up to 50 percent of these untrue profits on their very real payouts. As these discrepancies accumulate because of large sums of this stolen money and unfeasible liabilities, it is quite natural that we have an ordinary recession every decade on average and real catastrophes every 70–80 years. It is my belief that the arrest and conviction of the creator of the financial pyramid in the United States, "Bernie" Madoff is nothing more than the hypocrisy of US officials. Most of the individuals concerned with financial policy in the president's administration are fully aware that in reality all the banks and financial institutions, whether bankrupt or still in business, are operating according to the principles of such a pyramid. And the financial system of the United States (and the European Union as well) is functioning more or less according to Madoff's principle. "Bernie's" mistake, or his "crime," was that he was the only and full beneficiary of his fraudulent activities. If he had shared his take with a wide circle of market participants, then the collapse of his pyramid would have looked like an ordinary ostentatious, but legitimate, bankruptcy.

But apart from the imminent defects of the existing financial system, the more dangerous contribution to the instability is created by the politicians. And the more ambitious and successful (in their opinion) is the project that they carry out, the bigger the outcome. But more often than not, it has the opposite result of what they wanted to achieve. If we look at the history of the last century, we see that this thesis has been confirmed on numerous occasions. An outstanding example (according to many historians) is the Germans' idea to infiltrate Russian politics with Vladimir

Lenin & Co. in 1915–1917. From the perspective of the short-sighted interests of the German government, the communist project led to the Treaty of Brest-Litovsk under German terms. But the long-term effect, terrifying for the Germans, is the fall of Berlin in 1945. The cost to humanity as a whole was 50 million lives, not to even mention the horrors of the Holocaust. I would like to think that even the thick-headed military staff of the Kaiser's Germany would not dare to take such action, if God—seething in compassion for humanity—had decided to demonstrate to them what "successful" execution of their plans would mean. It must be said that some historians believe that the 1917 revolution in Russia was financed by Anglo-American sources. And even in this case, it is impossible to say that it was money well spent by those politicians. Because what America and the West got was the Soviet Union, which was such a severe threat to Western civilization during the Cold War that enormous amounts of money had to be spent on defence; also, the division of Europe for half a century was no blessing.

Another example is the American financing and support of Soviet enemies in Afghanistan in the 1980s, which ended up with 9/11, Al Qaeda, Islamic fundamentalism flourishing in the whole world, and the need to deploy troops into Afghanistan—with the same expected zero or even negative effect as was the case with the USSR, but with greater spending and unpredictable long-term consequences from the possible return to power of the Taliban. Yet another of President Reagan's contributions to this litany of schemes was his grand conspiracy against the USSR to decrease oil prices in the 1980s. The result of his efforts (threatening to deny military support) was that the Arab sheikhs flooded the market with oil, causing prices to fall. The Soviet Union could not handle this and fell apart. But the oil sheikhs, in order to justify the volume of sales and extraction of this cheap oil, started to exaggerate the volume of oil reserves. Nobody can really imagine the degree to which the numbers were "massaged" as far as reports of reserves are concerned. But it is known that starting from the '80s, the sheikhs liked to do it to avoid spending a great deal of money on exploration drilling. This is why many experts believe that anywhere from a third to a half of all the "existing reserves" of Saudi Arabia and of other Arab oil sultans are blatant false additions. Incidentally, despite the alleged large reserves, in the last few years Oman's oil extraction has been falling, with no reasons given.

Furthermore, not so long ago the representatives of the International Energy Agency (IEA) confirmed that there are constant exaggerations as far as the available oil reserves are concerned [8]. One of the main reasons for this is that releasing real information may cause panic on the world financial markets. The market is being deceived due to the pressure from the United States, who started this spiral of lies in the '80s as a tool to destroy the USSR, but did not care much about the ways out of this situation. As the *Guardian* says, the key role of the IEA is to not annoy the

Americans [8]. In reality the peak of oil extraction has passed. In 2005 the IEA predicted the growth of world oil extraction to be 120 million barrels a day, then gradually reducing the prediction to 116, and then to 105 million. The original figure of 120 million is complete nonsense, but the current figures are also exaggerated. The employees of the organization believe that in the foreseeable future (4–7 years), it will be challenging to keep up the daily supply volumes even at the level of 90–95 million barrels. It is also important to remember that even in 2007, demand reached 88 million barrels a day. In 2011 it is expected to pass 90 million. During the peak of the 2008 crisis, demand fell slightly to 86 million barrels a day. Catastrophes such as the recent disaster on the BP rig in the Gulf of Mexico will cause the cost to soar and will partly reduce extraction from the deep marine shelf in the near future. We must consider the fact that the IEA reports are the documents upon which most governments base their decisions regarding energy. Looking back over the agency's activity, we can see that from the moment of its establishment in 1974, those most informed and interested in this question are the Western oil corporations, who never followed the agency's advice. A fact indicative of this point is that for the duration of the agency's activity—that is, more than thirty-five years—no construction has been initiated for a single oil refining plant in the United States, even of an average power capacity.

With the decline of the world economy following the 2008 crisis, there has been some reduction in oil consumption. Some experts immediately started saying that there is a lot of oil in the world and so there will be no oil shock. But the economy is gathering momentum as it is coming out of the crisis, and I am sure that these optimistic speculations will be forgotten in the next three to five years as energy consumption escalates. Even though there is a trend in the reduction of energy expenditure per unit of output, the population continues to grow at the same time. When it stops there will be a possible increase in costs of providing for pensioners on a global scale. It is difficult to imagine, according to the opinion of many experts, what will happen in the next five to ten years. If the Saudis & Co. ceases to maintain the level of extraction, what kind of a shock will it cause on the oil markets? Then the prices in the range of US$200–300 (in today's purchasing power) per barrel will "bury" the world economy as we know it today. And the US economy will be the first one to be affected. The effect of the forthcoming oil crisis will be much greater than the effect of the collapse of the USSR. Many believe that the fear of a possible oil shock was the reason for America's invasion of Iraq. The final intention of the United States was not so much to occupy it for itself, but to guarantee that the Iraqi oil, which is still quite plentiful, will be supplied to the world market and dampen the anticipated oil shock in 2012-2015.

It is really true: *"For they have sown the wind and they shall reap the whirlwind"* (Hosea 8:7). And the irony of the result of Reagan's actions is that the beneficiaries of the rise in oil prices will be (and already are to a considerable extent) the few countries that appeared out of the rubble of the USSR—most of all Russia, which to this day is the biggest military nuclear threat to the United States. The rise of oil prices even to US$200 per barrel will give Russia a level of power and independence from the United States in decision making in the global arena that the Soviet Union never had from the time of the Nazi's fall, trapped as it was within the frames of the "Cold War."

So, what will the approaching oil shock lead to? Even if we leave out the catastrophic consequences of possible new wars and outbreaks of terrorism, food production will definitely suffer. Alternative energy sources are not readily compensating for a lack of energy from traditional sources (fossil fuels) at any practical price level. It is possible that because of the fall of energy production, the role of alternative energy will grow, but the agricultural tractors will still run on increasingly expensive fossil fuels. Furthermore, as the population becomes poorer, the demand for the cheapest agricultural produce will rise. This will intensify the use of fertilizer, pesticides, hormones, antibiotics, and GMO crops, which will reduce the final anthropy of food.

# 2
# The Nonlinear universe of Food and Body Interaction

## 2.1  Evolutionary disanthropy

In this section, I will lightly touch on the questions concerning the evolution of man's food habits and the processes of transferral in the organism to understand the influence of anthropic food. Investigating the principals of anthropic nutrition will help us to explain why today people feel illness and fatigue before they even reach old age. It will also help uI don't follow.s to understand what kind of changes in nutrizzzzzztion will help return us to a healthy state of existence that is based on the healthy metabolism determined by evolution. That is, it will lead to what I call an anthropic existence.

As I mentioned earlier, the life of hunter-gatherers was essentially healthier than that of modern people, with a high mortality rate from "industrial" traumatism and from social reasons (wars) rather than disease. Even those members of the population who managed to reach older age did not have such conditions as cancer, diabetes, and stroke. Anthropic food in natural habitats provided the early people with the whole range of necessary bioorganic and mineral substances that were ideal for the demands of the metabolism. The transition to agricultural development, which undoubtedly was a solution to the challenges that society was facing as a whole (partly during the Neolithic revolution in the nooevolution of humanity), nevertheless brought about poorer living standards for the majority, disease, and a lower life expectancy for many generations. The development of medicine undoubtedly led to increased life expectancy to a certain extent, but medicine is mainly aimed at fighting diseases that spread across the human population as a price of social development in the course of nooevolution. If this progress could be possible without the negative effect of disanthropic factors of our existence, then the life expectancy in many developed countries would be one-and-a-half to two times greater. The disanthropy of food is one of the crucial factors that is destroying our biological potential and vitality.

The carbohydrate- or starch-based diet that has become the norm in the millennia of human development has led to exponential population growth, human domination of the planet, and on the whole to great social, scientific, and techno-logical progress. But it has not had such a positive effect on health. The modern commercial, unnatural methods of protein production, which are designed to feed the seven billion people of the planet, are in conflict with the individual's health requirements. This dialectic contradiction between the interests of each human and humanity will undoubtedly soar in the foreseeable future before it will be solved, as I hope, in the future prospects of nooevolution of humanity. However, this will not happen for several generations. The process has gone so far that the nutritionally complete food habits of the modern Bushmen, the poorest inhabitants of this planet who have inherited the lifestyle of hunter-gatherers, is an unattainable luxury for most people living in the richest countries today. In such green and luscious coun-tries as Ireland during the nineteenth and twentieth centuries, millions of people died of famine, whereas the Bushmen would have been thriving.

If we look at a whole range of non-infectious diseases that have become pan-demic, we find it possible that many of them—such as heart disease, cancer, and cirrhosis of the liver—are caused by an unnatural disanthropic diet, especially in the Western countries, that was inherited by humanity from the times of Neolithic agricultural revolution. This diet is high in carbohydrates, processed plant and ani-mal fats, chemical and high-temperature processing. All of this is made worse by the barbaric methods of growing crops, meat, and seafood that I mentioned earlier [18], and is characterized by the destruction of the vitalizing anthropic qualities of food.

When food ingredients get into the human organism, they go through a trans-formation in the digestive tract. The biochemistry of the processes, or of the reaction pathways, of the human metabolism has mostly to do with the subject of colloidal chemistry involving a particularly large selection of chemical compounds in mul-tiphase colloidal solutions, which are most of the fluids in our organism including blood, lymph and tissue fluids, which have different compositions in different or-gans. The variety of chemical compounds and their variations in the colloidal liquids of the organism determine the complicated nature of chemical and phasic transfor-mations. Most of these processes are not very well studied by modern biochemistry. The available scientific knowledge of the reactions' sequences is very fragmented; this means that the hope of controlling such processes by using, for instance, medi-cine is almost always doomed to a limited success rate.

By changing a trajectory of one of the reactions in order to fulfil a particu-lar aim, the action of the medicine inevitably changes the trajectory of many reac-tions, and this leads to side effects. When the organism is primarily getting other substances instead of anthropic substances from food that has been evolutionarily adapted, there is a breakdown of metabolic pathways. The result is that chemical compounds that are toxic for the organism are formed.

The properties of colloidal systems are particularly sensitive to the combination of the bioorganic molecules that they are made of. They are also sensitive to the stereometry of proteins and to other variations of qualities of food components. This can be demonstrated quite easily using the examples that have to do with animal feed. A few years ago, a British company that I worked for supplied the dry milk protein casein for use in certain colloid solutions by the paper industry. We faced a mystery: some batches of casein destabilized the colloid solutions so much that they were rendered useless. Chemical analysis did not yield any clues as to the cause of this as all of the different batches of casein were chemically identical. In the end we realised that the difference in the batches of casein was caused by the difference in cow feed—dry feed or green grass. This shows that protein substances with an identical chemical composition behave differently in colloidal solution depending on whether the cows ate green grass or dry grass. We are shown that in the real metabolism, the resulting qualities of substances, in this case milk, depend on the phasic states of substances in the ingested food.

Because a multitude of bioorganic substances in addition to proteins make up the human organism, it is obvious that food other than anthropic food causes considerable metabolic breakdowns of the organism's systems. This manifest as a constant feeling of light fatigue: we get tired easily after exercise and our external appearance suffers due to inner intoxication of our bodies from damaged metabolic processes. Gradually the precursors to many vital hormones stop being produced. With age, disanthropic food habits cause the production of some vital hormones to drop up to twentyfold, maybe even more. These factors, along with the changes they cause in the physical and chemical conditions of structures within the organism, lead to premature aging and often—due to epigenetic effects—are irreversible. Because the trajectories of the cascades of reactions of the metabolism change, then because of inverse relationships, many anabolic and catabolic chain reactions also change. This in turn shifts the organism away from its optimal state even further.

There is a quite well-founded suspicion that after a long disanthropic influence, the distorted pathways of metabolic reactions are epigenetically stored, and the organism keeps to them for some time even when there is a satisfactory intake of the anthropic substances. This mechanism is embedded in the evolutionary nature of the organism like an adaptation in case food habits change according to the change of the environment. I want to underline here that the stable changes in the metabolic processes caused by a change in diet are not extraordinary, but a natural mode of existence of the organism. These especially stable deviations of the metabolic processes generally happen because of a change in gene expression (activation). That is why it is difficult to reverse the damage caused by disanthropic influences, and to heal, it may take as long as it took to damage the organism. It is conceivable that the organism may not have the time or the capacity to change back due to some irreversible changes in gene expression. In part, an attempt to shorten this time needed is the use of drastic control signals in the form of medicine to correct the state of the metabolism. However, as was mentioned earlier, at the current level of knowledge, this is mostly ineffective.

It is important to add that is highly desirable to keep up anthropic food habits during any individual's whole life, starting from childhood. Otherwise, prolonged exposure to disanthropic states—the degree of influence of which varies with the individual and depends on many factors, including genetic factors—will lead the metabolism into a critical state and an inevitably premature death. A cancer study [73] was carried out on a large sample of immigrants in Sweden (more than six hundred thousand) in the first and second generations. The incidence of cancer in the first generation that arrived in Sweden at the age of twenty to thirty was exactly the same as in the same population group in the countries of origin. However, the incidence of cancer in the second generation and the structure and frequency of the disease corresponded to the rates of the disease of the Swedish population (which, by the way, was decreased). We can therefore conclude that the incidence of different types of cancer is determined by lifestyle factors, and most importantly by nutrition in the developing years.

Genes-diet or so called nature-nurture controversy has never been more moot in all of human history than in last 100-plus years. The main source of growth of this contradiction lies in modern food's loss of originally inherent anthropic properties. There is only a relatively small number of people whose metabolisms can return to the genetically (evolutionary) determined phasic state with the return of anthropic nutrition after a long-term disanthropic diet. Most people are quite vulnerable to disanthropic states that cause irreversible damage. This phenomenon is analogous to morphogenesis—organ regeneration in animals. For example, lizards can regrow lost tails. It is as if the organism remembers the state of the organ, including the size of the animal, and regenerates the lost tail to the exact size as it was before. In the same way our organism, after relatively short and small variations in nutrition returns to its original metabolism by itself. But prolonged exposure to disanthropic food causes irreversible changes in the cascades of metabolic reaction pathways in most people. One of the simplest examples of reversible versus irreversible metabolic changes is drug abuse. In most cases the changes in metabolism are fast and irreversible. However, a small number of people after prolonged hard drug use can give up and reinstate the normal metabolism of endorphins and other relevant substances in the cells and organs. The level of the deviation of the phasic state of the metabolism where morphogenesis is impossible is different in all of us. But the best way not to depend on this is to provide the maximally possible anthropic character of nutrition to every individual.

Another example of a disanthropic state, which in a way is similar to drug abuse, is when people who are severely overweight cannot give up food which is high in calories and is unsuitable for the human metabolism. The reasons for this will also be discussed. To conclude, I can say that it is likely that disanthropic food habits act as an early initiator of metabolic changes as well as the promoter of existing damage at later stages. Abstaining from disanthropic food can lower the intensity of later-stage damage.

## 2.2    Thermodynamics of anthropy

There are three different types of living systems—organisms, organs of organisms, and communities of organisms—each one of which is a whole structure, and its qualities are determined by inner interaction and metabolism. Living systems of higher animals and human beings on all levels are complicated hierarchic structures of relatively independent parts that cannot efficiently function independently for prolonged periods of time. For example, in a human there is a relatively independent community of cells and intermediary media such as blood, groups of cells in the forms of organs and tissues or their parts, and organs with complicated structures with a large combination of different cell communities (including the bacterial environment of the organism). All the cells in the tissues have structural, functional and informational connections. From this standpoint a human being is an integral being, the characteristics of which cannot be determined solely by the properties of its components. A human being as an entity is a product of some "ordered" interaction of its parts, which is mainly determined by metabolic processes. A stable state of the organism exists because of the processes of metabolism—its continuous exchange of energy and matter with the environment. These types of systems belong to the class of nonlinear open systems with self-organisation and are characterized by continuous dissipation of energy.

Schrödinger suggested describing these processes as the production of $-E$ – negative entropy based on

$$-E = k\ln(1/D)$$

Where the value $1/D = f$ - this characterizes the degree of order in the system, $D$ – the degree of disorder, and $k$ – Boltsman's constant.

Let's assume that when the organism is consuming anthropic food, the negative entropy production is

$$-Ea = k\ln(1/Da)$$

Where 'a' stands for the anthropic state.

When there is disanthropic food at the same conditions the production of negative entropy is

$$-Ed = k\ln(1/Dd)$$

Where 'd' stands for the disanthropic state.

The difference between $-E_a$ and $-E_d$ is the degree of disanthropy of nutrition

$$-dE = -(E_a - E_d) = k\ln(D_d/D_a) - k\ln(f_a/f_d)$$

When nutrition is anthropic this is equal to zero. The formula shows that the degree of disanthropy of nutrition is determined by the ratio of variables $f_a$ and $f_d$, $f_a$—the evolutionarily determined order of the metabolism and $f_d$—the order of the metabolism with disanthropic nutrition.

At any degree of disanthropic nutrition, by definition, the amount of the produced negative entropy is always less than when nutrition is absolutely anthropic. The decrease of production of negative entropy due to disanthropic nutrition leads to the increase of overall entropy of the organism, which in medical terms is called disease. Even though limits exist on how to describe such processes based on Schrödinger's entropy implicitly defined by a set of microstates of the system [222,223], the stated limits do not influence the course of formal discussions, unless we suppose that there exists a large number stationary stable states of the system.

The metabolism is a sum total of a wealth of biochemical processes of metamorphosis of matter and energy from the environment in the form of nutrition and oxygen in living organisms. Metabolism and metamorphosis of matter and energy is the process of absorbing/receiving different substances from the environment and using them in the living processes, growth, and (ideally) complete excretion of the resulting product. In the cells and intercellular space, these processes consist of very complex chains of reactions that include many cascades of chemical metamorphosis that develop into quite varied consequences called the trajectories of the metabolism. Unimaginably complex chaotic movements of separate particles—molecules or bioorganic clusters of molecules—are further complicated by simultaneously on-going chemical reactions.

The details of physical and chemical reactions of external influences on the organism and food assimilation demands a detailed analysis down to a quantum level. A very interesting example is a study [80] which illustrates an exclusive importance of quantum effects, such as an entanglement on metabolic processes in the cells. The influence of quantum entanglement of electrons was studied in the reaction of photosynthesis in bacteria. It has been noticed that nature uses effective mechanisms on many different occasions, and that's why we can confidently assume that the same happens in some other metabolic processes. The properties of molecules for quantum entanglement, a rather delicate process, are crucial. When food metabolites are absorbed in a disanthropic quantum state, this can change the character of quantum entanglement and the transfer of energy in metabolic processes. This theoretically could lead to a breakdown of the quantum "computer" of the cell and cause it to give out abnormal control signals. This changes the trajectories of reactions in the phasic space of the metabolism, increase entropy, and lead to a state of disease of the cell.

In terms of negative entropy obtained by the body from food, it is divided into compositional and configurational parts. Reductionism prevailing in modern nutrition is based mainly on the primitive assessment of some part of the composite of negative entropy, and considering the balance of the diet in terms of content of protein, carbohydrates, vitamins, minerals, etc. However, the configuration state of the molecules entering the food that is dependent on many factors—such as the origin of food (the wild, or the cultural, feeding or breeding conditions), the conditions of processing, and chemical additives—determines the fine-tuning of the metabolic processes that are not taken into account. Depending on these conditions, the proteins, for example, can have different lengths or different amino acid sequences that determine their status in molecular clusters. Or the food may contain or not contain the genetic information [572] that is necessary for the management of processes in our body. The difference between these characteristics of the proteins and other chemical compounds in processed food and in food from their natural forms determine the overall extent of the anthropy of food.

At the same time, it should be noted that the composition and magnitude of the negative entropy also depends on the conditions of processing of our food and plays an important role in the processes at all levels of metabolism. For example, a very fine process in the cell nucleus depends on the availability of "nuclear localization signal" molecules in proteins. But these signals depend little on the sequence of amino acids in the protein and are highly dependent on the magnitude of the positive charge in small groups of amino acids in proteins. This means that these processes are determined by the negative entropy of the protein composition, which may vary considerably during the processes of cultivation and processing of a particular protein. Thus, for the fine-tuning of metabolism, which determines such important characteristics as the longevity of the body, both types of negative entropy are important. At the same time, maximizing its absolute value is achieved when using the most anthropic food.

## 2.3   Strange attractor of metabolism

The sequences, directions, and frequency of branching of the cascades of the metabolic chemical reactions are regulated by genetic programming and by chemical composition and nature of the reactions themselves, which are best adapted by evolution to taking place in the organisms. It is a result of evolutionary self-organization of living systems. The emergence and stability of such self-consistent structures of biochemical reactions are determined by positive and negative feedback loops in the system.

The existence of a positive feedback loop—direct chemical interaction of incoming substances from food—causes instability to the spatial-uniform state of the system and makes it possible for the complex dissipative structures to emerge. The existence of negative feedback gives rise to the self-consistent, non-linear, open non-equilibrium system of reactions and determines the stationary phase state of the metabolism depending on the incoming substances.

Any biological system has limits to its stable existence. However, the variations in external governing factors, in our case the changes in the composition and phasic parameters of incoming food, leads to a change in the state of processes in the metabolism. Drastic change leads to the biological death of the organism. The weaker changes lead to degraded states that do not provide minimal entropy. In normal life it is not the choice between life and death, but between a greater or lesser evil, between different stable stationary states of the phase space of the reactions of the metabolism. These stable states have different levels of entropy. The states of higher entropy on the macroscopic levels manifest themselves as a higher level of degradation of cell and organ function.

The behaviour of complex systems such as the metabolic processes is fairly unstable in relation to small changes in external controlling factors such as changes in incoming food composition. In complex biochemical systems, sometimes changes in reaction constants, which also include the concentration of components, even by a millionth of a per cent can dramatically alter the type of chemical processes. For the purpose of description of such complicated (dissipative) non-linear systems during a long time span, a wonderful mathematical concept was introduced: strange attractor.

Here the strange attractor of the biological system is the state of the system in which it is stable, or, to put it another way, the point (area) in which it is stable is the macro state of the system with minimal entropy.

The strange attractor of the metabolic processes in the organism is represented by the area of phasic space, filled continuously with trajectories of biochemical reactions which are determined by genetic factors and control signals from the environment. What corresponds to the evolution of the strange attractor in the time of the life cycle of the organism? It is the change in size and/or the shape of the phasic space (change in number and types of ongoing reactions) under the influence of internally determined genetic signals as well as external control factors—in our case, diet. In terms of the metabolism self-consistent of the organism, this means that there is a certain set of chemical reactions in a chaotic system, which regenerates itself in the state of dissipative equilibrium with the environment with minimal entropy. In addition, it depends on external control factors in the form of the composition of incoming food.

The strange attractors of the metabolism of living systems organize a chaotic set of chemical reactions and interactions of substances into a metabolic system. This system has certain stability to the influence of external signals. In the phase space of the strange attractor of the metabolism the set of different chaotic and simultaneous reactions gives rise to complicated phase structures with unique properties. We call it life. From this stance, it is possible that life itself did not start from not-so-impossible events like the accidental appearance of DNA or RNA molecules. Rather it began with the appearance of the initial chemical metabolism in the primordial soup of low molecular weight chemicals. This topic, being an important one when it comes to nutrition, will be discussed in the forthcoming chapters. The stable phasic areas in the attractor of the process of primitive metabolism caused the synthesis of complicated structures. With this approach the emergence of live matter from non-living matter cannot be explained by the appearance of the first nucleotides. Instead we describe it by the widespread chemical reactions of amino acids, minerals, and alkalis with the emergence of oligomer (short polymer) molecules in the primordial ocean of our planet. This ocean was probably full of the simplest amino acids which exist, if we believe the radio astronomers, even in interstellar space.

If we look at the process of the emergence of life from the point of view of the famous paradox of the chicken or the egg, then it is my belief that the primordial set of chemical reactions, structured in phases, is in fact that definite "egg" from which the first oligomers of nucleic acids came and consequently transformed themselves into primitive RNA. From this perspective the first embryo of life was a set of some very simple auto-catalytic, auto-oscillating chemical reactions with replication, similar to the Belousov-Zhabotinskiy (BZ) reaction. It means that life itself should be considered as continuous metabolic processes—i.e., a set of structured cascades of chemical reactions. From this point of view, even the material structures of the organism—cells and their colonies such as tissue and organs—are metastable formations which only exist as cascades of continuously overlapping chemical reactions with differing intensities in different parts and which are stationary only due to the balance of existing feedback loops. The speed of these reactions is different in different parts of the organism. For example, in bone tissue of higher animals these speeds are minimal. Every organism can be looked at as a constantly working open chemical reactor that interacts with the environment. Even though the metabolism has a great number of degrees of freedom, because of a huge number of reactions with negative feedback loops, these kinds of systems give rise to a hierarchical system of control and relations, which a physiologist treats as cells, organs, tissues. Here, certain special properties of live matter appear which do not exist in the non-live world—the viscoelastic properties of living media on all levels of hierarchy of their structures. These aspects of structure of live systems will be discussed in the following parts of this book.

To a certain extent the metabolism of live beings and biocenosis can be viewed as a special state of matter which behaves as a whole and is determined by a practically infinite set of unceasing reactions. Even though the process is obviously chaotic, the interpenetrating cascades of metabolic reactions lead to a certain cooperative self-consistent state of the phase space of the organism. This cooperative state is life itself. It is theoretically known that such systems as the metabolism, being macro-determined systems, are not very stable to changes in initial data or border (external) conditions because they are based on a huge chaotic set of chemical reactions. From this approach the importance for metabolism of such external control signals as food is obvious. We assume that the process of spontaneous organization of such systems and their transition to the state of the strange attractor plays an exclusively important role in biological evolution. It has been observed that on the border of organized behaviour and chaos—"the edge of chaos"—certain processes take place that remind us of evolution or cybernetic methods of information processing. It can be said that evolution itself provides partly spontaneous organization simultaneously with natural selection. In this case only a certain set of strange attractors arise and exist, and the necessary number of iterations [14] is reduced in order for one or another biological species of organs or systems of livelihood to appear.

Maybe such a mechanism led to a relatively fast evolution of the primordial chemical soup, the appearance of the first cells, and of other more complex strange attractors of the metabolism of organisms all the way to the highest forms. The point of attraction (short term) of all chemical reactions in an anthropic metabolism is a state of full or relatively full health of the organism. Theoretically, we can assume that we can express the strange attractor in the form of a system of N-equations that describe all the metabolic processes. The metabolism in itself is a dissipative process and its representation in the N-dimensional phase space, assuming the existence of attraction, has a "contracting" character. M. Feigenbaum, in his work [43], shows that the iterative transformations for such equations, after N-iterations (at any value of N), converge to a one-dimensional equation with the solution being a point in N-dimensional space. This idea, when applied to the strange attractor of metabolism, allows us to view it as a process of consecutive iterations during the transformation from an endless number of chemical reactions to integral organized states of cells in the organism. Further iterations determine organ function and the metabolism of the organism on the whole. In this case the solution to the initial N-dimensional system will be one single point which represents theoretically absolute health. The significance of the existence of one single solution to the initial system of equations is the fact that they prove the existence of one single state of absolute health. Of course, in reality there are a large number of states with relatively small deviations.

If we view the organism as an analogue computer which is constantly solving this set of N equations, then the deep connection between the concepts of mathematics and biophysics in the organism becomes very clear. It is impossible to formulate the whole system of equations for the metabolism due to a huge number of processes (the dimension of the attractor). However, even the general analysis of such systems allows us to reach certain conclusions. We can also safely assert that a certain set of disanthropic factors will inevitably lead the metabolism to a certain state different from anthropic. Deviations from an anthropic diet influence the strange attractor of the metabolism and, by definition, lead to a state of the phase space different from the evolutionarily determined minimal level of entropy. In order for the sum total of chemical reactions to be suitable to the evolutionarily determined genetic programming, it is important that the incoming food nutritional elements have an initial set (spectrum) of phase states that conform to the set (spectrum) of phase transformations in the organism. When these nutritional substances are missing from the anthropic spectrum or are replaced by other disanthropic substances, then there are breakdowns and changes in metabolic pathways. This causes the organism to function differently, and if prolonged or in extreme cases, this may lead to genetic or epigenetic mutations.

The application of the strange attractor theory provides us with certain heuristic possibilities to describe and understand processes in biological, ecological, and social systems. Let us take an example of the unification of the internal metabolism with the sum total of chemical and phase properties of the environment. This is also known as the biotope, and within the frame of a sole strange attractor we can analyse the influence of ecology (here the term ecology is used to describe biofeedback between the organism and the environment) and nutritional composition of food on the functioning of the whole system. In some cases it allows us to theoretically analyse, for instance, the influence of the content of endogenous and extraneous essential amino acids on the metabolic state of the organism. There are an almost infinite number of other examples. All the organs and the cells of the organism can be viewed as strange attractors in the frame of analysis of their livelihood independently or in their interactions—paired and hierarchical. By considering the phase states of the organic food components and minerals in complex bioorganic food substances, their interaction in the metabolism and chemical transformations are also subject to formalization by applying the concept of the strange attractor. The necessity to take into account the phase states of organic substances and minerals for the process of their assimilation into the organism becomes more obvious.

We also have to bear in mind that scientists are limited as far as being able to study many properties of pharmaceutical drugs on human beings, and in vitro studies are not always productive. So intuitively they have been using the strange attractors of animal metabolisms with somewhat similar biochemical characteristics as a model. Thus they can research the changes in the phase space of the strange attractor of the human metabolism. A good example is when studying the influence of steroids. In the human organism, steroids are responsible for many cascades of metabolic reactions that are in charge of immune system function, bone tissue formation, brain function, etc. Not enough or too many steroids in the organism are equally harmful. For example, too many steroids give rise to Type 2 diabetes and memory breakdown. One study [168] shows that in the endocrine system of zebrafish (Danio Rerio), cortisol is metabolised by the same enzymes as in the human organism. This makes the strange attractor of fish metabolism an excellent experimental model to study the strange attractor of the human metabolism. Using the fish metabolism as a model for the human metabolism provides an understanding of individual cascades and trajectories of metabolic reactions without having the knowledge of the exact details of chemical transformations in fish. In such an integral fashion, we can potentially view any organism or biotope as a united strange attractor. In such an attractor, the internal metabolism is joined with the metabolism of the biotope and the principle of anthropic eating for the human being transforms into the principle of minimal possible entropy of the phase state of the united attractor.

It is true that by definition, due to the existence of an area of attraction with minimal entropy, or in other words, with the maximal production of negative entropy, there should be trajectories in phase space which could provide for that—i.e., the principle of anthropic eating is as unnatural as the separation of the human metabolism from the biosphere and society. Today the degree of separation of the human being from the natural biotope is quite significant. The most extreme case of such a separation is living in urbanized areas with air pollution, lack of physical exercise, and lack of natural food. From this point of view, the anthropic principle of nutrition and its application will be discussed below. It is the conceptual base recognition of the right nutrition for maximum health and longevity of the human organism. The study of the strange attractor also opens the possibility to look at the disanthropy of the phase space from a mathematical point of view—i.e., the deviation from the anthropic state of complete health that depends on the incoming food substances. In theory, for mathematical analysis it is possible to build multi-dimensional spaces with physiological parameters—for example, pulse, temperature, speed of metabolism, etc. The divergence from the point of true health—i.e., the point in phase space where these parameters show health—will show the true state of disanthropy. Here a perspective for the use of powerful topological methods and computer analysis opens up. Similar studies of heart function from the point of view of the assumption of the strange attractor are given in study [14]. Also the advantages of such an approach based on the study of chaotic processes of the strange attractor in the non-linear dynamic system during the fibrillation of the heart muscle are demonstrated in study [88].

On a larger scale, our civilization is also a strange attractor that is united with the environment. Unfortunately, at our current stage of nooevolution, the feedback is in the form of environmental pollution and agricultural engineering for the sake of higher productivity, at the cost of losing the anthropic qualities of the final product. I hope that this will not last forever. The strange attractor of civilization—or to be more correct, the strange super attractor "civilization"—has unique attributes compared to other such systems in the universe: it is able to change its own phase state and the phase state of the surrounding universe, as if "conquering" it. This conquering of the universe happens due to generation of knowledge and skills by humanity. In terms of the strange attractor, it can be viewed as the growth of the fractal dimension of the attractor. If the hypothesis about indefinite existence of human civilization holds the dimension of the super attractor, it will increase infinitely; the area of the phase space will increase due to the inclusion of an ever-increasing number of processes and volume of the surrounding universe into the "metabolism" of this attractor.

## 2.4   Evolution of the strange attractor

As I mentioned earlier, I believe that the process of bio-evolution of man is not actual now because long-term biological processes have been replaced by more acute processes that are much quicker—the processes of deliberate environmental change and of man himself in the process of nooevolution. However, the bifurcations of evolutionary trajectories, and maybe also in the case of nooevolution, especially in the early stages, can have dramatic and not very well researched qualities.

A very interesting and important overview of the evolutionary scenarios is described in one study [15], which is based on the concept of K/R selection and supplemented by the concept of L selection in [16]. The evolutionary scenarios are studied in limited and intermediate types K, I, and R at which the maximum limit for a population fluctuates between the maximum limit K and the minimum limit L, beyond which the species disappears. At K, the population aims to reach the numbers K due to taking up territorial niches mainly due to competition. Populations with the R scenario are characterized by quick fluctuations in numbers between the levels K and L. Growth in numbers can sometimes have an explosive character, but the population numbers at the R-scenario is exclusively unstable, the species are maladapted to the negative conditions of the environment and competition. The species which are in the L-scenario—i.e., the species on the lowest level of survival potential under the conditions of natural selection manifest a greater ability to survive in hostile conditions of competition and adverse environmental influences. Also, they do not multiply very quickly.

In the nooevolution of the human species the common laws of evolution can also act and manifest themselves. From this point of view, due to the action of nonlinear laws inherent to complex open systems, with the addition of economic and geographic differentials of moral and ethical norms in the evolution of human society, bifurcations can exist within the system that separate it into parts that evolve according to separate scenarios. At present, humanity on the whole finds itself in the R scenario of development with an explosive growth in population, heading towards the K-level, which, taking into account that resources are finite, will lead to a relative drop in population numbers due to a lower life span of the species. Such a scenario is happening in Africa right now. To a certain degree, the countries of South East Asia are also evolving according to such a scenario.

The "Golden Billion" from the developed nations is pulled towards the L-scenario, though in some parts on the periphery it may evolve according to the R-scenario. The sum total of these situations in our globalized economy generates many demographic and economic challenges. This is particularly reflected in the availability and quality of food products. The deterioration of the quality of the mass food supply is predetermined, at least in the medium term (twenty to thirty years). This assumes the independent care of individuals and families about the quality of nutrition for optimum health and longevity. The concept of anthropic nutrition in my opinion provides a solution to this challenge. Natural evolution in nature assumes a choice of one of the scenarios without our influence, which is determined by external factors, genes, and social structure. However, in the process of nooevolution, the meaning of free will of the individual becomes critically important. It is necessary to understand that the modern evolution of medicine is generally aimed at the symptoms of the nooevolutionary disanthropy of our existence; and it switches humanity from the K-scenario over to the R-scenario of development. Within certain time limits, this creates certain advantages if we are considering the average life span of an individual and fast population growth. But in the end, it is contradictory to the individual aim of achieving a genetically predetermined life span. Here the capabilities of modern medicine are quite limited.

If we are to view the existence of a population of organisms and the environment from the standpoint of a united strange attractor, then it is possible to find an area of attraction where the population cannot exist. It is possible that change of the attraction point by such factors as population growth and environmental changes may lead to a situation in which the minimal entropy of the united strange attractor is achieved by the survival of other than human species. For the human being, this scenario is an ecological catastrophe. Humanity has gone through such catastrophes many times during its history. The first disaster of this type led to the Neolithic revolution. This means that the process of nooevolution has something to do with thermodynamic principles, which exist everywhere else in nature. Instead of dying, humanity changed the rules of the game—it started farming and growing animals for food. As was mentioned earlier, the level of minimal entropy for the strange sub-attractor of the united attractor, the individual, has increased; life span has decreased; and the number of diseases, as compared to hunter gatherers, has increased. However, the common level of entropy of the united strange attractor decreased, and a certain growth in ordering took place due to new knowledge, skills, social structures, and other changes. I am afraid that the L, K and R scenarios that were mentioned are not some particularities, but are general laws imminently inherent to any evolution of species, which possibly also include nooevolution. The existence of one or another scenario depends on the existence of feedback mechanisms that change the character of evolution and changes in scenarios. For example, one of the many types of feedback mechanisms to the growth in population and the general disanthropy of urbanization is the widespread use of fast food. It adds to and increases the general disanthropy of urban existence and could speed up the change in scenario. This is also true for many other aspects of life that we are accustomed to call "quality of life." To be fair, it is important to note that we cannot insist that changing to the R-scenario of evolution is necessarily bad for all species. Theoretically, during this change, certain members of the population become more active and dominate in the short term (a good example of this is the large number of field commanders in war zones). This allows for population growth and survival of the species. However, it is bad, or very bad, or outright awful for the individual member of the population, which is in essence every one of us. A very good example of this point is the demographic processes in some of the third world countries: high population growth with a high mortality rate and lower life expectancy in the conditions of utter poverty in Bangladesh or some African countries like Sudan, for example. To a certain extent this scenario can take place in any part of the world as the result of war, economic crises, etc. The above scenarios are a manifestation of some nonlinear evolution laws by which nature manages to decrease total number of iterations and achieves, because of it, truly amazing speeds in the evolutionary process.

## 2.5   Genesis of human metabolism

I apologize once more to the reader that a book about nutrition contains information which seem to be irrelevant. But it is my opinion that the presented concept will be more convincing by observing the mechanisms of interaction of humans with the environment (which also includes the global society), which influence the nutrition intake of the individual as a biological and social animal.

Living nature has a huge reserve of strength and endurance for survival in hostile conditions. But changes and breakdowns in the genetically-evolutionarily determined optimal trajectories of the metabolism considerably worsen its state in the thermodynamic sense, without necessarily leading to death. The emerging instability of the state of the strange attractor of the organism on the whole, as an open, nonlinear dissipative structure which involves matter and energy exchange, leads to an absolute and/or relative degradation of the organism—i.e., a thermodynamic death. In living organisms, including humans, this manifests itself as a gradual breakdown of cell function and internal organs and, ultimately, death of the organism. This takes place in a much shorter period of time than the normal evolutionarily determined life span of the organism.

I have already argued that when it comes to nutrition, in some approximation we can look at the human being as a synergistic structure in a state far from equilibrium. Such structures, known from the simplest examples of inanimate nature, remain in equilibrium with a small range of changes in parameters. That's why the disanthropic nutrition of modern man, along with other disanthropic factors of life, worsens the thermodynamic stability of the organism, which is expressed in real life as disease.

The process of metabolism in itself includes interrelated but inverse processes: anabolism and catabolism. Anabolism is a sum of processes of the biosynthesis of organic substances (the components of a cell and other organs and tissues) which provides growth and development of biological structures and energy reserves in the organism. The core of the anabolic processes is the chemical restructuring of bioorganic substances which are ingested with food into exclusively complex and varied biochemical substances that are involved in the organism's function.

Catabolism is a sum of the processes of breaking down biochemical substances into relatively simple compounds that are suitable for use in the processes of biosynthesis or for the final products of metabolism releasing energy. Catabolic processes extract the chemical energies from the substances in food and use this energy in the anabolic processes. The processes of anabolism and catabolism are in a state close to dynamic equilibrium in the organism. In childhood, as the organism grows, the anabolic processes dominate, and the catabolic processes dominate as we age. Which processes are more dominant in the organism is also determined by the state of our health and physical, intellectual, and psychological activity.

There are more than 200 types of cells in the organism (skin, nerve, liver, etc.) and in the vital functions of the organism, a large variety of proteins and nucleic acids are synthesised. Amino acids are the building blocks of proteins, which create oligomer chains of 30 to 20,000 amino acids in the formation of molecules. Each protein, from the thousands of them in living cells, has its own unique sequence of amino acids which determines its function in the organism. The function of proteins happens in the way of physical and chemical interactions with other molecules. They change the form and structure of these molecules in the body of cells and their membranes or in the extracellular space. Some proteins take part in many reactions, but thousands of them interact only in a narrow band with specific molecules. The slightest change in, for example, stereometry (the shape of molecule), radically changes the character of interaction and the intermediate and final results.

Modern disanthropic nutrition content of the food limits the delivery of substances suitable for the human organism. However, if the wrong molecules take part in the metabolic processes, or there were not enough right molecules, the biosynthesis lacks certain biochemical compounds, and/or compounds differing in chemical formula and/or the stereometric and conformational state—i.e., to change in the phase space of a strange attractor of metabolism of our body. If it is the case, in turn, the catabolism of the products of anabolism of these substances, formed from the remnants of wrong molecules, lead to formation of substances improper to the body metabolism, that the body sometimes does not have the capacity for removal to outside through the existing metabolic pathways. So part of the metabolic waste is left in intracellular space and can neither take part in the processes of the metabolism nor be expelled fast enough. The slow expulsion of these substances due to the change in equilibrium leads to their accumulation; they are poisonous. This has consequences for all the trajectories chemical reactions and the phase state of the strange attractor of the living system.

These unsuitable products of the metabolic breakdown are those infamous toxins that are being cleaned out of the organism in detox clinics worldwide. Many specialists in the biochemistry of the human organism believe that these clinics have limited effectiveness.

## 2.6   Flicker-noise from our food

The interaction of metabolism and food can be analysed using the generalization of the study of the influence of 1/f, or flicker-noise, on open non-linear systems in the form of strange attractors. Flicker-noise of food could be presented as the spectrum of nutritional content of the food.

Flicker-noise is one of the characteristics of dynamic chaotic processes. The power spectrum of flicker-noise it is the distribution of energy over frequency. The dynamics of primitive systems described by a single characteristic frequency, through which can be expressed the characteristic time, length, etc. In complex systems, the characteristic frequency is replaced by the frequency spectrum. The spectral density $S(f) \sim f^{-1}$ is characteristic of flicker-noise. In general, the phenomenon of flicker-noise is widely represented in nature: for example, the earth's crust vibrations, temperature and pressure of the atmosphere, the photosphere of the sun and stars, and the fluctuations in velocity and other parameters of the biochemical and biophysical processes. There is a point of view according to which flicker-noise is inherent to all systems that are far from equilibrium. These systems include the metabolism of living things from cells to higher organisms. Therefore, application of the idea of flicker-noise, in my opinion, is well suited to describe such processes.

Besides, the metabolic system is a so-called stochastic processes in which auto-oscillations are possible (self-supported oscillations) caused by the presence of feedback. At certain ratios of the parameters of stochastic systems auto-oscillations occur regularly, with a constant frequency, and this oscillational state is only possible steady state system. In the case of such a complex system as a strange attractor metabolism analogue of such self-oscillational state is the existence of a stationary system of metabolic reactions. In such a system, flicker-noise is the result of external action spectrum of food nutrition's on metabolism.

Power of flicker-noise is inversely proportional to the frequency. This means that the "slow" changes in the characteristics of the process are more likely than the "fast," especially at low frequencies. In the case of a metabolism, instead of the frequency, can be used, for example, the rate of the entropy production per unit of time by nutritional element of food in its metabolic changes. That is, analogue of low-frequency will be low rate of entropy production by the organism during metabolism of food element, regarded as a single isolated flicker disturbance—a "single line" of flicker-noise spectrum. "Slow" changes of parameters of the process in this case correspond to the minimal changes of metabolism structure or by other words—metabolism in its maximum stable state. If we will omit some small variations of metabolism due to periodicity of eating food, the "slow" changes in the ideal case (fully anthropic food) are responsible only for aging, which is normally a slow enough process.

But how does the stability of structures of metabolism, for example, in the cell emerge? More or less we can understand the stability of the elephant metabolism, which can theoretically occur due to the great mass and inertia associated with this process. But in the smallest cells, inertial processes play almost no role at all. Nevertheless, their livelihood is very active and durable. As will be shown in subsequent sections of the book, the stability of biological systems is determined by the existence of the viscoelastic properties of organisms on all levels of the constituent structures.

It is known that self-organizing non-equilibrium processes, for which the 1/f-noise is characteristic, can be easily adjusted to external influence. And this indicates that organisms react to change in nutrition with relative ease and flexibility. If we take as an example the human organism, then we can see that it adapts to changes in food relatively easily.

Flicker-noise, or an element of it—a small perturbation—can be represented as a transition over trajectories in the phase space, when the result of the reactions has probabilistic dependence from a certain set of deviations of some of the preceding reactions [13]. Flicker-noises represent correlated noises, but the correlation has a probabilistic nature—i.e., there is a quite certain, but not an absolutely firm, connection between the preceding and succeeding reactions.

If we look at it from a more general approach, we can imagine the 1/f system as a sum of components, in each of which there is an ongoing production of negative entropy. At some point the state of the element drastically changes (bifurcation) and there is a sharp change in entropy, but after that a new cycle begins. That is why systems with flicker-noise can respond with a powerful reaction to a small (infinitesimal) disturbance.

A necessary condition to support the high sensitivity of the 1/f system is a fairly large number of active elements, in our case a large number of structurally arranged cascades of the reactions of the metabolism in the cells and the media of the organism. As is mentioned in one study [279] the nature of flicker-noise remains unclear. One of the hypotheses links it to the existence of processes with relaxation times (fading memory). It is known that the self-organizing non-equilibrium processes, which are characterized by the 1/f noise, easily adapt to external influences. It will be shown below, in part 3 of this book, that the source of the relaxation processes for the metabolism of live organisms are the viscoelastic properties of their biopolymer structures.

In the classic case, the spectrum of flicker-noise is qualitatively different from the Gaussian distribution. The main difference is that the flicker-noise distribution is deformed towards the lower frequencies. In the case of complicated organisms and their nutritional processes, we can use the speed of the metabolism of nutritional elements as a frequency of the flicker-noise. In this case in the joint evolution of organisms and the environment, deformation of flicker-noise distribution to the lower part of the noise spectrum does mean that the organisms adapt to the potentially available sources of food. In the case of the human being and his predecessors, this evolutionary adaptation attempts to transform the largest number of elements in the biotope into maximally anthropic compounds with optimum speed and efficiency of metabolism.

At the same time, for different representatives of closely related species (for example, primates), the spectrum of these elements of biotope used as food may be close, but at the same time different, as the result of separate evolution in the last few million years. Humans and chimps are a good example of this difference. One study [94] found that the sialic sugars molecules  suitable for chimps, but absent in humans, have a serious disanthropic effect when they make their way into the human organism: they cause inflammation. There are two main versions of these sugars: Neu5Gc and Neu5Ac, which have a difference of only one atom of oxygen. Humans have only the Ac version, whereas monkeys and other primates have the Gc version. This separation happened three million years ago, and since then the foods that contain these sugars are disanthropic for humans. This was discovered during an attempt to use biotechnological drugs made from animal tissue. The date when this difference was born coincides with the dates when the sharp stone tools emerged.

It is quite possible that this not a random coincidence—i.e., the first evolutionary bifurcation might have brought about this biochemical division of the metabolism. However, it is possible that in reality it was the other way around.

More evidence of the influence of food on the evolution of the human metabolism is presented by the well-known phenomenon in which modern humans broke through the traditional barriers of body weight and brain size that exist in the animal kingdom. The case in most animals is that the bigger the body, the smaller the size of the brain in relative terms. Explanation of that phenomenon is perhaps an increase in the variation of food sources inclining towards a seafood diet. Developing a large brain size depends on the availability of some fatty acids which are not found in terrestrial foodstuffs. However, these fatty acids are common in marine animals [179].

Experience from different areas of science where non-linear processes in far-from-stable states can be seen shows that the power of single flicker-disturbances turns out to be inversely proportional to the frequency variation. In our case this will mean the inverse proportionality of the degree of deviation from anthropic properties. This is obvious because if the active element in food has maximally anthropic qualities, then its influence on the metabolism is maximally anthropic by definition. In my opinion this confirms once more the rationality of using the flicker-noise approach in order to look at the processes of food assimilation. Small deviations from anthropic properties make the penetration of food elements and their participation in the trajectories of reactions in the phase space of the metabolism more probable.

As the flicker-noises of food (nutritional elements of food) enter the organism they change the phase space of the metabolism, and the natural selection provides for survival and vitality of the organisms with that metabolism, as they rely on a larger spectrum of available food. The evolution of the combined strange attractor of the human being and the nutritional environment in this case provide minimal entropy, or in other words, the maximal generation of Schrödinger's negative entropy of this strange attractor. All the nooevolutionary activity of man in nutrition, production of medicine, and many other things can be seen as an attempt to achieve minimal local entropy in the strange attractor of the human metabolism in separation from the common attractor. This includes the natural nutritional environment. It mostly has to do with the fact that as nooevolution progressed the natural food environment has been lost or cannot support the increased population density locally and globally.

In simplified terms, the $1/f$ system of the metabolism represents in itself the sum total of reactions. In each reaction there is an accumulation of some relatively small deviations from the standard trajectory. At some point, when these deviations add up and the phase state of the strange attractor experiences a bifurcation—a leap into a different state happened which can be reversible or not. Biologically this often manifests as a change in the epigenetic function of the genome.

With the so-called stochastic approach to the systems that demonstrate the dynamic chaos (metabolism), it can be assumed that at the absence of the distortion of the external noise—new nutritional signals—the trajectory of the system in phase space will aim towards one or the other point of attraction, depending on the properties of the strange attractor of the system. The influence of the change in external flicker-noise will give rise to random splits between the trajectories, the statistics of which will be determined by the properties of the noise spectrum and the non-linear dynamics of the whole system. In this case the stochastic set of bifurcations can be associated with the information flow through the system. In evolution, live organisms adapted to using the inherent internal noise of biochemical reactions and the signal noise of incoming food to process information—the biochemical transformation of anthropic food in the metabolic processes according to the evolutionarily determined scenario. When an external noise affects the stochastic systems, a stochastic synchronization occurs—i.e., a takeover by the system of the external frequency of the signal. In the practice of eating, this corresponds to evolutionary adaptations of the organism to some nutritional element of food and to their spectrum. At the wide spectrum of flicker-noises from incoming food, it is possible for a synchronization of an ensemble of stochastic resonators (an ensemble of metabolic reactions) with the external noises of food. Such a process is called stochastic resonance. The role of the noise spectrum necessary for the realization of the stochastic resonance is carried out by the inner noise of the biological system—the evolutionarily formed ensemble of metabolic reactions in the organism.

As I mentioned above, the systems with flicker-noises are unstable in the face of small disturbances. A necessary and sufficient condition for the high sensitivity of the strange attractor to flicker-noises is the presence of a large quantity of active elements. It is fulfilled in the strange attractor of the metabolism by the huge number of trajectories in the phase space (the ensemble of metabolic reactions) and by the variation of substances in the available food. Change in this variation, or in other words, the nutritional deviations for similar organisms, can lead to dramatic consequences for evolution of their populations. We all know the fates of the two human races, the Neanderthals and the modern humans, based on the anthropological research [190]. Specific chemicals locked into bone suggest the Neanderthals got most of their protein from large animals, such as mammoths, bison, and reindeer. The modern humans that were living at the same time had more diverse tastes. As well as large mammals, they also consumed smaller mammals, fish, and seafood. Humans had a much broader diet than the Neanderthals. Such dietary differences could have played a major role in the extinction of the Neanderthals roughly 24,000 years ago [190]. The chemical measurements from the bone collagen protein from Neanderthals and humans are very different due to a different diet. High levels of some carbon isotopes in bones from Italy and France serve as evidence that humans ate some seafood. Also, it looks like the Neanderthals did not eat as many vegetables as the humans. A study of ancient DNA offers preliminary support for that conclusion. The Neanderthals had a mutated gene that prevented them from tasting bitter chemicals found in many plants. This gene encodes a receptor that detects a chemical called phenylthiocarbamide, which is closely related to the compounds in plants. Relatively small and at first sight insignificant changes in genes, related to new signals from food consumption (flicker-noise) changed the fate of the Neanderthal race. This shows the significance of food and gene interactions for the survival of an entire population. Such flexibility of metabolism may explain why modern humans thrived in ancient Europe while the Neanderthals perished.

At least the Neanderthals and humans ate natural food. But from the point of view of nutrition, the disturbance of the metabolic system is caused by deviations from the anthropic composition of food. These deviations lead to changes in trajectories of metabolic reactions in some organs, cells, and other subsystems of biochemical transformations in the organism. When, as a consequence of disanthropic nutrition, such changes accumulate, they redirect the metabolic processes to a new stationary state. Due to the modern rate of change in food habits, especially in the last 100 years, the mechanisms of natural biological evolution cannot keep up. On the whole, this is reflected in the state of the systems of the organism in the most disadvantageous manner. In particular, a change occurs in the production of hormones, essential amino acids, peptides, proteins, etc. However, it is possible to return the disturbed metabolism to the normal state by the cooperative flicker-noise of anthropic elements. Moreover, the presence of anthropic particles in food stabilizes the phase space of the metabolism in the state which provides minimum entropy. A particular mechanism of returning a changed metabolism to its natural state is provided by the expression and the repression of specific genes under the influence of anthropic diet. This is due to the fact that the natural metabolism is encoded in our DNA and can epigenetically manifest itself only under anthropic dietary conditions. Of course the situation is far from linear and is not so simple. Sometimes gene expression, especially if it happened at a young age, can be very stable. Many believe that they can be passed on for three to four generations. The reinstatement of a changed metabolism to its normal state will be very slow at best. But the ability to reverse the metabolism induced by food fills us with a certain level of hope.

A typical example of a dramatic negative effect of flicker-noises of nutrition is the mass obesity in some populations of the United States. Scientists connect this to nutritional choices which include large amounts of bad-quality fats present in the fast-food industry, high sugar, and low mineral content. The influence of these flicker deviations leads to unbalanced hormonal systems in the organism and ir-reversibly alters the trajectories of cascades of reactions of the metabolism. This concerns children especially. There is a lot of data in the media that shows that these problems occur mainly in the lower income sectors of the population, especially single mothers. Because of their economic condition, lack of education, and certain cultural influences allow their children to eat fast food from a very young age.

Also, food habits that lead to obesity change the trajectory of reactions to such an extent that it can cause diseases that have nothing to do with genetic factors. The genetic predisposition to diabetes in general is not correlated to the frequency of leukemia cases; however, diabetes induced by diet leads to an increase in leukemia [178]. The causes of both diseases are not genetically connected. This means that dietary induced metabolic changes affect a huge amount of cascades of chemical re-actions in our bodies that are not directly related. Such disanthropic disturbances in the diet of modern man, in all their huge variations, have long-term effects on the metabolic processes. This leads to a deviation from the normal of the point of attrac-tion in the phase space of the strange attractor to a point of attraction (a state) which is characterized by an early degradation of the metabolism. In such case the state of the population is characterized by a pandemic level of spread of non-infectious diseases, leading to lowered life expectancy. The "deformed" state of the metabolism can be "remembered" by the organism to such an extent that even if the disanthropic factors are totally removed, the metabolism continues to function along the wrong trajectories of reactions [227]. This can be seen very clearly with diabetes. The syn-drome of metabolic memory means that the diabetic vascular stresses continue even after blood glucose level normalization. This point of view is supported by labora-tory and clinical studies for both types of diabetes.

The existence of this metabolic memory suggests the need for the earliest pos-sible treatment in life to restore the metabolism to its maximal health in order to keep long-term complications from the disease at bay. Metabolic memory can also play a positive role: if, for example, after a prolonged disanthropic diet the organism begins to receive anthropic food, it can use the memory to help restore the healthy functioning of the metabolism. This is what happens when the states of deformed metabolism are corrected by an anthropic diet. In general, metabolic memory is connected to reversible and irreversible (almost irreversible) epigenetic changes that will be discussed below.

## 2.7    The cumulative interactions

The interaction of the strange attractors of the organism and food take place during the metabolism of substances, energies, and negative entropy. The last value can also be seen as information exchange or a measure of order. The studies [28, 29] are about complex non-linear open systems—strange attractors as cumulatively dissipating structures, which is in fact what complex organisms are. It is shown that such multiphase structures that have a lot of phase borders have the ability to focus and accumulate mass and energy and to generate negative entropy by the means of inner order. The higher the hierarchy of inhomogeneity in such a system, the more mass and energy it can accumulate, and the more complex the system itself. That's why such systems have the ability to achieve higher concentrations of energy at a given set of conditions. Interaction on the phase borders gives rise to boundary-cumulative streams-subsystems. Due to the dynamics of inner processes with feedback (in our case, food metabolism), these subsystems undergo a structuring of the processes of phase interactions, and surplus energy is dissipated from this area.

The dynamic unstable state gives rise to extreme dissipative processes. The extreme state in the phase space of digestive reactions is achieved due to the formation of areas of focusing energy flows and masses. They have considerably higher intensities of transformations of anthropic compounds of foods in the boundary-layer of the strange attractor of the human metabolism—the digestive system. What happens in this zone is the initialization of new degrees of freedom (transmutation) of the accumulating particles in the phase space—the biochemical transformations of chemical and bioorganic elements of food in the course of digestion. In the area of the attraction of energy and mass flows in the phase space of the reaction trajectories within such systems, "cumulative jets" arise, which are hyper conductive of processes in which the "activated" particles pass along stable trajectories of metabolic reactions with the least energy expenditure. It is important to note that the processes described here are in the phase space of metabolic reactions and that is why the terms "jets" and "particles" have a considerably different meaning from the normal interpretation of these words. "Hyper conductivity" is understood as the chemical transformations of food elements along the cascades of metabolic reaction pathways with the lowest possible energy expenditure.

This way, the transformation of substances along the cascades of metabolic reactions is energetically fed by the flows of energy and substances into the area of attraction within the phase space, and by generation of negative entropy. In practice this means that all the structural and hierarchical levels of the organism behave as strange sub-attractors, in which a certain minimal level of entropy is reached due to the organism's metabolism exchanging substances with the environment. The best example of this is a cell with a developed metabolism and in a state of isolation: a bacterium, a completely independent organism. This approach allows us to define the bioactivity of food particles and microelements to the human organism—their inherent ability to be part of the cumulative jets of chemical interactions in the phase space of metabolic reactions.

We must note that in such systems, the movement of inactive particles happens according to a completely different scenario. In the case of digestion, these particles are inert, dead weight (indigestible) nutritional food elements. It is very dangerous when disanthropic food particles become activated. These compounds take part in the "wrong" chemical reactions. They skew the trajectories of metabolic reactions and direction of cumulative jets. Also, the initially non-uniform chemical composition of the incoming food can destabilize the emerging dissipative structures of anthropic cumulative jets and break down the organizing hierarchical systems of selection, filtration, and local attraction in the phase transformations along the metabolic pathways.

According to the general properties of such non-linear systems, in such strange sub-attractors of the metabolic processes, the rates of reaction in the cumulative jets increase at a great rate, while the characteristic scales of processes decrease. This activates previously dormant areas of the phase space—i.e., a greater intake of various compounds and structural systems of metabolism into the processes of cumulative hyper conductivity on different hierarchical levels. All of this is accompanied by the dispersal of cumulative jets in the form of bifurcations of metabolic pathways. Under normal conditions of anthropic nutrition and a healthy organism, all these processes are in accordance with genetically determined programs of metabolism. However, the presence of disanthropic food elements lowers the degree of order due to large scale fluctuations. Consequently, the directions of metabolic pathways change to trajectories unnatural to the organism. In this case the formation of the combined "big crack" in the phase space of the metabolism happens. This is the process of falling out from the active phase of a significant number of cascades of chemical reactions. The big cracks appear when the "small cracks" in the chemical processes happen according to a hyperbolic scenario, or in other words, an "explosion." This explosion represents disease in one of the vital organs.

When there is a lack of anthropic food elements in the organism, there is a rise in entropy, first because of the disappearance of evolutionarily determined metabolic pathways (due to a decrease in production of negative entropy); and second, because of a deformation in the evolutionarily determined order in the phase space of the metabolism as there are disanthropic compounds present, which also, by definition, lead to a decrease in production of negative entropy. And vice versa—the presence and influence of anthropic food compounds causes an ordering of cumulative jets along the metabolic pathways which increases the production of negative entropy. The disintegration of these cumulative jets in the phase space due to disanthropic food consumption can lead to two outcomes. One possibility is that it causes changes in the point of attraction within the phase space of the metabolism's strange attractor. Another potential consequence is full or partial disintegration of this phase space as a result of the "cracks" that were mentioned earlier. In medical terms the disintegration of the phase space of the strange attractor looks like a gradual break-down of organ function in the late stages of disease.

Some disanthropic nutritional factors described here as disanthropic fluctuations have more significance than others. A greater fluctuation, with a relatively long existence can interact with other variations of the same scale and then something called the non-random states correlation occurs. This is the emergence of the alternative metabolic pathway that was discussed earlier. On its own or combined with other alternative trajectories, it leads to a large scale change in the set of states, which is equivalent to the appearance of a new large "branch" or "crack" in the phase space of the metabolism. This change slowly takes over the whole phase space or its larger part and commands all the other fluctuations (the "slaving" principle)—i.e., it evolves into a dominating regime. This means that due to the "slaving" principle in non-linear systems, certain separate fluctuations escalate (micro changes) into a correlated, coherent interaction of a new type on the macro level of the organism. In clinical practice it manifests as one of the diseases of the metabolic syndrome.

Very interesting research of the mathematical modeling of this subject is carried out at The Translational Genomics Research Institute (TGen) in Arizona [45]. S. Kim's study uses the Boolean modelling (probabilistic Boolean network) to look at the interconnections of the strange attractors in the organism of the final state (aging) of the phenotypes of the cells with reservoirs (basins). This is the state of the external metabolism, which leads to the attraction in the state of the old cell. It has been shown that certain key influences exist which determine the final transition to the final state of the cell independent of the existence of the particular genetic mechanism. This mathematical study shows again that such external influences as nutritional deviations are one of the most important factors which determine the state of the organism.

# 3
# The Nonlinear Biochemistry of Eating and Living

## 3.1 Biochemical aspects of the strange attractor of the metabolism

Essentially, the non-linear approach opens the understanding of metabolic processes as hierarchical structures. For such structures, it is natural to have the properties that we see in metabolic processes in the organism—i.e., the formation of cascades of reactions that have long- and short-distance dynamic orders. The influence of external signals, in our case the disanthropic nutritional factors, can give rise to completely new elements of structure in such systems. They arise due to the existence of many fluctuations, or in other words, a set of control signals that bring the system to new equilibrium. Unfortunately, these new states are almost always bad news for the human organism.

Every part of the organism and of its metabolism has many interwoven functions with other parts. This interaction often happens through metabolic components, which are generated on a certain cascade of reactions and take part in many more cascades in different organs. Also, the production of metabolites by different organs or different parts of the body participates in commanding many functions in other tissues and liquid media of the organism. The same proteins which are generated in the metabolic reactions play a key role in the emergence of different degenerative diseases [84]. This proves the opinion of the book you are reading right now that the signals of the broken-down metabolism are manifested on different levels of the hierarchy and in different parts of the phase state of the metabolism. As shown in the study [86] the protein that is the so called factor of transcription studies [84, 85] show that the bone system of the organism, in addition to its two commonly accepted functions (mechanical and generation of blood), also plays vital metabolic roles as an endocrine organ in the production of the hormone osteocalcin, which in its active form controls the metabolism of glucose in the organism. If we look at the production of osteocalcin as a negative feedback reaction to the incoming nutritional elements and their metabolites in the reaction cascades, we can conclude that along with the direct reactions of the metabolism, negative feedback chemical reactions are generated according to the same cascade principle. Of course this division in the real metabolism is relative, but it allows us to have a better understanding of the strange attractor of the metabolism as a defined system of reactions built according to the network principle, where changes in one point cause changes in all the other points.

Lack of certain elements in the diet can cause tough consequences for the human metabolism. This is especially true for essential amino acids. As an example to prove this thesis we can look at the results of one study [200] of the function of the amino acid methionine in the cell defence system. The study explains that when cells are confronted with an invading virus or bacteria or exposed to an irritating chemical, they protect themselves by going off their DNA recipe and inserting the wrong amino acid into new proteins to defend them against damage. Thus, the more errors, the better. These "regulated errors" comprise a novel non-genetic mechanism by which cells can rapidly make important proteins more resistant to attack when stressed. The normal error rate in constructing proteins by placing amino acids incorrectly is around 1 in 10,000. But one amino acid, methionine, has an error rate at a much higher level of 1 in 100 and therfore plays a crucial role in cell protection against invading viruses and other threats. This mechanism is very important to our wellbeing. But methionine is not synthesized in humans, so we must ingest it from methionine-containing foods. Of course it has many other important functions in our organism too.

Even clearer evidence to prove the function of the metabolism's strange attractor according to the network principle is given in a study [45] examining thWhat is?e interaction of proteins in cells. Research has shown that cells in the strange attractor of the metabolism contain the most complicated network of hundreds and thousands of proteins interacting in the cell, and each one (protein) is interconnected with hundreds or maybe thousands of others simultaneously. The change in the state of one protein due to a certain reaction is reflected in the change of state of all the others. The study of individual interactions in such a case gives researchers some additional information, but it is almost useless. If we can imagine that scientists got data for trillions of different states of this network of proteins, then the volume of the acquired information will not allow the whole picture of cell function to be seen because the quantity of information exceeds the possibility of its cognitive processing in effigo (my vandalisation of Latin: effigia means imagination) even by using supercomputers. Sciences such as molecular biology, biochemistry, genetic biology, and evolutionary biology contain too much data for scientists to parse. The best way to deal with this obstacle in my opinion is the phenomenological approach to the organisms as to non-linear structures. But that will be discussed later.

In the last twenty to thirty years, metabolic issues have had a tendency to be explained in terms of free radicals and antioxidants. This is why it is interesting to view this interaction from the perspective of the non-linear concepts discussed earlier and the biochemistry of the cell. Also, the theory of antioxidants itself needs certain clarifications. On study [27] shows that a surplus of antioxidants itself does not really lead to good health. This conclusion is based on the analysis of the level of antioxidants in cells of mammals, including humans, with different rates of metabolic reactions. However, it is important to note that the level of free radicals in the organism plays a very important role in the metabolic processes and cell function as far as aging and degenerative diseases are concerned. The level of "leakage" of free radicals from the mitochondria determines the condition of the whole cell. However, it is not possible to regulate this level by primitively changing the amount of antioxidants in the cell or its environment. The nature of this process is much more complicated.

During the process of metabolism in the organism, the inner energy of chemical food compounds transforms into chemical, heat, mechanical, and electrical types of energy. The vital processes are supplied with energy from respiration. Production of energy without the participation of oxygen, as in glycolysis (breakdown of glucose into lactic acid), for example, is called anaerobic respiration. The energy released in the anaerobic processes is normally not sufficient for the functioning of most organisms. Aerobic respiration, which takes place in the presence of oxygen, is energetically more efficient. When the organic compounds are oxidized, the breakdown of chemical bonds leads to a specific release of a relatively large amount of energy. Essentially, the process of oxidation in the organism is the same as the process of burning in an open bonfire, but it takes place at a much slower speed—i.e., the process of chemical oxidation of substances in the organism is the energetic basis for its livelihood. However, pure oxygen is quite dangerous for live organisms. In small doses its effects are medicinal, but in larger quantities it is poisonous. For higher animals, the optimal concentration of oxygen in the air is around 21 percent.

The oxidation of food takes place in the mitochondria. Most of the oxygen which comes into the organism through breathing (about 99 percent of it) is used up in the processes of energy production due to oxidation. However, a small percentage of the oxygen is transformed into the so-called active forms of oxygen that exist in the form of free radicals or hydrogen peroxide molecules. Many scientists believe that these substances during some biochemical breakdowns in the organism lead to the formation of cancer cells and the breakdown of the natural process of apoptosis through DNA damage. So what is the evolutionary concept behind the production of these free radicals in the organism? Many scientists believe that thanks to the existence of free radicals tied up by proteins on the outer surface of cell membranes, the cells acquire the capability to destroy bacteria and even viruses. During the long process of evolution here on Earth, they learned to uphold a certain stable balance of radicals. This balance is supported by the function of the cell mitochondria and by the antioxidant content of the incoming food, or by the molecules produced from the nutritional compounds of food which have precursor properties for the production of endogenous antioxidants. It is thought that the natural antioxidants from anthropic food of endogenous origin partly recombine the surplus of free radicals. By this action they provide the necessary level of free radicals so the organism functions at its best.

The disanthropic nutrition of modern man does not provide the supply of either "good" antioxidants or their precursors—the raw materials for endogenous antioxidant production. This leads to the destruction of natural balance, DNA damage, and breakdowns in cell and mitochondria function. It seems that a minute imbalance triggers explosive growth of damaged cells, which is expressed as organ tissue degeneration or cancer. It is known that up to a certain age, a child's organism endogenously produces the necessary antioxidants. However, as the child grows, so does the need for external sources of raw materials to produce antioxidants and the antioxidants themselves. All of these external compounds in modern literature are known as vitamins and essential amino acids, which are critically needed by adults. This need increases with age. Some people, due to their genetic makeup, are able to endogenously produce these antioxidants throughout their lives. As shown in one study [494], people with exceptional longevity did not have healthier habits overall than the comparison group in terms of body mass index, smoking, physical activity, or alcohol consumption. While longevity genes may protect centenarians from results of bad habits, healthy lifestyle choices and especially reasonably anthropic diet remain critical for the vast majority of the population. These are the people who are genetically predisposed for a long life—the centenarians. For everyone else a considerable increase in longevity can be achieved by a balanced supply of the necessary precursors for the endogenous production of antioxidants from food with anthropic qualities. However, it is important to note that the prevalent factor is the concentration of free radicals, which depends on the rate of "leakage" during the functioning of the mitochondria [27]. The existence of antioxidants in the organism has a positive effect if it's the right type in the right concentration found in the right place. Even an unnatural increase in natural antioxidants stops having the necessary effect after it reaches certain concentrations. That is why it is important to stimulate their endogenous production by supplying the right precursors through nutrition. Optimal mitochondrial function with the lowest possible leakage of free radicals can be partially provided by the supply of all the necessary nutritional elements or their metabolites to the cells and the mitochondria and by maintaining it at the evolutionarily determined level by consuming anthropic food. On the whole, as far as natural antioxidants are concerned, I believe that their main beneficial quality for us is as a natural preservative for food during storage.

Eating incorrect food that is offered by the modern agriculture and food industry causes many natural processes in the organism to fade out. Hormone production falls, other vital compounds are also reduced, and the function of many organs, glands, and cells is suppressed at the molecular level. Prolonged exposure to disanthropic food can lead, as we saw earlier in many publications, to irreversible changes in metabolic pathways. Thus we conclude that it is necessary to maintain the anthropic balance of the incoming nutritional substances from an early age. In this case the organism continuously produces substances that are determined genetically to maintain anthropic balance and in a way constantly trains the organism to do that. Life expectancy increases and the aging process slows down. However, it is important to stress the point that all of the above has to do with only natural food from natural sources, and not to supplements in tablet or powder form, which only have a negative effect on metabolic processes. When the free radicals are produced in the cell mitochondria, there are certain levels of equilibrium of oxidative-regenerative reactions. Deviation from this leads to a break down in cell function. Antioxidants have to be present in necessary amounts in the necessary places. We are talking about natural antioxidants from nature, but mostly about endogenously produced antioxidants made in a body from the precursors that are supplied by incoming food.

In any cell of the organism eukaryote, there are two types of the inheritance devices: in the cell nucleus and in the mitochondria – the organelle which generate almost all the energy in the organism in the form of ATP. ATP is a complex chemical compound that in essence is a nucleotide which also provides for the production of negative entropy in the process of hydrolysis—i.e., during the attachment of the water molecule. The water molecules come in normally during the processes of synthesis and fusion of organic compounds in the cells—i.e., of the complication of the cell structure and in turn of increase in order, or in other words, reduction in entropy. The nature of eukaryotes is such that the mutations in the genetic apparatus of the nucleus are slowly gathered over its whole lifetime and influence a relatively small part. However, it is impossible to deny the connection between genetic mutations in the cell nucleus and such diseases as cancer. The genetic apparatus of the mitochondria is a totally different story. In the course of the organism's lifetime, there is a significant build-up of defects in their genetic apparatus. The sources of these defects are the normal genetically predetermined accumulation of defect as the organism ages, as well as the additional epigenetic mutations which are connected to the adverse factors of the environment: radiation, chemical substances in food, and other influences. If the mutations affect the so-called transcription factor, which controls the synthesis of the proteins in the mitochondria and determines epigenetic expression—rate of activity under the influence of some chemicals—then the influence of the mutations manifests in certain organs and tissues of the organism. This leads to the degradation of the processes of the respiratory chain in the mitochondria and organ function breakdown in the form of deformed trajectories of chemical reactions of protein synthesis. If, for example, there is a breakdown in enzyme synthesis, then this strongly affects the metabolism of minerals in cells and organs. This has to do with the fact that many enzymes have such a subtle phase-molecular structure, that they are capable of identifying minerals not only by their chemical and valence composition, but at the quantum level also by their isotope composition. The breakdown of this function leads to organ function breakdown via changes in metabolic pathways. But apart from that, enzymes play an extremely important role in the endless interactive metabolic processes which include proteins and other vital substances. That is why changes in their synthesis cause many changes in the biochemistry of the cell and the degradation of its function.

As is mentioned earlier, the number of mutations in the mitochondria in a lifetime greatly exceeds the number of mutations of the cell nucleus. For example, in the simplest eukaryotes the number of mutations in the mitochondria can exceed the number of mutations in the nucleus a thousand times. The same applies to the cells of the human organism. In the so-called control region of the human cell mitochondria, the speed of mutations exceeds the speed of mutations in the nucleus tenfold or more. That is why the control region is also called the "hyper variable" genome region. The genome of the mitochondria also has a "coding" region, which is responsible for the coding of certain proteins that participate in the main functions of the mitochondria. The mutations that have been accumulated in the control region during the aging process and from external influences are manifested in almost all the cells of the particular organ or the tissues of the organism. The mutations in the coding region genome manifest themselves in separate cells, rarely affecting more than 1 percent of organ or tissue cells.

In such a way, the influence of external factors on the control region of the genome of the mitochondria leads to very significant consequences for the function of the mitochondria. As a result the mitochondria become more "excitable" to external signals (the example of activation of a medium in a non-linear approach as given in 2.7), and in particular, can increase the division speed in the cell as compared to non-mutated cells (the cumulative jet from 2.7). As a result there is an accumulation and prevalence of the defective mutant mitochondria in the cell. At every cell division, they are multiplied and their number greatly increases as compared to the number of such mitochondria in normal conditions, and in this way they take over the whole organ or tissue of the organism (the change of the point of attraction of the phase space from part 2). Even though the cells have mechanisms to destroy the defective mitochondria, the general level of "semi"-defective mitochondria in the organ increases anyway.

Under the influence of the external chemical stress during mutations, the spectrum of the active genes of the mitochondria changes relative to the normal distribution. As a result cells cross over to new quasi-stable states that answer to a new flow of the sum total of the metabolic reactions, or into the new phase state of the strange attractor of the metabolism of the cell. However, this new phase state is different from normal because the whole set of reactions and resources of the cell are directed toward survival and not toward carrying out its genetically determined function; or in terms of thermodynamics, the genetically determined level of generation of negative entropy is not provided for. In real life, this faulty functioning of the cells is manifested as low energy, slow recovery after illness, general weakness, and depression. Prolonged and significant changes in the respiratory functions of the mitochondria lead to serious breakdowns, from rare genetic diseases to diabetes, cancer, Parkinson's disease, and premature aging.

The fact that a high frequency of mutations in the mitochondria plays a significant role in tumour formation is proven by the data from one study [271], which looked at the influence of metabolic rate on the development of stem cells. It was shown that cells with high metabolic rates provided much higher numbers of tumours than cells with lower metabolism. But then researchers administered the mitochondrial inhibitor rapamycin to stem cells with high metabolism, which significantly decreased their tumour inducing capacity. The ability of stem cells to form tumours remains the biggest obstacle to their clinical use. The importance of the study [271] is also in the fact that all the processes in the mitochondria of normal and stem cells are very closely connected to signals from the ingested food. It is the mitochondria that convert the nutrients in our food, with the help of oxygen, to a form of energy that can be used by the body. And that is why the function of the mitochondria and the accumulation of mutations in their genetic apparatus are very much dependent on our daily diet.

In addition, according to another study [525], the presence of a functional angiotensin system in human mitochondria provides a foundation for understanding the interaction between mitochondria and chronic disease states and reveals potential therapeutic targets for optimizing mitochondrial function and decreasing chronic disease burden with aging. Angiotensin is a peptide hormone that also stimulates the release of other hormones. Changing the angiotensin level in the body's cells had affected mitochondrial energy production. To keep energy production in aging cells' mitochondria on an acceptable level, we have to inhibit angiotensin converting enzyme (ACE) activity. But how to achieve that? The study [526] shows that the inhibition of ACE activity was associated with both phenolic and flavanol content in the foods. It does mean that superfoods in an anthropic diet rich in both ingredients—such as cacao, red wine, and green tea leafs—are very useful for inhibition of ACE and thus keeping the level of angiotensin in an old body's cells more or less the same as in a young body.

## 3.2   The fabric of life

If we look at the set of chemical transformations in organisms solely in terms of the strange attractor of the metabolism, a certain difficulty arises. If we stop supplying the system (which in itself is a chemical reactor) with nutrition containing the necessary elements, then the reactor will soon stop working. But live organisms have a much higher stability than any chemical reactor. So what's going on?

When we were viewing the metabolic processes using the abstraction of the strange attractor, it was based on the non-linear interaction of linear media. However, everything is much more complicated in nature. The strange attractor of the metabolism of live organisms has an additional considerable non-linearity—the mechanical non-linearity of a medium. Practically all media of the organism are mechanically non-linear; they are the so-called viscoelastic media which are characterized by a certain spectrum of times of relaxation from deformations and tension. This manifests on all the levels of structure of the organism, from tissues to individual molecules, which in essence are biopolymers or are situated in biopolymer environment. Feel yourself—you will find viscoelasticity in everything: cells, skin, bones, eyes, etc. Everything in your body is viscoelastic with capabilities of reversible deformities and relaxation of tension in a certain period of time—the relaxation time. In this part of the book, my intention is to demonstrate that the viscoelastic qualities of the macromolecules of biopolymers (mainly proteins) influence not only the mechanical properties of tissues of the organism, but also all the processes of energy and matter transference, the function of chemical transformations, and the function of the genome.

Viscoelastic materials can have the viscous behaviour of a fluid and the elastic behaviour of a solid simultaneously. This gives rise to many particularities of their mechanical behaviour. The existence of relaxation processes leads to a determined delay in feedback in coordination with their characteristic times. If, for example, as one study [48] notes, we talk about a network of protein molecules in the cytoplasm of a cell, then the change in state of one macromolecule is reflected in the change in state of the nearby molecules with a characteristic delay. And the movement of the signal of the changed state of the network, the time of delay in each area, varies according to the characteristic relaxation time for each intermediate macromolecule. The same is true for the mechanical and chemical feedback from the change in states of macromolecules in the network of proteins. To use an acoustic analogy, our level of knowledge and understanding of live organisms on the micro level is like an unimaginable cacophony to our ears. But it is the same cacophony as Bach's music would be for a Neanderthal. In reality it is a powerful symphony of life.

The influence of the viscoelastic, as they sometimes call liquids with non-Newtonian properties, manifests itself in the additional non-linearity of processes. Once upon a time, I theoretically and experimentally studied the simplest dissipative system—the process of Bernard's convection in the thin horizontal layers of the non-linear viscoelastic liquid of the second order [97,98] which was experimentally modelled by the solutions of polyoxyethylene, a polymer with prominent non-Newtonian qualities. The subject that was studied was the evolution of the wave package of perturbations of finitesimal size at the point of bifurcation of heat transmission in the non-linear viscoelastic medium. It was theoretically found that the bifurcation which is responsible for the start of convection process and its fading, because of additional non-linearity of the equation of the state of the liquid, has a hysteresis property. This means that convection arises abruptly by jump, and fades out monotonously. And both processes happen at different normalized temperature difference (Rayleigh numbers). Because the phenomenon of starting convection, in essence, is the bifurcation of the process of energy transferal by heat conduction at small temperature differences—at the emergence of the simplest dissipative structure—then undoubtedly, similar hysteresis processes can take place in other more complicated systems as the physical and chemical reactions of the metabolism in the viscoelastic media of the organism. Viscoelasticity is readily evident in all tissues and cells of all the organisms where energy dissipation or hysteresis takes place, between the loading and unloading of any part under mechanical forces of any nature and origin. By the way, in one study [233] at the generalization of Prigogine's theorem for non-linear systems far from equilibrium, the influence of processes of relaxation on physicochemical processes has been studied. It was shown that the phase transitions of the second type—analogues of "jumping" bifurcations, which emerges and fades out by hysteresis trajectory—exist in such systems. The bifurcation of this type can be viewed as moving through a barrier that has a hole with a one-way valve. The return to the original state, even if possible, then only goes through the loop of hysteresis—i.e., through a different valve, or strictly speaking, along a different trajectory.

It is true that in gel-type media of live organisms, the frame of the biopolymer matrices holds a certain stable structure. Physicochemical reactions that take place in the low-viscous phase of the gel are accompanied by a movement of groups of atoms. The low-viscous phase in which conformational relaxation of the elements of the matrix takes place, is not an undefined obstacle for the movement of certain groups of atoms of biopolymers within certain boundaries, during some kind of a physicochemical reaction. But the change in conformation of the elements of the biopolymer matrix will affect the resilient qualities of these elements and of the matrix in general. And this applies to the viscoelastic properties of the gel also. However, the movement of molecules or group of molecules—the "geometry" of which is not confined to the parameters of the conformational changes of the elastic properties at given tensions which are determined by the elements of the given matrix—could be difficult. This way the physical or chemical bases of the conformational are created, and this also applies to the viscoelastic specifics of biochemical reactions in the space of the biopolymer matrix. From the general point of view, the reason for such specifics is that the redistribution of the chemical bonds during a chemical reaction changes the distribution of electron density in the reacting biopolymer molecules, which leads to their conformation to be different before and after the reaction. Often such a conformational transference is reversible. The reversal at the end of the reaction can be viewed as the conformational viscoelastic relaxation of uncompensated tensions that arise in the biopolymer molecules after the act of redistribution of electron density or other changes in the course of the reaction. An example of a direct conformational transfer of such a type is the attachment of the oxygen molecule to haemoglobin. The joining of a very small molecule of oxygen to haemoglobin causes considerable changes in the conformation of some proteins—a turn of some groups of atoms in the molecule compositions by 10 degrees.

Viscoelasticity plays an important role in molecular processes of signaling in the cell. Research [303] in cellular mechanotransduction (the process by which cells convert mechanical stresses into chemical activity) focuses on how extracellular physical forces are converted into chemical signals at the cell surface. This study shows that the mechanical forces exerted on the surface-adhesion receptors, such as integrins and cadherins, are also channelled along the cytoskeletal filaments matrix and concentrated at distant sites in the cytoplasm and the nucleus. In general, cells sense their surroundings through mechanotransduction—that is, by translating physical forces and deformations into chemical signals such as, for example, changes in intracellular chemical concentration or by activating diverse chemical pathways. In turn, these signals can adjust cellular and extracellular structure. This mechanosensitive feedback modulates cellular functions as diverse as migration, proliferation, differentiation and apoptosis, and is crucial for organ development and homeostasis. In addition, protein-formed ion channels that detect mechanical stresses participate in such processes as sensing or heartbeat. Consequently, defects in mechanotransduction—mainly caused by mutations or by changes in viscoelasticity of the protein matrix that disturb cellular or extracellular mechanics—are implicated in the development of various diseases, including cancer and metastasis progressions. These molecular mechanisms might act at some distance to induce mechanochemical conversion in the nucleus and alter gene expression. Evidently, viscoelasticity in molecular processes of mechanotransduction plays a key role in the determination of qualities and functioning of cells, organelles, cell media, and matrices.

It is obvious from the above examples that the reactions in gels with protein matrices depend on the viscoelastic properties of the matrices, which in the course of the reactions are also subject to change. Also, it is absolutely obvious that the excitation and fading out of processes that are connected to chemical reactions, which depend on the viscoelastic qualities of the matrices or the changing of the viscoelastic properties of matrices, will take place along different trajectories. This is hysteresis. It can manifest itself in different ways across the cell. For example, in one study [305] using a method developed to watch moment to moment as a molecule is moved to precise sites inside live human cells, scientists found that the proteins at one location may signal division and growth of cells, and the same proteins at another location may signal death. In general, many proteins inside of cells move in both directions—to the periphery, a locale known as the plasma membrane—and from the periphery back to different organelles. Any single protein plays multiple roles in the cell by changing its location. Chemical signalling inside cells regulates the movement of protein molecules through complex feedback loops defined also by viscoelastic properties of the cellular protein matrix. In this manner, by influencing the speed of diffusion of proteins and the hysteresis character of excitation and fading out of chemical reactions, the mechanical properties of different polymer protein and lipid matrices inside the cells and on their membrane have a significant influence on the vital cellular processes. It is very important from this point of view that almost any protein that fails to be delivered to the target area in the cell due to some malfunctioning in viscoelastic matrixes could disturb the cell's proteins' homeostasis that could lead to inappropriate protein aggregation in cell parts and to the degradation of cell functions.

However, the hysteresis type of excitation and fading out of reactions or phase transitions affects the character of influence of feedback, the main function of which in the cell is in controlling concentration of different chemical compounds in different places. This is an additional non-linearity. The existence of viscoelastic non-linearity is a substantial element that affects all the metabolic processes, and as it seems, is responsible for stability of the metabolism on all the levels of the hierarchy of the organism. As proof of this discussion, let us look at the theoretical approach of one study [222], which shows that in the physical-chemical systems, if the instability to finitesimal perturbations (bifurcation) appears-disappears (direct-inverse transition) according to the mechanism of hysteresis type (viscoelastic systems), then the macro states of the system determine its micro states. The situation is in a way the opposite of what exists in the classical linear medium, where the ensemble of micro states determines the macro states of systems. For the viscoelastic systems, the macro state provides the realization of corresponding microstates through renormalization of probabilities, as the result of which the ensemble of possible micro states narrows [222]. Such a renormalization of probabilities means that the probabilities of the states of the ensembles of particles which do not lead to the macro state of the system become negligible. From the physical and chemical point of view, this means that all the possible physicochemical configurations of molecules can participate in the function of processes of the structures which emerge as the result.

In this way, the viscoelastic properties of a medium determine the main feedback mechanism, which in turn determines its properties on the micro level. They also determine the variety of emerging structures and their stability due to having relaxation times. For such a system, during the calculation of statistic weight of each macro state (a state, the logarithm for which, according to Boltsman's formula, is exactly equal to the constant of entropy), according to [223], must not add up the probabilities of alternative microstates, but should multiply the relative probabilities of all micro states. This leads to the negative meaning of entropy, which strictly speaking is not the entropy meant by Boltzmann-Schrödinger [233]. However, the viscoelastic properties of live systems in any case are one of the sources of negative entropy. The exceptional thing in the analysis of the study [222] is the fact that all the above mentioned is realized for the systems with a limited cycle or with Feigenbaums convergence—i.e., the existence of a system in the state of a strange attractor. It is possible that this discussion and mathematical analysis [222,223,233] are the basis for the necessity and sufficiency of definition of live systems as nonlinear open structures with viscoelasticity. Anyway, I adhere to this approach in further discussions.

It should be noted also, as shown in work [607], that the evolution of physical non-Markovian systems with fading memory—i.e., having the relaxation times is described by equations with fractional derivatives describing some kind of fractal structure with channels. Channels structure may be different and generated by a specific fractal structure of the medium. Such processes are classified as a process with the "remaining" memory. In our case, a channel refers to a different rate of diffusion through a gel formed of molecules of different "size."

In statistical physics, one of the criteria of the irreversibility of the process is the change of sign of the time. As stated in the specific process described in the fractional derivatives, is that with this change, some states of the particles are preserved, while the other part corresponds to an irreversible loss. According to this work, in the relatively slow relaxation process (relaxation time is substantially greater than the characteristic time of the process), when the magnitude of stress changes more slowly than its first derivative does, irreversible loss of some of the system states occur. Loss of the states in such a description naturally takes into account the irreversibility of nonlinear processes. This is somewhat analogous to what is proposed in [222], a renormalization of ensemble probabilities of possible states of particles in phase space with hysteretic type of direct and inverse transitions (bifurcation) observed in nonlinear viscoelastic media in the processes of heat transfer [98].

## 3.3   Cybernetics of living molecules

An dominant opinion exists that the individualization of the first cells of the primordial organic soup and of the simplest isolated metabolism happened thanks to the forces of surface tension. It is possible that this is not quite the case. The viscoelastic properties have also probably played a role in this as well. It is thanks to these qualities that the disintegration of areas of reactions involved in developing the first polymers or oligomers (short polymers) is accompanied by fragmentation into macro droplets. The breakdown of threads of viscoelastic polymer solutions is easily observed experimentally and the photographs of this process are published in many books on the rheology of polymers. Droplets which look like beads are formed on the threads of the solution that retain their stability. The conditions that form these beads, according to {101}, are certain interrelations between the inertial forces and viscoelastic characteristics of the liquid that is stretched into a thread. But what could be a source of viscoelasticity in primordial soup?

For reasoning the existence of viscoelasticity in a soup, I have suggested my own hypothesis for the emergence of life. According to this theory, when life was born out of the primordial soup, the simplest self-exciting oscillation reactions led to the formation of matrices from the first oligomers (relatively long molecules). This most likely happened in an isolated volume with the participation of amino acids, nucleotides, and other substances. Long oligomer molecules spontaneously and irregularly formed a primitive matrix structure of the primordial gel. When a new long molecule is formed based on the existing molecule, there is a local decrease in entropy in the solution. Also, because of the so-called dynamic asymmetry of diffusion of the solvent and the polymer, viscoelastic separation of phases occurs as well. This viscoelastic phase separation help to convert the negative entropy associated with energy flow and composition of the system into negative entropy associated with configuration entropy of macromolecules.

It should also to be noted that the basis of a viscoelastic gel could be not only peptides based on amino acids, but on lipid molecules as well. As shown by many researchers, the abiotic synthesis of lipids is possible in primordial conditions. Lipid bilayers have viscoelastic properties and can exist in both phases as liquid and as gel. Additionally, the lipid layers typically have a porous structure with the existence of channels for molecular transport. Theoretically, the formation of lipid gels can provide the appearance of viscoelastic properties of the medium required in the concept of life origin I have advanced. However, this does not change anything practically in the subsequent arguments. Moreover, lipid molecules and oligomers might create intersecting matrices or lipid molecules may cover filaments of oligomer matrices and establish certain conditions for the reactions and mass exchange in the regions of the viscoelastic phase separation required for the evolution of molecules and structures.

As well, authors of the study [608] proved that not only lipids but peptides also can organize themselves as bi-layers, and they have generated the first real-time imaging of the self-assembly process. But there are some other possibilities as well. Duplicating the harsh conditions of cold interstellar space in their laboratory, NASA scientists have created primitive membranous structures found in all living things [575]. These chemical compounds may have played a part in the origin of life. In the lab, the scientists recreated the conditions found in space—which is a cold vacuum—zapping a series of simple ices with the ultraviolet radiation found everywhere. They created solid materials which, when immersed in water, spontaneously created soap-bubble-like membranous structures that contained both an "inside" and an "outside" layer. Those bubbles theoretically may have viscoelastic properties for beginning the process as was described above.

It seems like the replication of molecules happens chaotically in the beginning. That is, almost every newly formed oligomer molecule can be considered a mutation. It is because random replication is forming different oligomers each time. The more stable mutations are inherited, repeated multiple times, and create new mutations. This is the beginning of the biological evolution and self-assembly of the primordial gel, which is generally observed for such systems—for example, in the processes of polymerization. The appearance of following mutations is a branching process, and the influence of preceding mutations on this has a characteristic similar to the flicker-noise. Due to this, a macrostructure of the system that is created is the strange attractor. However the structure of molecular replication must be stationary for the state of the strange attractor to exist. In the primary state, the feedback mechanisms have a linear or a "mildly" non-linear character, but because of this they do not provide enough stability for emerging structures. An example of such "mildly" non-linear structures are Bernard convection cells in a thin layer of liquid losing its structure immediately after the heating cut.

The existence of such a system in the form of the strange attractor is initially provided by the viscoelastic properties of the medium (the gel), which provide strong non-linear feedback mechanisms. To put it another way, the emerging macromolecular structures have a certain range of relaxation times of their mechanical states, which stabilizes the system at large fluctuations of external conditions. The mechanical viscoelastic properties act as a special feedback mechanism by regulating the mechanical stabilization of the bi-phasic gel and of molecular replication through the asymmetry diffusion processes. To a certain extent, the viscoelastic phase of such a system is like a "reservoir" of saved negative configurational entropy. That is, in unfavourable conditions of the external environment, the processes and structures function more or less normally for a certain period of time due to the relaxation time of this negative configurational entropy of macromolecules. Configurational entropy has also the meaning of "informational" entropy, which is responsible for encoding in the system. The emergence of encoding is the essential difference between  living and nonliving systems. The existence of negative configurational entropy in such systems narrow the ensemble of probability of particle states in phase space; that is, it ensures the flow of physicochemical transformations in the most effective way. The effectiveness of the process is determined by the fact that most of the reactions occur in a fairly narrow spectrum. Although the system remains chaotic, exclusion of part of the reaction spectrum in the system reduces energy dissipation and hence increases the energy conversion efficiency. The accumulation of negative configurational entropy in the system contributes to the emergence and evolution of fractal hierarchical structures. The changes (evolution) on the higher levels of the hierarchy affect (in different degrees) the ensembles of probabilities of states of the particles on all the lower rungs of the hierarchy. This in turn provides a unified evolution of the system as a whole at all levels of the fractal hierarchy. Viscoelastic feedback loops in nano-scale, micro-scale, and the scale of the whole system (different relaxation times) are the main contributing factor to the functioning and evolution of such systems.

The basis of evolution is mutations and competition. In such relatively simple systems, these happen on the molecular level. The rate of replication, as in any evolutionary process, provides a larger number of individual molecules and improves survival. Competition between molecules for matter, energy, and space begins at their construction stage. The amount of substances and energy is regulated by the laws of conservation. However, the exchange of energy and matter in live systems is also determined by the second law of thermodynamics, according to which in the process of exchange there is a certain irreversible loss of free energy as a result of increasing entropy. This is probably the reason for the birth of competition at the molecular level. Emerging competition leads to the transformation of negative entropy associated with the flow of mass and energy to negative configurational entropy of macromolecules' structure. We can assume that in the early processes of metabolism, the main role was not played by the existence of some long sequences of reactions, but more likely by the branching of trajectories in the form of a high frequency of mutations when molecules replicate. This mechanism provided a very fast evolution in the primitive stage because the stability of such a system was determined, at least partially, by a continuous changeability. In other words the first stable live systems were very different from the modern structure of a stable system of metabolic reactions. They existed as a constantly evolving structure: the necessary sources of bifurcations were mainly mutations and the non-linear feedback mechanisms provided by the viscoelastic properties of systems. It seems that the evolution of viscoelasticity during the gelation reactions consisted of modifications of the hydrophilic/hydrophobic balance. This allowed the foundation of hydrophobic interactions and hydrogen bonding. So why is non-linear feedback so critical? The thing is, just the existence of long molecules in such a system does not fully differentiate live matter from non-live matter. What is crucial is the evolution of molecules and the hierarchical structures which are based on them—and that only appears when there is non-linear feedback. The mechanochemical non-linearity of the medium determines, for example, the existence of the so-called non-linear signal enhancer and various forms of resonance. In such a mix, more complicated structures can evolve.

It seems that the "live" medium of replicating and interacting molecules was in itself quite an extensive structure of a three-dimensional gel, similar to mucous in modern organisms. Through the mesh of the gel matrix, thanks to low viscosity on the nano scale, there was a free delivery of new low-molecular elements (e.g., amino acids). Based on the chemical or physical interaction with the molecules of the matrix, these elements built new oligomers on the gel structure, some of which were integral to its configuration. This process is very similar to the normal process of gel formation during the polymerization of monomers. Such systems belong to the class of systems in which, along with the replication reactions and diffusion, there are intermolecular Van der Waals forces at work. A distinctive specialty of such systems is that the structures emerging within them can be smaller than the characteristic length of diffusion (the typical distance that can be covered by a molecule from the point of its birth to the point where it will take part in some reaction). That is, microscopically and nanoscopically sized structures that are characteristic of biological systems [348] may materialise.

As a matter of fact, the phenomenon of microscopic self-organization and ordering in the systems of interacting molecules is a well-known fact in the modern physical chemistry of polymers. Periodic structures in such a system could emerge in many ways. For example, we can consider a macromolecule that is composed of two or more chemically incompatible monomers, chemically bound at their ends. Such macromolecules are called copolymers. Chromosomes are an example of a copolymer, exhibiting rigid regions alternating with semi-flexible regions [496]. In a solution or a gel of such copolymers, we would expect a phenomenon: a viscoelastic phase separation in polymers [102], but occurring locally and producing complex periodic structures. Alternatively, in other research [483] it is found that in the initial state of viscoelastic phase separation, an increase in the relaxation time tends to increase the rate of growth of the structure factor, and tends to decrease the wave number of the peak in the structure factor. This means that the sizes of spontaneously emerging structures of gel during its evolution—i.e., with increasing elasticity—decrease but appear faster.

It seems as if early life was in the form of layers of constantly evolving viscoelastic gel on the bottom, or the surface of the oceans or the surface of "fresh water" ponds, or in porous volcanic rocks in water. Whether it was the bottom or the surface depended on the source of energy, either the sun or the thermal springs from the mantle. The source of chemical elements is also important: this could be the waters of the primordial ocean, or the highly concentrated mineral springs, or fresh water minerals dissolved in porous rocks. Moreover, the morphological and rheological behaviour of the self-assembled molecular network formed in this gel is presumably related to the thermodynamic character of the gelation in given external conditions. We can also view such systems in terms of cybernetics. A fundamental phenomenon which is common to polymer molecules is the ability to build new signalling molecules. The links of the gel matrix are a network of non-linearly connected oscillators with the capability to transfer signals. Modulation of the signals by the properties of the system (iteration) is a change in the frequency of mutation. That is, the polymer matrix is a system which records, transforms, and translates information. But it is a quality of some cybernetic systems in which primitive structures have appeared, thanks to molecular evolution in which there are constant mutations. Such a gel or matrix can be viewed as a lot of very simple cybernetic elements, where each one is capable of performing a small number of operations and keeping data about few proceedings. In mathematics such structures are called cellular automata, which carry out analogue processing of information. Analogue devices are not programmable in the general meaning of this word. The programming of their function is determined by the physical-chemical processes that take place in them. The preferential assembly of some types of molecules by replication was happening in these analogue structures, which corresponds to the appearance of the primitive genetic code. Such open self-organizing systems, in cybernetics, belong to the class of adaptive systems. In the process of evolution, they have the ability to adapt to changes in their internal conditions. They can continuously alter their structures and the algorithms of their function as well. Besides, this information processing on a molecular level is a well-known fact in biochemical systems. As an example of similar molecular computing, we can consider the process of the regulation of messenger RNA (mRNA) synthesis by special proteins called transcription factors. This makes it possible to theoretically implement an arbitrary logic relation between the transcription factors and the result of synthesis. Implementation of logic actually establishes a platform for in vivo molecular computing.

The probability of the normal evolutionary emergence of, say, the relatively long RNA molecule, as was mentioned by many researchers of the subject, is close to zero. But thanks to "cybernetic evolution," the process was going quickly relative to any other scenario due to the "explosive" mechanism. It is possible that the results of a study [316] are a distant illustration of such a type of fast evolution on the molecular level. In that study researchers developed a special method to artificially induce accelerated evolution of enzymes in a test tube, which enabled them to

engineer "tailor-made" enzymes. The method is based on introducing many mutations to an enzyme, and the mutated versions were scanned to select for those exhibiting improved efficiency. These improved enzymes then repeatedly underwent
further rounds of mutation and selection for better efficiency. This method can make
enzymes more productive by factors of hundreds and even thousands. In the study
mentioned above, the selection was carried out artificially. However, another study
[222] reveals that for non-linear, non-equilibrium systems with a limited cycle (in
whichthe iteration processes converge) and a hysteretic character of bifurcations (viscoelasticity), we can determine the sequence of microstates (process of "calculation")
by the "future" macro-state of the system. When forming long molecules (oligomers), they can create matrices with physical or weak chemical bonds.

In order for the cybernetic analogue system to function adequately and quickly,
there needs to be a certain directionality of the processes. Otherwise the function of
such a "computer" will boil down to the banal processing of variations that even at
top speed can take too long ompared to the time of the existence of the visible universe. In the study mentioned above [316], the selection was carried out artificially.
However, a different study [222] reveals that for non-linear, non-equilibrium systems
with a limited cycle (the iteration processes converge) and a hysteretic character of
bifurcations (viscoelasticity), we can determine the sequence of microstates (process
of "calculation") by the "future" macro-state of the system. When forming long molecules (oligomers), they can create matrices with physical or weak chemical bonds.

The formation of the first matrices means the formation of the first structures.
The conditions of existence and interactions in such a matrix will be different
from the initial conditions when it did not yet exist. The matrix will only accept
molecules that, in one way or another, conform to the conditions of their stay in
the matrix. Other molecules either stay in the medium, slowly saturating it, or
leave the matrix. This is the emergence of a simple feedback mechanism from the
rejected molecules. In turn, this changes the primary process of oligomer formation
itself due to the altered chemical conditions. This phenomenon is mathematically
called the renormalization of probabilities of ensembles of microstates of molecules
by the final macro-state of the system [223]. So the method of renormalization of
polymeric gels could explain the appearance of periodic structures, as a study [484]
conducted by H.C. Öttinger and his colleagues confirms. In recent theoretical pro-
posals for the process of critical dynamics at polymer gelation [484], H.C. Öttinger
suggested a dynamical renormalization approach within a general framework of
non-equilibrium thermodynamics. By this method he could demonstrate that the
self-similarity of the structure close to gelation can be directly connected to a critically high relaxation time of viscoelastic gel arising.

A good analogy is the autocatalytic reactions, when the result of the reaction
depends on its products. And this is not only an analogy; there are many reasons

to assume that the reactions of the described type of molecular replication happen along the autocatalytic pathway [387]. This is what gives our "computer" such fast processing speed and the renormalization of probabilities that is mentioned in [224]. In the autocatalytic reactions, the set of possible ensembles from molecular replication is quite narrow due to the nature of the process itself. Moreover, in autocatalytic processes the reactions distribute themselves at a much greater speed than in other, similar systems. So, according to [387], for a general chemical process with a reaction rate of 1x10-5s/molecule it would require around 20 billion years to generate a mole of reaction product. In contrast, it only takes 79 µs in an autocatalytic process to produce the same amount of product. But even so, autocatalytic processes alone are not enough to support the conditions for the emergence of life. The viscoelastic feedback is also a key factor.

But what chemicals in primordial soup can catalyze the autocatalytic process of replication? The possible answer is some small organic molecules. The smallest RNA enzyme—ribozyme, a form of RNA with only five nucleotides that can catalyze chemical reactions—is described in an article [506] published in the *Proceedings of the National Academy of Sciences*. According to the author of the work, "It appears that the first catalytic macromolecules could have been RNA molecules, since they are somewhat simpler, were likely to exist [or could be formed] early in the formation of the first life forms, and are capable of catalyzing chemical reactions without proteins being present." The existence of that kind of small-molecule organic catalysts could support initial autocatalytic reactions in primordial soup leading to the formation of gel from relatively short oligomer protein or RNA molecules according to the viscoelastic scenario of pre-life and life origin presented above.

Let's try to explain the meaning of renormalization by using an example not far off the described physicochemical system. Suppose that in the primordial soup there is an area where various chemical reagents are delivered. These reagents can take part in reactions between themselves and/or with products of reactions. We should also assume that these reactions result in the spontaneous formation of oligomer (relatively long) molecules. During the course of the reaction, oligomers accumulate where the reactions take place. Physical and chemical bonds form randomly between them, and when a certain concentration is reached, a primitive matrix develops.

Imagine as well that some oligomers, due to their special properties (chemical composition, stereometry, etc.) can leave the reaction zone through the cellular mesh of the matrix (by diffusion, for example). Let us also suppose that there are molecules that cannot leave the reaction zone. As the concentration of such molecules increases, the reactions that create them are inhibited. At the same time, the concentration of molecules that can leave does not inhibit their production, as their concentration always stays low. In this case the set of ensemble of the possible states of the system (participating reagents, types of reactions, and the types of oligomers being formed) is significantly smaller than the set of ensembles of the possible microstates of the system at the primary stage. This is because relatively few variations of oligomers can be created, and the diversity of chemical reagents and reactions these molecules could be involved in is also reduced. It does mean that in such systems the emerging molecular organization is not only a direct consequence of the reactions pathways involved in the assembly process, but also that reactions pathways are changing from the influence of viscoelastic feedback of evolving supramolecular structure. Thus we can say that there is a renormalization of probabilities of the ensembles of the microstates existing by the macro-state of the system. In terms of non-linear analysis, this can be called the first iteration. I want to note that I have not made any improbable assumptions and that this is quite possible to have been happening in the early evolution of life on Earth. If there is some sort of autocatalysis, which can appear for many reasons, then the speeds of reactions in such a system can increase considerably and narrow the spectrum of ensembles of microstates.

It is also important to note that the initial gel-matrix formed by oligomers will perform the function of a parametrical filter because the diffusion processes within the matrix are dynamically asymmetric. That parametric filter regulates diffusion in the solvent phase depending on the effective size of the molecules. This means that at the exit point of such a reaction area we have a modulated distribution of molecular concentrations according to their physical and chemical properties. This probably leads to the next iteration: the formation of more complex structures. For example, it could lead to a denser packing of oligomers, or to matrices with more regular configurations, or more cross-linking bonds, etc. As the system becomes more complicated, the variety of ensembles of possible microstates narrows. This brings about the materialization of structures similar to those structures formed from a class of amorphous oligomer molecules known as "intrinsically disordered proteins" [357]. Unlike typical proteins in the cell, intrinsically disordered proteins do not adopt a stable globular form in isolation. Rather, they are like a messy, unfolded string of yarn, whereas typical globular proteins more closely resemble yarn neatly knit into complicated and functional shapes like that of a glove. The equilibrium states of molecules in such an ordered structure are known as an alpha helix (like the coil of a phone cord) called the F state. Researchers [357] have seen how the protein shape changes soon after binding to its partner molecules. This means that in our dynamic conditions of a continuous supply of new molecules through the initial matrix, similar new states might provide the system with many more modes of regulation and selection of incoming molecules. Each molecule aggregates differently within different structures, and thus has unique properties and modes of interaction with other long- and small-molecule partners. The rise of viscoelastic properties in oligomer matrices considerably complicates the character of molecular interactions in such a system. In terms of the dynamics of non-linear processes, there is an addition of viscoelastic non-linearity of media in the equations describing the processes.

As a result of viscoelastic phase separation [102], the stability of the matrix structure will grow. Even when the process of generation of new molecules is halted, the matrix can theoretically exist indefinitely if the external environment and the stability of phase separation will permit it. The existence of viscoelasticity is an important feature of processes that distinguish them from "normal" inertial non-linear processes, which generally cause dissipative structures to rapidly disappear when no more mass and energy are fed into the system. Because of viscoelastic separation of phases, there will be different areas in the matrix mesh with different diffusion speeds for particles of varying sizes. The tension condition of the matrix will influence the "mesh size" (molecular interactions) of the matrix structure, and will modulate the diffusion processes accordingly.

It seems that we have to view diffusion and reactions in such a system as ongoing in a fractal phase space due to the range of relaxation times in the viscoelastic-gel-forming phase. One study [354] demonstrates the possibility of stable auto-oscillating processes in such a medium. In the intricate viscoelastic media containing reacting particles, it will seem as if the emergence of lasting periodical (in space and time) sets of physicochemical reactions are leading to the production of long and complicated aperiodic molecules. These have many cooperative spatial modes that could be described by the set of relaxation times for each mode. Theoretically spatial-temporal patterns can be generated due to the reaction and diffusion of a number of chemicals in Newtonian (not viscoelastic) fluid in so called reaction-diffusion driven instability. The reaction kinetics in such a system are stabilizing and diffusion—a homogenizing process. But in the case of nonlinearity of media (viscoelasticity) where the processes are taking place, some additional spatial-temporal pattern linked with relaxation times of viscoelastic media (instead of homogenization of processes) emerge, because hysteresis type of bifurcations.

On every level of the hierarchy in such systems, if the process of multiplication and mutation has started and there is strong non-linear feedback, the system will go towards the closest stationary state with minimum entropy. The level of entropy for these systems is determined by the macro-state of the system [223]. This directed drift towards minimal entropy of the stationary macro-state will be the aim of the cybernetic system to allow "channelling" of the iteration process. When molecules multiply, this channelling will expel those with extremely exotic mutations that will not bring the system to a stationary state. If this is the case we don't have to rely on such an improbable event as the accidental appearance of an RNA or DNA molecules and the four-letter code. With our approach, they appear as the result of the cybernetic iterative process and the mutation of complex molecules with the appearance of complex hierarchical structures with viscoelastic feedback loops. These structures that appear as the result of this evolution change the non-linear dynamic properties of the system. As the system is changing, at each stage it comes to some stationary state—a hierarchical structure, which in itself is a "building block" for creating the next structures. The appearance of these hierarchical structures in itself is an iterative procedure during the work of such an analogue computer. It seems that the four-letter code of the nucleotides appeared gradually, through the appearance and evolution of two- and three-letter codes.

Probably, at the initial stationary stage, there was a two-letter code. The following iterations created three- and four-letter codes. In a way this is reminiscent of the synergetic structuring of energy and matter flows in dissipative processes and the appearance of physical structures. But in this particular case, the emergence of physical regularity has the properties of the informational cybernetic structure in the form of the genetic code. However, we should bear in mind that the stationary macro-state does not completely determine the set of microstates during the channelling of the evolution of systems. Theoretically, the "channel" can have an infinite ensemble of these microstates or evolutionary trajectories that lead to the final stationary state. Also, the data from a study [233] where similar systems were researched shows that they have the ability to independently generate "chaos," which is determined by the final stage [222]. It does not depend on external noises and fluctuations, but is determined only by the non-linear dynamics of the system—i.e., by the imminently inherent properties in conjunction with the external boundary conditions.

Also, this approach pushes back the time of appearance of life in the form of the current four-letter code, to the appearance of the preceding two-letter code. As there are no principal differences in the primitive life forms which are based on the two-, three-, or four-letter codes, the physicochemical mechanisms that take place in such a gel molecular computer need to be studied. We can only note that a big role is played by hydrophilic and hydrophobic interactions that are inherent to amino acids, as everything happens in water. In such gel structures, self-assembly is possible; this is proved by study [327] which examined the movement of nanoparticles in narrow channels. The study shows a mechanism that leads to nanoparticles displaying self-assembling behaviour in a series of experiments in vivo and in silico. The researchers found that the disturbances of the fluid induced by each flowing and rotating particle drives neighbouring particles away, while the migration of particles to localized streams due to the momentum of the fluid acts to stabilize the spacing between particles at a certain distance.

In essence, the combination of repulsion and localization leads to an organized structure. The researchers also found that by simply adding short regions of expanded channel width, the particles set could be re-self-assembled into different structures in a controllable and potentially programmable way. Even more complicated structures should appear due to the deformation of channels happening according to the viscoelastic mechanism, if we view the gel as a porous-elastic medium with relaxation times of the wall. In such a system, the frequent positioning of mutant molecules and their self-organization always lead to the continuous modifications of gel properties—to the evolutionary transitions that lead to the emergence of higher-level systems hierarchy through the assembly of components. The emergence of the first small fragments which resemble the modern micro RNA and peptides complicated the structure of the gel, and the structure of these fragments and peptides to the point of the appearance of first proteins and primitive RNA. As indirect confirmation of the approach described above, it could be very interesting to look at the results of a study [497] where the engineered scaffolding from RNA molecules in the cell has boosted output of enzyme biosynthesis production in two orders of magnitude. RNA scaffolds in the cell organize and concentrate the enzymes, interim products, and final products in the reaction zone exactly as described above in viscoelastic gel.

It is important to note again that on the physicochemical level of such a cybernetic system the above "channelling of calculations" happens because during the complication of structures and selection, not all the mutant molecules can take part in the evolution of the gel structure. Also, it seems that in complicated structures, because of structuring of chemical and physical interactions, the spectrum of created mutants must be narrowing. All of this leads to the narrowing of the set of ensembles of particle states for selection and the acceleration of evolution within each iterative "procedure."

The above approach is based on the assumption that the process of molecular evolution in the appearance of live matter is determined by the emergence of oligomer molecules capable of creating viscoelastic matrices. The structural self-organization in time and space of such processes—i.e., polymerization of monomers into polymer systems—was studied in [328]. It turns out that in such systems, as the words of John A.Pojman describe, spatial pattern formation can occur on the microscopic level to form structures reminiscent of those seen in the Belousov-Zhabotinsky reaction and biological systems. Regular spirals and bubbles can be formed in the self-propagating polymerization process. According to another of his publications [506], "if spatially bi-stable reaction systems are operated in size responsive chemo sensitive gels, the size changes can provide a feedback which beyond plain reaction diffusion instabilities can be the source of new self-organizing phenomena, referred to as chemo mechanical structures." In relation to spatial pattern formation in gelation processes in [506] Pojman made an interesting observation: "Typical phase separation leads to a two-phase disordered morphology. Multiphase polymeric materials with a variety of co-continuous structures can be prepared by controlling the kinetics of phase separation via spinodal decomposition using appropriate chemical reactions. By taking advantages of photo-crosslinking and photo isomerization of one polymer component in a binary miscible blend, researchers have been able to prepare materials, known as semi-interpenetrating polymer networks, and polymers with co-continuous structures in the micrometre range." But these are the hierarchical systems we talked about earlier, which occur thanks to the non-linear viscoelastic properties of the macromolecular gel during polymerization both from synthetic and natural monomers—e.g., the amino acids in the primordial soup. In my opinion Pojman's researches [328, 506] are an excellent illustration of my suggestion that life appeared through the emergence and evolution of the viscoelastic gel.

Theoretically, such an approach could lead to new methods of polymerization in viscoelastic gel. As I have suggested, the emergence of self-assembled molecular and supramolecular structures of polymers is inevitable. That polymer, via some controlling mechanical and chemical factors of the gel matrix and polymerizing monomers, could be designed to have sophisticated complex properties and functions. As the first step in our in effigo experiment, we can prepare a gel with certain viscoelastic and molecular properties of a matrix and with low-molecular phase by making a solution of monomers (possibly more than one type). Next we can initiate a polymerization reaction in the monomer phase by any known method. In such a case, the properties and structure of the new polymer or co-polymer will be dependent on the viscoelasticity and matrix structure of the initial gel, where both variables can be theoretically quite easily controlled. It means that by applying the concept of processes leading to the origin of life to industrial technology, we could develop methods to design new, complex polymer systems with new material and functional properties that will be radically different from any ordinary plastics. By this method, if we will use two or more layers of different gels with different monomers, we can produce extremely complicated multilayer polymeric or co-polymeric materials with some predetermined functionality. If we add to the processes other variable parameters such as kinetics of reactions in different phases, diffusion rates, and viscosity, the possibilities to manage the structures of such materials are almost endless.

The equation [349] of the bi-phasic reactive physicochemical system is distinguished by two characteristic times responsible for the starting of the reaction t and for the influence of the feedback reaction T. The analysis shows that when t<T (i.e. when there is a relatively large relaxation time of the feedback mechanism [viscoelastic medium]), this gives rise to complicated structures in the system—oscillating dissipative structures with parameters that change in time. If we imagine that there are several relaxation times or they are represented by a continuous spectrum (which complies with the real biopolymer systems), then there is an even stronger complication of systems. And if we imagine that all this is taking place not in the bi-phasic model medium but in a multi-phasic medium, then this guarantees even more complicated structuring of the systems which gave birth to life on this planet.

The common feature between all living beings known as viscoelasticity could be the basis for the evolution of the chemicals systems on which life is based. It could, in fact, be the essential feature controlling evolution at the prebiotic level. Because all living beings are made from the same basic chemical molecules that have more than likely been around since the beginning of life on Earth, researchers are studying the life origin chemical reactions as, for example, small sugar (glucose) binding sequences that occur in all protein and peptides like insulin [622]. in the influence of emerging viscoelasticity of media could provide fundamental insights into diseases such as diabetes and cancer, and processes of the feeding. Another team of the researchers from the University of York [623] studying the origin of life found that by using common left-handed amino acids, it is possible to catalyze the formation of right-handed sugars. Such sugars might evolve due to the molecular complementarity in the formose type reaction. The chemical path known as the formose reaction, discovered by Aleksandr Butlerov in 1861, is a potential route from the simple molecules, which might have been present on Earth before life began, to the sugars essential to life. The formose reaction begins with formaldehyde, thought to be a plausible constituent of a prebiotic Earth, going through a series of chemical metamorphoses leading to many sugars, including ribose, which is a key building block in DNA and RNA. However chemists found that in general the formose reaction produces a very tiny amount of ribose but instead a lot of other sugars that lack any biological use. This means that this reaction must take place under special conditions favorable for the production of ribose. Such conditions could theoretically exist at the lattices phase of peptide viscoelastic gel formed in the prebiotic soup. The gel lattices might have an effect on selection by chirality and atomic composition of complementary short strands of sugar molecules incoming from the reaction zone of different chemicals. Theoretically it could explain how essential-for-life carbohydrates originated and why the right-handed form of them dominates in nature. For life to have evolved, we have to have a moment when non-living things become living. Everything up to that point, according to common beliefs, is chemistry. A main point of this chapter that is not pure chemistry, but rather mechanochemistry in viscoelastic phase, was and still is the controlling factor of the first emergence, existence, and functioning of all living beings. Researchers in the field of life origin are still a long way from being able to assemble living cells from scratch in the laboratory. According to biochemist David Deamer of the University of California, Santa Cruz [576], life began with complex systems of molecules that came together through the self-assembly of nonliving components. A useful analogy for understanding how this came about can be found in combinatorial chemistry, an approach in which thousands of experiments are carried out in parallel by robotic devices [576]. But the question arises, what worked as the "robotic devices" in nature around 3.5 billion years ago.

Another researcher, New York University chemistry professor Robert Shapiro, published a book called *Planetary Dreams* in 1999 in which he argues that the simplest kind of life may arise as a result of organic chemistry reactions and the physical processes of self-organizing systems whenever the right constituents and conditions exist: a liquid or dense gas medium (not necessarily water), a suitable energy source, and a system of matter capable of using the energy to organize itself. This case also raises the question: what is specifically "a system of matter" that is "capable of using the energy to organize itself"? Without specifying the type and properties of the "robotic devices" of D. Deamer or "a system of matter" of R. Shapiro, these approaches in my opinion do not have the heuristic value.

The approach I propose indicates that emerging viscoelasticity in a system of organic chemistry reactions with diffusion in prebiotic systems leading to the appearance and evolution of hierarchical self-organized structures is laying the foundation of life, is responsible for the work of the "robotic devices" of D. Deamer in nature, and is "a system of matter capable of using the energy to organize itself " of R. Shapiro.

In the literature there are quite a number of objections to the mechanism of life origin in the primordial soup from simple chemicals, such as peptides, by the polymerization reaction [307]. Let's consider some of them.

For example, stated that the difficulty of activating amino acids and forming long peptides under primordial conditions is one of the great obstacles to the origin of life. But in a concentrated solution of 1 M (mol/l) of each amino acid, the equilibrium dipeptide concentration would be only 0.007 M [307]. And first we must suppose that in primordial conditions reacting systems existed with diffusion that could greatly change dipeptide concentration due to different diffusion rates for different components of the system. Also, the formation of dipeptide gels is possible in the range of quite low concentration—less than 1 percent [308].

High temperatures on the surface of early Earth, as many researchers advocate, would accelerate the breakdown of gel. The famous pioneer of evolutionary origin-of-life experiments, Stanley Miller, points out that those oligomers are "too unstable to exist in a hot prebiotic environment." But again, according to [308] at a concentration of 2 mg/ml, the dipeptide forms a stable aqueous gel at 60 °C. We cannot say this is low temperature for primordial conditions.

Some researchers argue that the heat also destroys some vital amino acids and results in highly randomized polymers. But I am proposing in this book a model of life origin in only the very early first stage of the prebiotic system that as result of "viscoelastic evolution" may create something like very primitive proteins or RNA; only further evolution of them may lead to the creation of the first life and make it suitable for polymers. The same is generally related to a problem of chirality of the amino acids.

Also, some researchers argue that amino acids in the primordial soup would be impure and grossly contaminated with other organic chemicals that would destroy peptides. Another argument is that it is a chemical impossibility for the primordial soup to accumulate large quantities of "condensing agents" for absorbing excessive amounts of water, which could slow the process of oligomerization.

In the model proposed in this book, the emergence of peptide gel implied the existence of a viscoelastic phase separation (spinodal decomposition). In the peptide phase of gel lattices where the reactions take place, the water content and the concentration of dissolved chemicals will be significantly lower than the average for a system. That is, it probably at least partially eliminates the question of the "condensing agent" and of the influence of chemicals that are hazardous for the existence of peptides and amino acids. In addition, according to some researchers, vital amino acids (for example, cytosine) are too unstable to have existed on a hypothetical prebiotic earth for long, and amino acids would be too diluted in primordial soup to actually interact with each other. And even if the amino acids could have formed dipeptides, they would soon hydrolyze. The existence of the dipeptide matrix skeleton phase could help to overcome all of these problems because processes will take place in concentrated phase, forming some kind of compartment physicochemically isolated in some degree from surrounding diluted phase. Also, viscoelasticity due to the evolution of the complexity of gel structures theoretically could support the tendency to form the coded oligomers/polymers required for life as opposed to random ones. During molecular and supramolecular evolution of gel structures, the randomness of forming polymers will go down. But that is not everything! Even the existence of randomness is playing an important role because the randomness is actually a source of mutations, which is the driving force of evolution in the viscoelastic gel model of life origin proposed.

Moreover, my proposed approach could possibly explain the emergence and domination of "left-handedness" in chirality for amino acids in life forms. This can occur because only some by the chirality of the amino acids preferable for liquid-crystal structures of supramolecular peptide gel matrix phase. Most scientists believe that the choice of chirality in life on Earth was purely accidental, possibly based on carbon; if an alien form of life exists somewhere in the universe, there will likely be another form of chirality. But some scientists are looking for fundamental reasons for the choice of chirality on Earth, such as the weak interaction, which might violate the symmetry of matter on a very fundamental level. If so, then the weak interaction is directly related to the existence of Higgs boson, which itself has never been observed but still is a major goal of the Large Hadron Collider at CERN. In this case, the emergence of life from the observed predominant chirality is based on the most fundamental properties of our universe.

The viscoelastic approach helps eliminate a contradiction highly discussed among scientists between the deterministic scenario of an initial prebiotic evolution and the contingency theory. The emergence of viscoelasticity is deterministic, but the emergence of something like RNA is still a probabilistic event with highly enhanced probability in viscoelastic media. Change in sets of initial conditions on the Earth at the time of the prebiotic state theoretically may lead to different forms of life. It means that there may exist some spectrum of trajectories of many different life forms. Theoretically, the approach I suggest may help to specify a range of parameters in initial conditions. Physical-chemical interactions may limit environments that are suitable for life and the possible range of their variation, restricted by only one condition: the possibility of a viscoelastic state of matter within this range. The emergence of viscoelasticity inevitably will lead to life formation. All main events in life formation such as self-organization and others could happen, be sustained, and evolve in the presence of viscoelastic media. Thus, the proposed viscoelastic hypothesis of the initial phase of life, though not absolutely complete and in need of a great number of refinements, offers the potential to explain some of the effects associated with the transition of non-living chemicals to the prebiotic state. The essence of the viscoelastic model is that a prebiotic molecular evolution from the simple molecules that were widely available on the early Earth, reacting into more complex chemical structures of oligomers, eventually led the system to the state of viscoelastic gel, which helped to evolve the first RNAs like molecules. Viscoelastic feedback loops in such a prebiotic system are the major selective evolutionary factor. During the process viscoelastic phase separation in the some degree ordered structures of the gel lattices could have at least a partial effect on the selection of complementary molecular strands coming in from the reaction zone pools of relatively random sequences. Viscoelastic phase-separation could have been a first mechanism leading to finely non-random macromolecular synthesis.

It is also important to note that the above arguments and approaches do not guarantee the evolution of the primordial soup into RNA- or DNA-coded life. The approach that is described here only explains that the initial production of relatively long molecules that form gel matrices inevitably leads to evolution of structures based on them, provided there is enough energy and matter for chemical reactions. It seems the life that we see based on the four-letter code is in fact the result of such evolution. This approach explains that in order to form the first nucleotide, there is no need to wait for an accidental combination from all the forming monomers. In addition, the creation of an artificial biopolymer matrix of DNA molecules has led to interesting results, which I think to some extent confirm the participation of viscoelasticity in the cybernetics of life. In a study [443], the researcher showed that DNA-based neural networks—a viscoelastic soup of interacting molecules of biopolymers—demonstrate the ability to take an incomplete pattern and figure out what it might represent; this is one of the brain's unique features. Such biopolymer systems with basic, decision-making capabilities could have powerful applications in medicine, chemistry, and biological research. Theoretically in the future, they could operate within cells, helping to answer fundamental biological questions or diagnose a disease. Biochemical processes that can intelligently respond to the presence of other molecules could allow engineers to develop increasingly complex chemicals or build new kinds of structures, molecule by molecule. Intelligence of such systems in my opinion is determined, at least partially, by their viscoelastic properties.

From what we have discussed, we can conclude that the resulting life form may not be the only possibility. Of course, if we consider our cybernetic system of mutating molecules aiming towards the "closest" stationary state possible at each iteration, then theoretically other life forms can be created. That's why the resulting final stationary state determined by the four-letter genetic code is probably also affected by certain initial physical conditions: the chemical composition of the proto-planetary cloud out of which the solar system was born; the local composition of the proto-planetary disc out of which Earth was born; the level of solar radiation or radiation from the Milky Way in this region of the solar system; atmospheric pressure; mineral content and viscosity of water; and so on. If the external conditions were different, then this possibly could have led to a different final stationary state with some sort of N-letter genetic code with an N value other than 4. It may be that this alternative possibility—the fact that mutations and evolution are possible not only in RNA or DNA systems—can be illustrated by study [302]. This study showed that prions (large protein molecules) can have many mutations and that they can bring about evolutionary adaptations such as drug resistance, which was previously known to only occur in bacteria and viruses. The prions are able to adapt and survive in a new host environment, producing distinct self-perpetuating structural mutations that provide a clear evolutionary advantage. Because in the stated process one of the main roles is played by viscoelasticity, then theoretically it is possible to assume the existence of different kinds of life based on different chemical elements. The only necessary condition is that the emergence of viscoelastic matrices must be possible.

Nevertheless, it is also important to note that according to the data from study [329] the modern-type system for the double-stranded DNA replication evolved independently twice: in the bacterial and archaea/eukaryotic lineages of cells. In this case we must admit that the convergence of evolution to the four-letter coding system is very strong. However, it is quite possible that during evolution of any kind, a certain first stationary state is reached that allows for survival of the species, but not with the lowest possible level of entropy. This is probably why our life expectancy does not greatly exceed the age of sexual maturity. Here the evolutionary mechanism is not remotely interested in achieving the following, hypothetically possible stationary states, in which we could live much longer. Contrary to popular belief, maybe evolution does not select the fittest from the available, but stops its selection at the fittest from the accidental. Here the term "accidental" implies the stationary states that have not only been realized due to the inherent properties of live systems, but also due to chance factors of the external environment. That is, as soon as a new species appears it takes up all or most of the space and does not allow similar species to emerge. This assumption may not be that absurd. If a mutation has assumed its rightful position as a life form, then the rules of the game change for every consequent mutation that could potentially be even better. And the "consequent" mutation is in a less advantageous position by definition, as the "previous" mutation has had free space to allow it to spread and develop. Here the advantages of the new mutation may not be sufficient to outcompete the previous species and survive.

So in this hypothetical situation, evolution will resume when the external conditions change; however, it will move in a slightly different direction. We can say that the evolutionary trajectory has changed. The same can be said in the case of some species diverging. If one branch fortuitously gained an advantage and colonized the habitat, then potentially, the more advanced branch may not survive. In such a scenario, continued evolution may not translate into ever-increasing fitness. The evolutionary variety is very well represented by the unique properties of animals and plants which evolved in isolation—for example, in Australia or the Galapagos islands. Also this idea is supported by a recent study [458] and demonstrates that the struggle for survival is stronger between more closely related species than those distantly related. The authors found that species extinction is happening more frequently and more rapidly between species of microorganisms that were more closely related. While evolutionists generally accept the premise, the study contains the most direct experimental evidence yet to validate it. Darwin's idea of "survival of the fittest" has been called into question as well in an article published not long ago [467]. Researchers challenged this old paradigm by showing that biodiversity may evolve where previously thought impossible. Authors have been watching hundreds of generations of bacterial evolution, about 3,000 years in human terms. It had been believed that the genome of only the fittest bacteria would be left, but that wasn't their finding. The experiment generated unexpected genetic diversity in which both the fit and the unfit bacterial cultures coexist indefinitely. Also, in an another slightly older study[468] published in the Royal Society journal *Biology Letters*, researchers provide further evidence that random genetic mutations over millions of years may also play a powerful role. The authors conclude that evolution should not necessarily be called "survival of the fittest" but rather "survival of the luckiest," which is closer to the aforementioned principle of "survival of the fittest from the accidentals."

Researchers at the University of Texas at Austin, led by Dr. Matthew Cowperthwaite, show in their computer models that life may not always be optimal and natural selection may not produce the best organisms [462]. The team developed numerical models of RNA molecules evolving by mutation and natural selection. Their computer models show that the evolution of optimal organisms often requires more than one mutation, in fact a long chain of interacting mutations, each arising by chance and surviving natural selection. As the authors explain, "Some traits are easy to evolve—formed by many different combinations of mutations. Others are hard to evolve—made from an unlikely genetic recipe. Evolution gives us the easy ones, even when they are not the best."[463]. Indeed, what the authors of this study call "easy ones" I call "the fittest from the accidentals."

However, perhaps nature has found a method to circumvent the above problems of evolution through the mechanisms of simultaneous (or cooperative) multiple mutations [404]. The study already demonstrated such a possibility that if mutations can happen cooperatively in twos, threes, or even more, then cells could make large evolutionary leaps through "fitness valley" and reach a different "fitness state" by acquiring multiple mutations simultaneously. It is most likely that the evolutionary mechanism is "interested" in the maximal speed of evolution. If we assume in effigo that it is possible to continue mutations on every level of our evolutionary predecessors and somehow to stimulate the survival of those with longer life spans, then it is quite possible that the representatives of intelligent life would have lived much longer. I would personally have sacrificed a couple of billion years of evolution on Earth in order to reach this goal. From the other point of view, the situation could have worked out so that intelligent life would have not appeared at all. And this, you must agree, is not such an ideal outcome. So the increase in the life span of intelligent life here on Earth in the current conditions is not so much a challenge for evolution, but for nooevolution.

But if we are to return to evolution of the non-linear dynamics of our gel system, then the oscillating reactions of polymerization in continuously mutating systems may have stopped periodically, probably because of the disassociation of the gel, and started up again along new trajectories depending on the mutations of the components. The superposition of the continuous mutation of the chemical composition, viscoelastic properties, and the dynamic non-linear processes—pulsations, convections, inversions of reaction fronts, etc.—caused the creation of highly complicated hetero-phasic structures in the process of gelation. The potential complexity of such structures could have greatly exceeded the polymerization structures described in the study [328]. Because this is true for any biological system, then the changes in their metabolisms happen because, for instance, the incoming chemical signals change the non-linear characteristics of the system, and thus, the characteristics of the stationary state of the metabolism. The behaviour of such systems, because of the non-linear effects, is extremely complex and determines the growth in the variety of reaction cascades under the influence of external signals. The key factor of evolution of our primordial gel is the high frequency of mutations. One study [263] illustrates the fact that a high frequency of mutations provides for fast evolution speeds; the research was about cancer of the pancreas. Pancreatic cancer is one of the most malignant aggressive types. This is because the pancreatic cancer genomes often contain a distinct pattern of mutation that could reflect changes in the repair mechanisms of the cancer cells. The pattern of mutations is dramatically different from those found in many other cancers. Although genome instability is common in cancer, further study has revealed the high dynamic nature of this variability and its role in rapid progression of disease in the body. Instability is the driving force behind this evolution that allows the tumour to adapt to new organs and tissues. Metastases of this cancer are therefore like a family: the different deposits of the tumour are genetically related to one another, but also have distinguishing genetic features that make them unique. The scientific team of this research suggests that the "galloping" mutation rate that develops produces cells that, because of specific mutations they acquire, can colonize other organs very quickly. Different combinations of genes are needed to survive in different tissues. The study [263] shows that even in one person's cancer, intensive cell mutations can quickly evolve the genomes which are specific for different organs. In these situations doctors should not prescribe treatment for one tumour, but for several genetically distinct tumours. Probably the similar high frequencies of mutations in the primordial gel lead to its fast evolution. The final result of the cybernetic evolution of the live gel was possibly the creation of first primitive RNA molecules, which relatively quickly lead to the emergence of the simplest DNA in primitive cells. So the "live ocean" in Stanislaw Lem's "Solaris" is a theoretically possible scenario for the evolution of a pre-live gel in the ocean of a planet with suitable environmental conditions. It is important to note that it is most likely that the primordial gel will be localized on the bottom of the ocean or coastal area ponds, in both cases filling up cracks and porous rocks, of volcanic origin for example, for the best protection from external influences and the abundance of minerals and chemicals in young volcanic rocks.

As the live hetero-phasic gel was evolving, its viscoelastic properties of gel or/ and lipid bi-layers may have contributed to the formation of the prototypes of the first primitive cells. As the mutated molecules and border conditions were accumulating in the cells or the primordial gel, new chemical reactions arose that provided for the stability of structures and their gradual change to the modern state of the metabolism. This is characterized by a stable set of chemical reactions at a relatively low frequency of mutations in particular media. In order to maintain their own stability, these structures have to develop their own metabolism that is not connected to mutations. In such live systems, evolution of the internal metabolism manifests itself on a different hierarchic stage—on the level of populations of primitive cells, their organelles, or other structures.

To a certain degree, the work of this early cybernetic system of evolution continues even now. In any case, the unity of live matter on the planet is obvious and the process of biological evolution of live systems happens on a cooperative basis. If a mutation appears in any part of the biocoenosis of the Earth, it spreads to the whole of the planet. In the world of bacteria, this happens extremely fast compared to the higher animals. Today the anthropogenic influences affect the "super computer" of the global biocoenosis in a way that is not foreseen by its "programming." Therefore it is difficult to predict what the reaction is going to be. One of the possibilities is that the computer will attempt to self-correct, to remove the irregularity in programming or remove a malfunction of one of the parts—i.e., remove humanity or the larger part of it after having identified it as the source of the problem. What the principle of this removing will be is uncertain, but there are many possibilities. An example could be a lethal virus which spreads so rapidly there is no time to create a vaccine. Another example is the frequent unexplainable mass deaths of birds. It's quite challenging to explain the mechanism of this phenomenon and how thousands or tens of thousands of one species can die in the same moment.

We have to be on our toes so that the same does not happen to humans! The nooevolutionary activities of man are becoming deeply hostile to the interests of the planet's classical biocoenosis. Exponential population growth coupled with the unstoppable growth in consumption can lead to a catastrophe comparable in scale to a massive asteroid impact. This disaster could be the way the planet can deal with the hypertrophy of its separate parts—in this case, the intelligent primates and their cultural biosphere. It is quite possible that the trigger for this process of antagonism between live beings and the cybernetic system could be the extreme population growth of the intelligent species and a lowering of biodiversity on Earth. The fact that biological evolution is no longer the dominating factor in further development of humanity is obvious evidence for the existence of this antagonism. However, fundamentally we are and will remain products of biological evolution. This is why it is important to keep to our biological essence on this stage of transition as far as lifestyle and nutritional choices are concerned. This means that if we adhere to the principals of anthropic nutrition, then there is no need to dogmatically follow the principle "back to nature." Currently, the conditions in some regions of the planet such as Africa, Oceania, and South America are not suitable according to epidemiological and a few other characteristics for people from North America and Eurasia.

Apart from that, the human activity of genetically modifying live systems can hypothetically send signals into the biocoenosis that will dramatically change its evolutionary trajectory. It could turn out that there will be no place for Homo Sapiens in the new evolutionary process. For example, there is a lot of research and development being done in the area of genetic manipulation with microbes. The spectrum of these manipulations ranges from yogurts to biodiesel to who knows what else. If the genie of these genetic manipulations manages to get out of its bottle, out of DANONE, for example, then all the activity of Monsanto with genetic modifications of plants will seem like child's play. By the way, some biologists [384] now suggest that a sixth mass extinction may be under way, given the known species losses over the past millennia. The differences between fossil and modern data show the extent of the current extinction crisis—results confirm that the current extinction rates are much higher than would be expected from the fossil record.

The main purpose of this chapter is to stress the fundamentality of viscoelasticity of biological objects for evolution from molecules to structures. Currently, viscoelasticity of macromolecules in biology is basically seen as the trivial consequence of having long biomolecules. This work views viscoelasticity as a fundamentally necessary and sufficient property of biopolymers that is needed to start and maintain evolution and to keep matter in its live state. In a hypothetical situation, if macromolecules did not have viscoelasticity, then maybe the live state could not exist. These seemingly speculative assumptions about the emergence of life, in my opinion, have a certain heuristic value as far as understanding the function of live systems is concerned. Possibly, the "explosive" cybernetic evolution of live gel formation set in place a variety of molecular mechanisms that provide for the huge variety of life forms which we have here on Earth. The analogy of the Big Bang is appropriate here. Scientists also assume that the properties of particles, fields, and possibly laws of interaction that were born in that moment of the universe's creation determined its modern structure. Apart from that, within the framework of these assumptions there is no need to discuss exobiological reasons for the creation of life, such as panspermia or the idea of God. Furthermore, this approach allows for the possibility of "other" evolution with different external conditions based on some other chemical elements, and possibly new unexpected metabolic pathways and life forms. And also we must bear in mind that this approach opens the possibility for life emerging per se as the result of the existing physicochemical diversity of the universe, its laws, and non-linear mechanical properties which fill up its matter.

## 3.4   Life of viscoelastic beings

Viscoelastic properties possibly played a key role as a primary feedback mechanism with delays (relaxation times), which created homeostasis of the simplest organisms and gave it the stable visible form. Dynamic equilibrium in the homeostasis of organisms is strongly influenced by the existence of viscoelastic properties, which exist in all structures of any organism, starting from macromolecules, organelles, cytoplasm, and elastic cell membranes and ending with elasticity of organs, tissues, bones, and non-Newtonian rheology of blood and lymph—everything in the organism.

The viscosity of the cell media could have been determining some inertia of the biochemical processes. However, as the viscosity is the invariant, it does not depend on the rate of matter transferral. And apart from that, there are two types of viscosity or diffusional asymmetry in live systems, which will be discussed below. In addition, the macro viscosity of live systems is determined by their viscoelastic properties. The reversible self-sustaining reactions in linear viscous media have a self-oscillating character, such as in the Belousov-Zhabotinsky (BZ) reaction with the periodical change in the colour of the solution. In the viscoelastic polymer media, if we imagine that this kind of reaction depends on the viscoelastic state of the macromolecules, the situation will be somewhat different. Such a self-oscillating reaction will have the characteristics of waves distributed at a certain speed. The speed of distribution would have been determined by the characteristic relaxation times of the solution. Moreover, if in our BZ reaction in effigo we would have a reactive polymer solution with a spectrum of relaxation times, then the reaction waves would have had fractal structure: at the wave (zones of colour change) there would be areas with shorter waves, which correspond to shorter relaxation times in the relaxation time spectrum of the solution. Within these shorter waves, there are even shorter waves, and so on. This would be true for all the sets of relaxation times which exist in the solution. But such a situation is realized in all the systems of organisms, where there is a huge amount of polymer molecules in vivo in an even larger variety of states and network interactions with an almost infinite set of relaxation times.

The Belousov-Zhabotinsky reaction here is used here strictly to illustrate our in effigo experiment as it is quite simple to visualise. In the simplest case of this reaction there is an elementary non-linear (the model for all live systems) self-supporting process, which is characterized by the periodic colour change. However, in the live cells and media of the organism, there are many thousands of simultaneous reactions with proteins having different molecular lengths and shapes, containing other biomolecules, and thus the set of relaxation times from the practical point of view is infinite. This means that in a real cell, there exists simultaneously a huge fractal set of such wave reactions that are determined by the parameters of feedback function with the so-called fading memory (i.e., the influence of distant events is less than that of closer events). Due to boundary conditions and to the fact that the cell's existence is stable in time, we can assume that in the space of its metabolism a superposition of standing waves emerges, which in vivo manifest themselves in the form of a hetero-phasic cell structure. The spectrum of relaxation times within the system stabilizes the system of reactions and excludes the possibility of a resonating oscillating regime of chemical reactions emerging, which would have been very probable, if the properties of the cells were to be determined by linear viscosity.

Most people look at a living cell as a kind of thing with more or less a permanent structure. But they are badly mistaken. Nothing is permanent in the cell. All parts of the cell are the supramolecular structures formed from polymer or lipid molecules that are unstable and are constantly in the process of association-dissociation. The prevalence of the association process is facilitated by an existence of relaxation time of the viscoelastic phase in the process of the dissociation of the supramolecular structures. That is, while the processes of molecular association in supramolecular structures from biopolymers dissolved in the cytosol, nucleosol, in membranes, etc. occur relatively quickly, the dissociation of structures take more time. The reason for this hysteretic behavior is the difficulty in the movement of polymer molecules or their end caps in supramolecular ordered structures, similar to the liquid crystals, due to the interaction of neighboring molecules in the viscoelastic phase separation state. We have to also consider the structure of the cell as a superposition of many oscillatory chemical reactions, such as the BZ reactions in a viscoelastic medium with an exceptionally wide range of the relaxation times. Each relaxation time, or to be more precise, narrow parts of its spectrum, is responsible for all cellular structures and determines the fractal dimensional state of the cells in the phase space.

Such supramolecular structures are generally very small organelles, much smaller than any part of the functioning cell. The transitory process of micro-phase of supramolecular structures to the bigger macro-phase structures can take place if the local concentration of supramolecular structures exceeds certain limits necessary for phase transition. Concentration growth in this locality is a narrowing of the ensembles of the probabilities of particle states due to the high concentration of supramolecular organized micro-phase. If transitions from small supramolecular structures to organelles like supramolecular macro-phase and back have histersis character, then the micro-phase system may "jump" to the new macro-phase state.

Self-oscillating polymer gels can demonstrate such processes because they are materials that continuously change back and forth between different states—such as colour or size—without stimulation from external sources. As we can see from experiments {413} and computer modelling {414}, gradients in cross-link density in a polymer matrix—a system very similar to developing polymer/oligomer matrices with spatial distribution of viscoelasticity—lead to the bending and self-propelled motion of active gels. Oscillating polymer gels undergoing the BZ {413} reaction provide an ideal model for probing the interaction between chemical energy and mechanical action. The BZ reaction in gel is converting chemical oscillation to mechanical oscillation of the polymer matrix. You can watch in a video {415} how these self-sustained pulsations are caused by the BZ chemical reaction interacting with the polymer matrix. Without any external influence, oscillation from this chemical reaction can develop within the material or cause the entire gel itself to pulsate mechanically for several hours, provided it is small enough. In such systems mechanical stresses in a matrix could be converted to chemical changes in mechanotransduction processes. Similar results were obtained in the recent research [504,505] where authors introduce autonomous gel as an actuating device converting chemical energy into mechanical motion. The polymer gels prepared had cyclic chemical reaction networks, and it was shown that the polymer gels generate periodical motion sustained by the chemical energy of the oscillatory BZ reaction. The scientists successfully made synthetic polymer gel move autonomously like a living organism and even achieved conveying the object by using the peristaltic motion of the gel. Although the gel is completely composed of synthetic polymer, it behaves as if it were alive. Again, there is proof that the basis for the function of living organisms is the viscoelastic properties themselves.

But gel oscillations play a very important role in cells. For example, they are known to be associated with successful progress of embryo development. Professor Magdalena Zernicka-Goetz of the University of Cambridge led a team of researchers to look for ways to assess fertilized embryos more effectively, allowing fewer to be implanted. In her experiments on animals, she found that when a sperm enters an egg, the egg's jelly-like innards would start to pulsate soon afterwards [541]. The oscillations seen in the egg's cytoplasm are caused by the influx of calcium ions after an egg is fertilized [542]. This temporal control in recently fertilized embryos is executed by modulating the force balance between two states: the strength of the biopolymer matrix and feedback loops based on signaling cascades of cytosolic calcium concentration. This concentration inside the cell is probably increased initially by an influx of calcium ions with or in the presence of sperm and decreased by the binding of calcium ions with the biopolymer matrix of the cytoplasm in the zygote.

A feedback loop emerges in a signaling network. An initial fluctuation of one component will propagate through the loop until it feeds back into itself and amplifies or reduces this initial fluctuation. When the intensities of positive feedback (amplification) and negative feedback (suppression) are balanced, a temporal state such as regular oscillations of the parameters of the system can emerge. In the case of the embryo observed in [542], the binding of calcium ions with biopolymers of the matrix may change their viscoelastic properties and consequently the mechanochemical state of gel. This leads to a change in volume of the gel. When some polymer molecules or parts of them elongate, the bonds break between the calcium ions and polymer so the gel returns to its previous state, but the concentration of ions in the cytoplasm increases and the process starts again. Such a feedback loop, I believe, is involved in the emergence of embryo pulsations, as is presented in the research [542]. Obviously, pulsations in human embryos are similar to those seen in animal eggs because of similarities in the embryos' biochemical properties and size. Furthermore, a team led by Princeton Professor of Molecular Biology Ned Wingreen recently reported interesting results in the journal *PLoS Computational Biology* [543]: contrary to the observed fact that embryonic cells are developing in synchrony, they are in a computational model prone to descend into chaos. From the beginning embryonic cell cycles are initiated by a wave of calcium ions that emanates from the fertilization site and prompts the embryo's cells to divide and duplicate—to oscillate, in biological terms. Wingreen and his colleagues found, however, that the natural spread of oscillation in the computational model is unstable and would result in an erratic patchwork of missed and incomplete cell divisions. But then the question arises: if there is potential for chaos in silico, how does the embryonic system avoid it and get into a synchronized state in vivo? To observe an in vivo outcome, oscillations should be arranged throughout the embryo in the stable wave pattern. Computational simulation produced the picture that the wave of calcium ions sparks cellular changes that set embryonic cells onto the path of synchronized development. As it should be in a typical BZ reaction, the stable wave emerges with characteristics (amplitude and frequency) throughout the embryo depending solely on the viscoelastic mechano-chemical properties of the cytoplasm and maybe even cell membranes.

Viscoelastic approach is a good framework to explain the oscillations observed in life egg [543] and non-life systems with BZ reaction in artificial polymer gel in [415]. Of course, particular classes of feedback loops could present very distinctive behaviors such as color changing wave or rate of oscillations change in the gel upon influencing an endless number of parameters in real gels. But important that as we see from the results [541,542, 543], from the very first seconds of any life, including our own, its viscoelasticity plays a crucial role. A good illustration of the importance of BZ reactions is the cell mitosis. According to a description of mitosis in study [618], microtubules extend from one of two spindle poles (the microtubule organizing center) on either side of the cell and attempt to latch onto the duplicated chromosomes. In addition to microtubules from both spindle poles that attach to the chromosomes, astral microtubules that are connected to the cell cortex—a protein layer lining the cell membrane—pull the spindle poles back and forth within the cell until the spindle and chromosomes align down the center axis of the cell. Then the microtubules tear the duplicated chromosomes in half. The process of mitosis is extremely important for the cell. Gaining or losing a chromosome during mitosis may lead to cell disorders and to whole body diseases like diabetes or cancer. Authors [618] noticed that when the spindle oscillates toward the cell's center, a partial halo of the protein dynein lines the cell cortex on the side farther away from the spindle. As the spindle swings to the left, dynein appears on the right, but when the spindle swings to the right, dynein vanishes and reappears on the left side. In this process dynein is anchored to the cell cortex by a complex that includes the protein called leucine-glycine-asparagine-enriched (LGN) protein. The stationary dynein acts as a winch to pull the spindle pole, and the microtubules and chromosomes attached to it, toward the cell cortex.

Researchers found that when a spindle pole comes within close proximity to the cell cortex, a certain signal protein emanates from the spindle pole, knocking dynein off of LGN and the cell cortex, stopping the spindle pole's forward motion, and freeing dynein to move through cytosol to the opposite side of the cell. These oscillations continue with decreasing amplitude until the spindle settles along the cell's center axis. Researchers also noticed that a layer of LGN extends all around the cell cortex, except in the areas that are closest to the chromosomes. As the chromosomes swing back and forth, the cortex area cleared of LGN changes in response. Because dynein needs to anchor to LGN, this cleared area ensures that dynein can only attach and pull to the right and left of the aligning chromosomes, rather than from above and below. Thus, we see that the most fundamental process in the life of the cell: the mechanism of the spindle settle in mitosis is controlled through the mechanical forces originating from the BZ type chemo-mechanical reactions in macromolecular viscoelastic media of binding the dynein on/off of LGN and both of them on/off the cell cortex. Such systems generate negative entropy. In reviewing the non-Markovian processes characteristic of viscoelastic media, the pre-historic account naturally introduces an additional feature, negative entropy, to measure how the order and complexity of processes emerge in chemical systems.

The Markov process is a random process for which the further evolution of a known state of the system does not depend on any previous states. Of course, the Markov processes are not suitable to describe polymer pre-biologic systems. The non-Markovian nature of things exists in phenomena such as the dependence of stresses in viscoelastic liquids, not only on the current rate of deformation, but also on the spectrum of relaxation times of the system. The non-Markovian approach proved fruitful for the polymeric non-Newtonian fluids in particular, and to describe processes with hysteresis. When the temporal and spatial variations of viscoelastic properties of gel emerge in oscillatory chemical reactions, a huge variety of increasingly complicated structures appear. The characteristic features of non-Markovian processes are certain stable spatial and temporal cycles that manifest themselves as a BZ reaction in the gel, as sustained mechanical pulsations.

The fact that these mechanical pulsations take a dominant role in cell function also follows from the results of some experimental studies. For example, the reconstitution of the actin matrix that is actively put under stress by molecular motors has been studied in research [495]. It shows that in an active and cross-linked matrix pulsatile, collective modes develop, which depend critically on the interaction strength of the molecular motor and filaments. This type of interaction is in some sense very similar to the BZ reaction when the dynamic process of movement of molecular motors through the gel is locally changing its viscoelasticity. In turn, the feedback mechanism causes changes in the dynamics of the molecular motor. Indeed the force exertion inside of dynamically cross-linked actin matrix of cell are expected to induce a force-rate-dependent rupturing of the crosslinking points and thus the locally induced (through the dynamic matrix rearrangements) dynamic changes of the viscoelastic environment will also affect and modulate the molecular motors dynamic. Thereby, it would be conceivable that the dynamic viscoelasticity of the matrix enables the appearance of collective pulsatile modes.

These pulsations can theoretically both slow and accelerate the passage of a molecular motor with the cargo, due to the effect of viscoelastic peristaltic pump, which may be especially noticeable for long clusters such as the motor-signal protein.

One of very important processes of pulsations is the work of our heart. Abnormal heart rhythms—arrhythmias—are killers. They strike without warning, causing sudden cardiac death, which accounts for about 10 percent of all deaths in the United States. That is more than 250,000 people. "The current anti-arrhythmic drugs do not prolong life," said Björn Knollmann, M.D., Ph.D., associate professor of Medicine and Pharmacology and the senior author of the current report. "There's a large need for new approaches to anti arrhythmic therapy" [564]. In their quest to understand how irregular heart rhythms arise—as a way to find new molecular targets for treatment—Knollmann and his colleagues have focused on the role of calcium inside heart muscle cells. Calcium ions are central to the contractile cycle not only in embryos as we see above, but for heart tissue as well. Electrical impulses based in calcium ions regularly circulate through cardiac tissue and cause the heart's muscle fibers to contract. The calcium ions interact with proteins called troponins, part of the heart tissue contractile filament. The interaction of ions with troponins regulates filament contraction. In a healthy heart, these electrical impulses provide smoothly travel contraction-relaxation waves through tissue. But mutated troponin had been linked to inherited forms of hypertrophic cardiomyopathy (HCM), which carries a high risk of sudden cardiac death. HCM is perhaps most famous as a cause of sudden cardiac death in young athletes, but it can affect individuals of any age [564].

The researchers examined the heart rhythms of mice expressing various troponin mutants that cause HCM and showed that the mice develop ventricular tachycardia (a particular arrhythmia). The risk for this arrhythmia was directly related to the degree of calcium sensitization caused by the troponin mutation: the higher the calcium sensitivity, the greater the arrhythmia risk. Troponin mutations associated with HCM increase the sensitivity of the troponins to calcium—they bind calcium ions more readily, which activates the filaments more easily and results in an increase in mechano-responsiveness and stronger contractions. Increased calcium ion sensitivity has also been found in acquired heart diseases, such as heart failure, that have a high incidence of sudden cardiac death. Authors [564] proposed that increased filament calcium sensitivity contributes to arrhythmia susceptibility.

In another study [565] of the factors that lead to fatal cardiac rhythms, a team of Canadian researchers has shown that the importance of communication applies to the heart cells. But what does cell communication mean here? It is based on calcium ion signals getting from one cell to another under electrical forces through the cell and intercellular space. The heart cells are electrically excitable cells, maintaining voltage gradients across their membranes by means of metabolically driven ion pumps, which combine with ion channels embedded in the membrane to generate intracellular-versus-extracellular concentration differences of calcium ions. In the mechanosensitive ion channels, a signal can be produced when forces acting within the viscoelastic lipid bilayer [273] rise to a level sufficient to produce conformational changes in the channels, forming proteins and thereby altering their conductivity. In a young person, these adaptive properties of membrane ion channels big enough to pump large numbers of ions through the cell's surface. But with age and mainly with long-time disanthropic diets and lifestyle due to epigenetical changes of proteins and lipid matrices forming channels is losing their flexibility. And when there is a sudden increase of heart activity, ion channels cannot provide enough ions to the cell's communication pathways. Modulated by viscoelastic properties of ion channels, electrochemical signals travel from the cell membrane and activate pathways for electrical impulses carried by ions to other cells through intercellular space. The viscoelastic properties of heart cells are very important for adaptation of cells' ability to transfer electrochemical signals in a wide range of intensity of mass transfer.

Equally important is how the ions travel and bind in itercellular space. In one study [565], researchers examined chick-embryo cardiac cells grown as a sheet of tissue. In the first two days after this arrangement of cells was created, spiral waves often formed in the tissue. When the researchers sprinkled the sheet of cardiac tissue with a drug that impairs communication between the cells (by partially interrupting electric signal transfer by calcium ions), they observed that the rotating spiral waves broke up into multiple rotating spirals. This breakup of spiral waves in the two-dimensional sheet can sometimes develop into troublesome and is believed to be similar to the 3D electrical patterns that cause human hearts to undergo ventricular fibrillation, a potentially fatal cardiac rhythm that often occurs when the ion communication channel between cells is impaired. Impairment of signaling between cells in this experiment means that the drug changes the viscoelastic property of the intercellular matrix and conditions of ion transfer. As we see from the studies cited above, we have three reasons for failure of the heart waves: loss of adaptation of membrane ion channels, troponin mutations changing in binding processes of calcium, and changes in intracellular matrix. All three reasons are emerging solely by changing the viscoelastic properties of cells, intercellular matrix, and filaments.

Loss of elasticity of channels in the heart cells and properties of the intercellular matrix creating as it look – new heterogeneous (for electrical impulses) tissue with different viscoelastic properties in the heart tissue, namely with different relaxation

times. Two types of tissue coexisting with two different relaxation times of elec-
tromechanical impulses leads to the emergence of two propagating front waves in
the heart tissue, which in turn leads to observed disorder in the heart pulsations.
Additional heterogeneity in the tissue – appearance fragmented areas of diseased and
normal tissue giving as well new additional waves propagating quite independently
and destroying normal coherent pulsations. When neighboring cells have strong
signaling and mechanical interaction with one another, electrical waves quickly pass
through the tissue unobstructed; when interactions are weak, wave propagation is
completely blocked. At intermediate levels of interaction, electrical waves break up
into multiple spiral waves.

Troponin sensitivity to calcium also changes the relaxation time of the contraction-relaxation processes of filaments. That leads to stronger heart beats that ultimately lead to faster destruction of heart tissue. If some parts of cells in the tissue
have different troponins, then two and more waves in a heart might emerge.

How to avoid both of these nasty outcomes? Medical researchers are trying to
find drugs with calcium desensitizing activity. But inevitably that drug will have
huge side effects like almost all drugs because there may be a situation where some
part of the tissue needs desensitization for calcium, but another does not. It is even
more difficult to change ion channels' functioning. But understanding the effects of
calcium ions' communication between cells and influence of viscoelastic properties
of matrices provides insights into the electrical malfunctions that are suspected to
lead to heart disorders, and may ultimately suggest strategies for avoiding them. I
personally think that it is almost impossible to find effective drugs or drug combinations if two or all three reasons for arrhythmia are acting together. The best possibility for getting long lasting effects is to change your diet to the anthropic style
and to try to restore epigenetic changes if they have not gone so far. If we can restore
the ability of neighboring cells to strongly interact with one another and prevent
epimutations of troponin, electrical waves will pass through the tissue normally at
almost at any age.

It is also interesting to note that viscoelastic properties apparently play an important role in the work of our brains. The study [566] focuses on identifying the mechanisms by which cells sense mechanical stress and transduce it into a biochemical
signal, developing a three-dimensional viscoelastic model of mechanotransduction.
Cells are exquisitely sensitive to mechanotransduction and actively respond through
a variety of biological functions including migration, morphological changes, and
alterations in gene expression and protein synthesis. The study found that due to
viscoelasticity of the cell matrix, the time-dependence of the sinusoidal force manifests itself in the stress concentration within the cell. The faster loading rates lead to
higher stress concentrations in the cell. Stresses transmitted via cell surface receptors
and the intracellular proteins that connect them to the intracellular matrices can also

induce conformational changes in them and, as a result, potentially alter matrices' binding affinity to signaling and other molecules or their diffusion rates.

Neuroscientists have long believed that vision is processed in the brain along circuits made up of neurons, similar to the way telephone signals are transferred through separate wires from one station to another. But scientists at Georgetown University Medical Center [567] discovered that visual information is also processed in a different way, like propagating wave packets. They visualized wave-like patterns in the neuron network of the brain's cortex using a new method called voltage sensitive dye imaging. That collective pattern emerges from the activities of billions of neurons in the visual areas. The wave packet patterns obviously play an important role in initiating and organizing brain activity not only in visionary processing but also in the memory functioning in the brain. This means that brain cortex processing information by the wave packets spreads in the entire network of neurons in some areas of brain—some sort of parallel computing. It appears that the brain processes all sensory information this way.

The same team uncovered spiraling wave propagation, resembling the spiraling pattern in the heart tissue pulsation [564], in animal epilepsy models. The authors think that this multiple spiral pattern in an area of damaged neural tissue can generate disorderly waves that invade healthy brain areas and start a seizure attack. This means that disorders such as epilepsy could be viewed not just as mis-wiring in the brain, but as an abnormal electrochemical wave pattern that invades the entire brain network. The neurons are electrically excitable cells, maintaining voltage gradients across their membranes by means of metabolically driven ion pumps, which combine with ion channels embedded in the membrane to generate intracellular-versus-extracellular concentration differences of ions such as calcium. In the mechanosensitive ion channels, signals can be produced when forces acting within the viscoelastic lipid bilayer rise to a level sufficient to produce conformational changes in the channels forming proteins and thereby alter their conductivity. Modulated by the viscoelastic properties of the ion channels, electrochemical signals travel along the cell axon and activate synaptic connections with other cells when they arrive.

Viscoelastic properties of neurons are very important for neurons' ability to transfer electrochemical signals in a wide range of amplitudes. In a young person, these adaptive properties of intracellular matrices of neuron and ion channels in the membrane are big enough to pump large number of ions through the neuron. As we age, the rise of amyloid deposits in cells and/or of the rigidity of channels forming proteins due to the changes in their structure (chemical bonds or molecular modification), the diapason of the variation of the intensity of ion pumping through them decreases. In such cases, for example, fresh visual images which consist from the many wave packets with high amplitudes cannot pass through the neural network (or at least through the many elements of it) and cannot be remembered. This happens with people in old age. But at the same time, the old images of long-term memory still might easily circulate in the brain cortex network because of the low intensity (amplitude) of the wave packets of old images. That is perhaps why old people remember some events which happened a long time ago but forget those that happened, say, half an hour ago. Of course we are now quite far from fully understanding how the brain handles these waves. But even in the brain—the most complex system in the universe—shall act simple enough physiochemical mechanism. Viscoelastic properties of brain tissue are playing the most fundamental role in its work.

We must also bear in mind that the relaxation times in such a system are related to the characteristic diffusion rates of the reagent molecules or the reaction sites. A good way to illustrate how elastic macromolecules influence cellular reactions is the existence of a duality in viscosity of solutions susceptible to structuration. This dynamic asymmetry of diffusion manifests itself on the macro scale as a well-known phenomenon of non-linear dependence of macro viscosity of polymer solutions from the sheer rate. In nano-scale via much lower viscosity of polymer solutions than in Macro scale. Nano-viscosity was discovered more than half a century ago during the research of sedimentation of small particles in ultracentrifuges. To everybody's surprise, it was found that nano-sized objects move in polymer solutions in such a way as if the viscosity of the liquid was tens of thousands of times lower than what was registered on the viscometers. Studies [99,100] show that every heterogenic hydrodynamic medium, and especially polymer solutions, has a characteristic molecular size, where the transition from nano-viscosity to macro-viscosity takes place. This size is determined by the mesh scale of the polymer solutions and gels. So if the effective size of a molecular coil is, for example, 10 nanometres, then any other particle of larger size that moves within the solution experiences macro-viscosity. If the particle is smaller, then it experiences nano-viscosity. The transition from nano-viscosity to macro-viscosity in the vicinity of the characteristic size happens exponentially—i.e., very sharply. For example, it has been shown in the study [121] that the anomaly of viscosity (diffusion)—i.e., the difference between macro-viscosity (macro-diffusion) and nano-viscosity (nano-diffusion)—is considerably reduced if the intra-cellular medium undergoes an osmotic stress, as when it is dissolved in water.

The diffusion of "small" molecules (e.g., carbon monoxide) in the protein macromolecular clusters happens at relatively low activation energy [234]. This means that the carbon monoxide moves quickly through the deforming mesh of the matrix of coils. The high degree of deformation is caused by the elasticity of areas of polymer chains (conformational freedom) that create the mesh. However, as soon as the size of the small molecules increases, then the rate of diffusion drops considerably. In principle, when considering such a diffusion in a polymer gel matrix to account for the viscoelastic properties of the matrix, we can use a concept analogous to the concept of the Kuna segment (relative to the moveable parts of polymers) for polymer solutions and polymer melts. Then the number of Kuna segments in between the knots of the matrix will determine its flexibility, and this means also its capability for elastic deformation. This capability for elastic deformation of the matrix meshes will in turn determine the character of diffusion of signal protein molecules and clusters.

The existence of dynamic scale asymmetry of the diffusion processes has a key role in influencing many processes in organisms, such as the speed at which proteins are transported within the cells and organelles. Around the cell nuclei there is a highly viscous medium which is densely packed with macromolecules that have a very low diffusion rate. However, thanks to nano-viscosity, for the smaller protein molecules, the rate of diffusion does not differ much from their rate of diffusion in a clean solvent—in this case, water. Plus, the results of study [112] demonstrate that changes in the properties of proteins in the vicinity of the cell nucleus, by influencing the properties of the matrix, change the processes of cell growth and proliferation.

A similar phenomenon is already used in electrophoresis. Electrophoresis is a process which enables the sorting of molecules based on size, molecular mass, spatial configurations, secondary structures and charge. Using an electric field, molecules (such as proteins, DNA, and RNA) can be made to move through a gel usually made of polyacrylamide or a biopolymer called agar. The molecules being sorted are dispensed into a well in the gel matrix. When a voltage is applied, separation takes place, and the bigger molecules move more slowly through the gel matrix than the smaller ones. The viscoelastic properties of macromolecular matrices have a greater influence on the rate of diffusion of signal proteins, the size of which is close to the characteristic. In this case their molecules "squeeze" themselves through viscoelastic protein channels. In other words, we can imagine the cell cytoplasm as a three-dimensional matrix that is formed by protein macromolecules. The aforementioned channels are in fact the mesh of this matrix, which contain the nano-viscous phase of the cytoplasm. The characteristic scales of the matrix mesh vary towards the cell organelles. The changes in elastic properties of the macromolecular matrix under the influence of different factors (e.g., the chemical or physical bonding of macromolecules due to incoming disanthropic chemical compounds or ions) can considerably influence the rate of diffusion of proteins of characteristic size. As I have mentioned, the change in the rate of diffusion in this area has an exponential dependency, so the changes can be considerable. Mechanical tension generated within the protein matrices of living cells is emerging as a critical regulator of biochemical functions in diverse situations ranging from the control of chromosome movement to the morphogenesis of the cell.

The diffusional asymmetry greatly influences the character of function of the nucleus. For example, one study [175] of the mechanism of mRNA transport in the matrix of the nucleus shows that although the transport of mRNA-protein (mRNP) complexes from transcription sites to nuclear pores moves by diffusion, their mobility is curtailed upon depletion of adenosine triphosphate (ATP) from the cell. It has been observed that mRNP particles tend to get stalled when they are passing through a high-density chromatin matrix. Upon ATP depletion this tendency is accentuated, resulting in a larger population of stalled particles. A possible explanation is that ATP depletion alters the chromatin matrix in such a way that the larger mRNP molecules stall.

What kind of structural changes in chromatin may bring this about? Some researchers have observed that chromatin becomes less flexible upon ATP depletion. Therefore, the high chromatin flexibility in a matrix, or in other words the reduced elasticity modulus of a matrix, enables the frequent escape of mRNP particles from their stalled states. Thus, ATP depletion will result in an increase in the fraction of stalled particles without affecting the diffusion constant of mobile particles. ATP depletion results in reduced "flexibility of pore size" (change in viscoelasticity) in the matrix "mesh," so overall, mRNP particles are less mobile. This view is supported by known facts of reversible curdling in chromatin upon ATP depletion.

The scientists [611] discovered that certain proteins, called extremely long-lived proteins (ELLPs), which are found on the surface of the nucleus of neurons, have a remarkably long lifespan. Results of the research suggest the proteins last an entire lifetime, without being replaced. ELLPs make up the transport channels on the pores of the nucleus membrane. Their long lifespan might be an advantage if not for the wear-and-tear that these proteins experience over time. Unlike other proteins in the body, ELLPs are not replaced when they incur aberrant chemical modifications and other damage. It does mean that their viscoelastic properties change over the time. This weakens the ELLPs' ability to regulate the flow of ions, proteins, and so on through the transport channels due to the lessened elasticity of chemically modified ELLPs.

Breakdown of protein viscoelastic homeostasis in nucleus pores lead to declining cell function not only in the brain, but also in other organs tissues like the heart and muscles. Authors of the study [611] also suggest nuclear pore deterioration might be a general aging mechanism leading to age-related defects in nuclear function. Nuclear pore deterioration might alter DNA transcription and gene expression programs. Scientists [436] have deciphered the structure of an essential part of Mediator, a complex molecular machine that plays a vital role in regulating the transcription of DNA. Mediator is a gigantic molecular cluster composed of twenty-five proteins arranged in a polymer-like structure.

It is a well-known fact that all cellular function is controlled by the genetic information through transcription that is enabling the production of necessary proteins. As cellular operations proceed, signals are sent to the DNA asking that some genes be activated and others be shut down using the Mediator—the transcription regulator that accepts and interprets those instructions, telling the enzyme RNA polymerase II when and where in the nucleus to start the transcription process [437]. Motion (diffusion) Mediators in the matrix structures and cell nuclei, clearly defined by the diffusion coefficient due to the phenomenon of viscoelastic phase separation in the matrix of biopolymer gel—i.e., the mechanical condition of biopolymer matrix—provide critical control for the velocity and concentration of Mediators in the target areas of cells and organelles.

For a better understanding of how our genes are controlled by the viscoelastic properties of cell matrices, it would be useful to look in the findings of a study [421] which analysed how proteins work as teams to control genes in the cells. Around half of human DNA is devoted to regulating how the genes that make proteins carry out their tasks [420,421]. According to the study more than 11,000 transcriptional co-regulators in human cells—co-activators and co-repressors—control how, when, for how long, and to what degree genes are expressed. Proteins are the final functional units emanating from the genes. They carry out all the biochemical reactions needed for a cell to live, grow, and function. Co-regulators are the helper proteins that actually decode the information in our genes [420]. This immense number of co-regulators, with their cooperative movement and final distribution in target places of the cell, are at least partially regulated by the many viscoelastic matrices existing in organelles trough viscoelastic diffusion asymmetry. This indicates that viscoelastic properties have a paramount importance for precise regulation in decoding genes. It means that not only the composition of the proteome (the entire set of proteins produced by a genome) arranges how it all works in the cell, but the mechanical states of the different biopolymer matrices are extremely important as well. If a gene makes a few percent too much or too little of a protein, then the cells do not function well. But the spatial concentrations of proteins in target places are just as important. Thousands of genes must be cooperatively fine-tuned in the cell. The cell is a master that through the producing co-regulators and viscoelastic regulation of their spatial distributions.

The same is true for complex clusters of protein macromolecules of the mitochondrial respiratory chain. Each time when we take a breath, our blood cells transport oxygen to the mitochondria, where they convert the nutrients from food to energy. This process is respiration. Interruptions have been linked to a number health conditions: diabetes, cancer, etc. Despite the fact that respiration is so basic, there is much scientists have yet to understand about how it is regulated [507]. The clusters of respiratory chain on the surface of the mitochondria, the matrices of which contain up to forty protein molecules that are intricately intertwined, control energy transfer by the use of special proteins through the nano-sized channels [106]. The viscoelastic properties of the cluster, or to be more exact, of the channel walls, have to play a critical role for the processes of nano-diffusion as well. Respiration depends on proteins synthesized outside the mitochondrion and imported into the cell, and on proteins synthesized inside the mitochondrion from its own DNA [507].

It has been shown that in the case of diseases such as diabetes and cancer, there is an increase in nano-viscosity in the cells. However, the effect of the change in the state of viscoelastic elements on these changes in viscosity has not been researched. The importance of determining the mechanism that affects the changes in nano-viscosity in diseased cells is critical as it can govern the possibilities and methods of correction of the cell state and the search for medicine. Similar processes happen during lymph and blood flow. So the non-linear viscous properties of blood are also determined by the characteristic scales of the viscoelastic elements of blood—cells and cellular clusters. For example, when blood flows through vessels, the balance of energy expenditure in the flow includes the dissipation in less viscous layers of blood plasma, energy expended on elastic deformation (elongation in the direction of flow) of viscoelastic elements, and the lower energy expenditure due to the orientation of viscoelastic elements. Finally, with the increase in sheer rate, viscosity is reduced, which allows for more efficient organ function. The influence of viscoelastic properties is important for blood and lymph flow and the biochemistry of processes in the capillaries. Abnormal haemorrheological property changes in erythrocyte deformability and in blood viscoelasticity may play very important roles in the development of microangiopathies (increased permeability and fragility of capillaries) in diabetes [174]. Compared with diabetics, healthy people have a much higher retinal capillary blood flow rate because, to some extent, the blood viscosity is significantly lower at all sheer rates. The main reason for this is that healthy red blood cells have much higher deformability. Incidentally, red blood cells infected with malaria stiffen (have a bigger elasticity modulus) as much as fifty times more than healthy red blood cells [309]. As in diabetes, this drastically changes the flow of blood in the arteries and in the capillaries. Again, this example shows that changes in cell viscoelasticity alter the function of organs.

Another interesting point is the thermodynamic aspect of viscoelastic behavior of practically all substances in the organism, starting from the intracellular scale. In a live state, all the viscoelastic elements of macromolecules, the protein matrix of cell cytoplasm, membranes, blood cells during flow, and tissues of the organism on the whole are in a state of tension due to forces of various natures. In the cytoplasm this is determined by molecular forces that are responsible for the mass transfer of elements. In the membranes this is controlled by the cell's internal pressure and the mechanical influence of the environment; in blood cells, by sheer pressure as they flow in the vessels; in tissues, by forces of gravity and muscle contraction. However, during the intensification of processes within the cell, the increased inner osmotic pressure in the nano phase on protein matrices makes them stretch. The level of these deformations is supported by continuous processes in the organism and is determined by nutrition, one of the main sources of energy keeping us alive. The deformed tense state of macromolecules has lower entropy than the relaxed state. During the function of all the hierarchical structures of the organism there may be fluctuations in the delivery of energy and necessary substances. This could be due to a lack of incoming elements into the cell. Accordingly, the concentration and diffusional flow of proteins smaller than the average size is reduced. The change of the diffusional flow lowers the pressure exerted on the larger macromolecules. In this case the viscoelastic properties of the macromolecular protein matrices can play a stabilizing role by, for example, reducing the sizes of the deformed matrix meshes in order to maintain the rate of nano-diffusion within the cell.

Another possible mechanism is related to the so-called viscoelastic phase separation [100] as a result of the dynamic asymmetry of diffusion processes. The viscoelastic properties of the large protein macromolecules in the cells play a key role in structuring and organizing the nucleus [103]. The viscoelastic phase separation reduces entropy. By this we can appreciate the importance of the existence of viscoelastic properties in macromolecules in the production of organelles and the function of the cell cytoplasm. It is also known as spinodal decomposition in viscoelastic fluids [483]. The fact that the viscoelastic gels' mesh space are in a bi-phasic state allows us to view them as a porous-elastic medium, where the macromolecular protein matrices make up the walls containing pores through which small proteins and other components can diffuse [164]. When this porous-elastic material is deformed, there is a deformation of the elastic phase, which creates a local pressure and changes the characteristics of the pores, potentially inducing local convection currents of the cytosol, which is the nano-viscous phase. The distribution of the waves of deformation within the cell has shown that in the space of a few seconds there is a temporary storage of mechanical energy in the elastic state.

Cellular and intercellular liquids cannot be described by the simple model of a homogeneous viscoelastic liquid due to its multiple phases of the neighboring protein matrices interacting with each other within the cell. However, diffusion through the pores of this matrix is determined considerably by the elastic characteristics of the matrices. This is why from now on we will talk about the viscoelastic characteristics of intra- and intercellular liquids, taking into account their multiphasic structure and the dependency of diffusion and currents on their relative sizes. Study [164] also shows that "long-living" pressure gradients of elastic deformation of cell matrices can induce convection currents within the cytosol and influence the mobility of the cell and speed of molecular transference. Obviously, this will influence diffusion and flow of proteins of characteristic size the most. It has also been shown that the processes of mechanical relaxation in the intracellular medium are not determined by one relaxation time, but by a whole spectrum of relaxation times.

Thus, from the results of this study, we can conclude that the viscoelastic properties of cellular media arise from different matrices of the cell and influence its biochemical and mechanical behavior. Mechanically induced biochemical activation (a.k.a. mechanochemical transduction) in cells and intercellular space enables an extraordinary range of physiological processes, which include the senses of touch, hearing, and balance, as well as growth and remodeling of all tissues and bones.

Also, this approach allows us to view viscoelastic matrices of an organism as a reservoir of energy, which helps to dampen the intensity of fluctuations of the metabolism. Or in other words, we can call it the reservoir of negative entropy. Often, when the dominating role of entropy is stressed, the elasticity of polymers has an entropic nature, and each molecule is in itself an entropic spring. The influence of viscoelasticity is more significant for the diffusion of proteins of close to characteristic size, and there are many of them in the cell responsible for critically important functions like all the other cell components.

To look at the cell as a whole or to focus on the behavior of the cell clusters, on the basis of in silico modeling of the viscoelastic properties of cells, study [165] has shown that biomechanical viscoelastic properties determine the full range of interaction and communication between cells, and patterns of growth and development of cells and tissues. In study [416] experimentation on non-living model cells that contain no DNA could help point to clues explaining the mysterious process of abiogenesis— the formation of life from non-living matter, an event that happened at least once during Earth's history. Scientists have demonstrated on non-living model cells that the structure and mechanical properties of a cell's membrane and cytoplasm may be as important to cell division as the specialized molecules—such as enzymes, DNA, or RNA—which are found within living cells. In these "cells" they established that cytokinesis—the process by which a cell splits to become two distinct cells—is possible in the absence of complex cellular components, such as genes and enzymes. In nature, living cells split into two asymmetrical cells with very different compositions and different "fates." For this seemingly complex task to be accomplished, some internal chemomechanical and mechanochemical processes must regulate both the reorganization of cellular parts and the maintenance of polarity, which is the property of a cell to exhibit distinct front and back "sides" with specific placement and distribution of cellular components.

Researchers built model cells, allowing water, lipids, and polymers to assemble into mimics of the most basic constituents of real, living cells, such as a membrane and cytoplasm. They then altered the osmotic pressure outside of the "cells" by adding sugar, which forced them to divide in a way that is reminiscent of how living, biological cells split under natural conditions. Like a biological cell, the model cell was designed to exhibit asymmetry in both its membrane and its cellular interior. The membrane asymmetry was modeled using two distinct lipid domains, while the cell's interior mechanical properties were modeled using two viscoelastic solutions of distinct polymers called polyethylene glycol (PEG) and dextran. These polymer solutions formed distinct domains, or compartments, on the inside of the model cells, with the dextran-rich compartment containing a higher concentration of a particular protein. The researchers observed that when the asymmetric mother cell divided, one cell inherited one lipid domain surrounding the PEG-rich interior, and the other cell inherited the other membrane domain surrounding the dextran-rich interior, which contained the larger portion of the protein. They also found that when they varied the relative size of the two lipid domains, one cell got both types of membrane and the other got only one type. The researchers note that the modeling technique seems to suggests that simple chemical and physical interactions within cells—such as self-assembly, mechanical (viscoelastic) phase separation, and partitioning—can result in seemingly complex behaviors like asymmetric division, even when no usual cellular components are present.

Around sixty years ago, scientists simulated early-Earth conditions in laboratories and demonstrated that many amino acids, which are fundamental constituents of proteins, can form through natural chemical reactions. From the exciting results of these experiments [416], we could come to the conclusion that chemical and spatial organization leading to spontaneous formation of biopolymers may have contributed to the emergence of early life forms. And it looks as though in nature, enzymes, DNA, and RNA play a major role in triggering cell division, but mechanical viscoelastic properties and structural organization of cells' lipid barriers and cytoplasm play an extremely important role in following process. The study [416] supports the hypothesis of the imminent importance of viscoelasticity in the functioning of cells.

The interaction of the mechanical viscoelastic properties of cell media and the processes of substance transferal within the cells should be taken into more consideration than is done today. For instance, applying particle-tracking micro rheology to freestanding phospholipid bilayers of cell membranes, researchers in study [273] found that the membranes are not simply viscous but rather exhibit viscoelasticity, with an elastic modulus that dominates the response above a characteristic frequency that diverges at the fluid-viscoelastic gel phase-transition temperature. Although the cell membrane is a protective barrier, it also plays a role in letting some foreign material in via ion channels dotted around the cell's surface. And as it is shown in work [274] phospholipid bilayers in cell membranes control and regulate the ion channels. This means that the viscoelastic properties of phospholipid layers on membranes directly influence the transport of molecules through ion channels. I am sure that by taking this factor into consideration, depending on many external physicochemical stresses that influence the cell, it is possible to carry out a more refined regulation of the processes of substance transferal through the membrane—compared, that is, to if we only consider the density of the electric field in the ion channel.

Also, in relation to viscoelastic properties of phospholipids on membranes, the interesting findings in study [275] suggested that the long-range motion in phospholipid membranes on short-time scales (from pico- to nanoseconds) is not diffusive but has flow-like characteristics. This is probably connected to the fact that as a result of viscoelastic phase separation in the phospholipid bilayer, possibly metastable long-range molecular clusters form with a certain orientation of molecules like in liquid crystals, which can be relatively free to move but only in directions defined by their orientation. This cooperative movement of clusters of phospholipid molecules in the nano-viscous state appeared in the study [275] as something very similar to their flow. All animal cell membranes are a viscoelastic bilayer of phospholipids made up to various degrees of fatty acids that must be acquired from the diet. The essential polyunsaturated fatty acids have been shown to be important in resistance to a variety of diseases and in coping with changes in body temperature. It is generally believed that mammals are unable to alter the proportions of essential fatty acids in their cell membranes except by changing their diets. Furthermore, mammals are unlikely candidates for extensive temperature-induced alteration, known to occur in fish or reptiles, because they typically maintain high and rather constant body temperatures. And we are all encouraged by nutritionists to eat polyunsaturated fatty acids, as these are "good for us" and we are unfortunately unable to make them ourselves. The (relative) levels of particular classes of polyunsaturated fatty acids have been associated with a plethora of human illnesses. The latest findings of Walter Arnold and his group at the University of Veterinary Medicine, Vienna, suggest that changes in fatty acid concentration of inner organs might be largely independent from immediate diet composition [382]. This means that the processes taking place in the phospholipid layers of membranes are determined by the combined response of the metabolism to the changes in external conditions. That is, the overall epigenetic state of the organism and of its cells is what influences the process of effective assimilation of polyunsaturated fats. This in turn means that in vivo, the management of the process of phospholipid layer formation can be carried out by a considerably long period of anthropic nutrition due to the correction of the overall epigenetic state of the metabolism. This example shows that the trivial advice of the contemporary nutritionists to eat more "good" fats more often than not is useless.

The viscoelastic mechanism of the strange attractor of the metabolism demands, along with dissipation, to take into account the non-linear types of equations of the state of the organism's media and determines its adaptive possibilities as viscoelastic deviations of the point of attraction in the phase space. I think that understanding the importance of counting viscoelasticity as one of the fundamental manifestations of differences between live and non-live nature is missing in life sciences. It is true that all the dissipative structures of non-live nature known to us exist as continuous energy flows and are stable at small fluctuations. One of the differentiating qualities of dissipative structures of live matter is their stability during long final fluctuations—negative conditions of the external environment. If organisms existed only in the form of classical dissipative structures, then changes in the external environment would lead to almost immediate death, with only a small time delay due to the inertia of processes. In the absence of the viscoelastic phase—i.e., the polymerization of molecules and of polymer matrices—macro-viscosity of the intracellular liquid medium would be much lower. The characteristic times of many biochemical reactions in the cells are relatively small. Any fluctuations in the delivery of substances and energy into the cell due to the insignificance of inertia at low viscosity would lead to the destruction of its structure. The viscoelastic properties of the cell, because of high relaxation times of polymer molecules, allow the emergence and stability of the structure.

At the same time, the viscoelasticity of structural proteins (Actin, Tubulin, Vimentin) allows the mesh sizes of the three-dimensional matrix to be adjusted and, therefore, the viscosity and diffusion rates to be maintained in the presence of changing protein concentrations (osmotic pressure). This in turn helps to maintain the proportional composition of the proteins.

There are cases in which viscoelasticity directly influences cell function in the organs. For example, study [306] compared actin matrices cross-linked with mutant and with wild-type proteins in kidney cells. Researchers found that the resulting altered viscoelastic properties contribute to the phenotypic changes in diseased kidneys. Interestingly, the study also found that mutations have a drastic effect on the properties of in vitro networks at sheer frequencies around 1 Hz, close to the human heart rate. In vivo, the pathological consequences of the mutant cross-linker are most apparent in cells that wrap around the capillaries in the kidney, which are subject to substantial stresses resulting from dynamic capillary pressure. What is intriguing in the specific evolution of viscoelasticity of these cells is that they have a relaxation time associated with heart rate. Such fine tuning of these cells demonstrates the critical importance of their mechanical viscoelastic properties for the function of complex organisms such as humans.

One study [284] presents a numerical analysis of the dynamic viscoelastic behaviour of the cytoskeleton (CSK), which helps us to understand recent results on the cellular-dynamic response and allows us to reunify the scattered data reported for the viscoelastic properties of living adherent cells. The researchers describe the viscoelastic properties of a refined cellular-tensegrity model composed of six rigid bars connected to a continuous network of twenty-four viscoelastic pre-stretched cables (Voigt bodies) in order to analyse the role of the CSK spatial rearrangement on the viscoelastic response of living adherent cells. This structural contribution was determined from the relationships between the global viscoelastic properties of the tensegrity model—i.e., normalized viscosity modulus (*), normalized elasticity

modulus (E*)—and the physical properties of the constituent elements—their normalized length (L*) and normalized initial internal tension (T*). A numerical method was used to simulate the deformation of the structure in response to different
loads, while varying L* and T* by several orders of magnitude. The results reveal
that *remains almost independent of changes in T*, whereas E* increases proportionally to the square root of the internal tension T*. Moreover, structural viscosity*
and elasticity E* are both inversely proportional to the square of the size of the structure. These structural properties appear consistent with CSK mechanical properties
measured experimentally by various methods which are specific to the CSK micro-
manipulation in living adherent cells. Results of this study suggest "that the effect
of structural rearrangement of CSK elements on global CSK behavior is characterized by a faster cellular mechanical response relatively to the CSK element response,
which thus contributes to the solidification process observed in adherent cells."

Each of the structural proteins creates its own three-dimensional matrix with its
own specific role in cells. Thus, the different proteins are exposed to different influences. For a more delicate numerical analysis of the mechanical structures of cells, it
would be useful to consider the effect of the interaction of various three-dimensional
matrices. This is important to understand as the mechanical properties of the matrix affect basic metabolic processes in the cell, such as the intensity of production,
movement, and the spectrum of proteins in organelles, cell signalling, and so on.

Until now, in the study of the biochemical processes of protein transport and
chemical reactions in the cell, the contribution of mechanical viscoelasticity of the
cell and intercellular structures was largely ignored. However, the presence of viscoelastic properties of the cell constituents changes the character of the excitation
(bifurcation) of various reactions in a cascade of metabolic reactions. In the end,
viscoelastic properties are the only systematic difference between the physical and
mechanical characteristics of living and non-living matter (excluding synthetic
polymers). It is unlikely that Mother nature had not used such a key difference in all
the vital processes in the most fundamental way. Because of the continuous dissociation and recovery of protein matrices, the viscoelastic properties, both inside and
outside the cells constantly undergo dramatic changes. This suggests that, even on a
micro-scale, the cells cannot be regarded as a simplified biochemical reactor close to
equilibrium. In addition, it is well known that the boundary between the intracellular and extracellular space (i.e., the cell membrane) formed from structural protein
matrix and the lipid bilayer has pronounced viscoelastic properties.

A variety of non-linear effects of viscoelasticity change our understanding of
how the cell works. The viscoelastic properties of cells at all structural levels, from
the extracellular matrix and membranes to the intracellular matrix and organelles,
protect cells from external stresses that theoretically could destabilize the generation of negative entropy and, consequently, cell metabolism. The simplified view

favoured in structural biology has to shift to a greater emphasis on protein dynamics in the viscoelastic environment. For example, scientists have discovered [276] that vaccines can help the immune system fight cancer, but vaccines that mimic biological structures can still fail if they do not take into account the flexibility and dynamics of the cell's environment.

In ordinary dissipative systems (strange attractors), viscous dissipation prevents the independence of small perturbations of the environment, so that the functions that describe their behaviour depend on the functions describing the behaviour of finite (large) perturbations. In other words, the system's behaviour at the lowest hierarchical level depends on the properties of the system at the next higher hierarchical level. For example, the diffusion and reaction of proteins in the mesh (pore) of protein matrices are dependent on its properties. When taking the viscoelastic matrix into account, it is obvious that the viscoelastic properties have a significant impact on the phase-space of the strange attractor reactions of the cell's metabolism.

The fractal dimension of the strange attractor metabolism is practically the infinite number of generalized dimensions (due to the large number of reactions) and so it can be broken down into multiple hierarchically related strange sub-attractors. The generalized fractal dimension of such attractors counts only effective degrees of freedom organized according to the physics of the sub-system and thus quantifies the complexity of its structure. For example, the hetero-phase structure of a cell can be divided into many separate phases of the individual attractors. Also, each cellular organelle contains and is surrounded by its own protein matrix that regulates the rate of transport of signalling molecules. If we consider each matrix as a relatively low-dimensional strange attractor, and the cytosol-containing signalling molecules as a stochastic set in a multidimensional phase space, then their union (because in reality they are united in a cell) is a multi-dimensional strange attractor.

For medical applications in the study of a disease, it is often enough to consider the metabolic conditions of the fractals of the whole body—organs, blood parameters, liver, kidneys, etc.—without details of metabolism in them. Nevertheless, the viscoelastic properties are manifested in the body at all levels, and we cannot get around them. Even on the DNA level, viscoelasticity is playing a noticeable role. A study [264] observed the conformational state of the genome of dividing cells of yeast (fission yeast). This study, published online as a featured article in the *Journal of Nucleic Acids Research*, is the first to combine microscopy with advanced genomic sequencing techniques, enabling researchers to literally see gene interactions. It is also the first to determine the three-dimensional structure of the fission yeast genome. They found that chromosomes spend the majority of their time clumped together in these large, non-random structures, and these shapes reflect various nuclear processes such as transcription and translation signals. When the chromosomes come together, they fold into positions that bring genes from different chromosomes near each other. This positioning allows the processes that dictate how and when genes are read to operate efficiently on multiple genes at once. Such structure is not merely an accident of chemical attractions within and among the chromosomes but an arrangement guided by other molecules in order to create a mega-structure in the cell that dictates genetic function. But in such polymer structures, an important role is played by the mechanical properties of the chromosome matrices, which define, through viscoelastic phase separation and the asymmetry of the diffusion, the rate of molecular transport and transcription of the genome structure.

Due to the fact that modern molecular biology has focused on studying the molecular details of countless chemical reactions in the cells, the big picture eludes our understanding. In the process of evolution, the work of the cell on the molecular level became so complicated that scientists do not see the root cause of many processes and their driving forces. In my view, the work in the cells at the molecular and system levels can be reduced, for better understanding of general principles, to mechanical processes in the viscoelastic media combined with a certain relatively narrow set of chemical reactions inherent for living systems since the origin of life billions years ago. That is why the excursion into the distant point in time at which those processes of life originated is so important for understanding the work of our cells, and, of course, of interaction of the genes, body, and food. I do not argue with the need to study details of molecular processes, but a clearer understanding of cell work at the system level can, in my opinion, greatly contribute to the development of modern biology. By the way, the research [613] of how the switches work also provides insight into molecular evolution, from the world of RNA in the origin of life to the now-existing world of proteins.

For example, the cells have evolved complex mechanisms called quorum-sensing systems [615] that provide for cell-to-cell communication, an adaptation that allows them to wait until their population grows large enough for getting some level of cooperative functionality in the tissue. When the numbers of cells in tissue become dense enough and/or reaches the specific spatial distribution of the cells, the mechanical tension in the system through some types of mechanochemical transduction processes initiate production of signal molecules at the level that triggers cascades of chemical reactions, eventually causing certain genes to produce the specific proteins. Quorum sensing provides self-organizing network properties to the tissue. The entire tissue acts as a unit, and this rapid, all-or-nothing response to the mechanical or chemical stresses makes quorum-sensing systems suitable for use as genetic switches. The study [614] found that cells can make use of many thousands (probably even billions [616]) of switches to enhance production of a wide variety of molecules that are necessary for the cell biological functions in given conditions. Stem cells, for instance, use switches for determining what kind of cell to become in response to the external stress. It should be noted that after half a century of stem cell study, scientists are still unable to clearly identify the characteristic distinctive properties of stem cells at the molecular level. Almost every month scientific works appear which suggest that the major characteristics of stem cells—conversion to another type of cell—could be attributed to the already differentiated functional cell. Possibly, science has to look at the problem from the point of view of systems dynamics instead of "simple" universal molecular reactions, and of signaling in gene regulatory circuits defined by mechanical viscoelastic state of cell matrices.

Molecular switches exist in two states, on or off. When a molecule switches from on to off, or vice versa, its conformation changes. This change in structure is mainly triggered by the physical or chemical binding of a signaling molecule. This means that changes are occurring in the mechanical state of the molecule and structure in which it is involved. The study [612] provides a general conceptual framework to understand the molecular principles of switching mechanisms and presents direct evidence for an allosteric mechanism with specific conformational changes. Allosteric regulation is regulation by binding a signaling molecule at the protein's allosteric site (that is, a site other than the active site) in the switching process. Through this mechanism, switch molecules exist in one of two conformations, tensed (T) or relaxed (R), and the relaxed subunits of the switch bind to the signal molecule more readily than those in the tense state. When a signal molecule interacts with a protein or RNAs, it is changing their conformation from T-state to R-state in order for the signal to bind. This state transfer has a characteristic mechanical relaxation time. That relaxation time controls the ratio of binding to un-binding processes, which determines the switched on/off ratio for a specific switch in the cell and the cell population. The state of on/off ratio for switches determines the cell function and depend from incoming to the cell nutrition's and the genetic information from anthropic food [572]. On an example of this mechanism, we can see that even on such a deep molecular level, hidden from clear understanding by the shadow of numerous chemical processes, the mechanical relaxation processes play a key role in cell regulation. In cells, there are other similar processes, which depend on the mechanical properties of the medium (matrices) and the conformational states of macromolecules. We can imagine the work of the cell as huge set of simultaneously occurring BZ reactions where oscillating modes of fractal wave packets probabilistically determine the ensemble of the state of switches in the entire cell.

Taking into account elasticity changes in cells may enable us to create new diagnostic techniques and procedures. Theoretically, it is possible to design sensors that will detect in vivo elasticity changes in cells, their organelles, and extracellular space. Furthermore, the sensors could identify the concentration variation of certain molecules as a consequence of changing the viscoelasticity of the biopolymer matrix and could help to detect early stages of cancerous or other degenerative changes in the cell or tissue wellbeing, allowing doctors to make direct appropriate therapeutic responses. Of course, extensive research in many laboratories is required to achieve that by creating disease profiles of viscoelastic changes, but it could give researchers and doctors truly invaluable early diagnostic tools.

## 3.5   Aging of viscoelastic beings

The viscoelastic properties of the protein matrix in the cell may also play a critical role in the aging process. There are many theories of ageing in cells, which do not take into account the effects of critical changes in viscoelastic properties. But even the aging process appears to bring about easily noticeable changes in the viscoelastic properties of body tissues: the loss of elasticity of the skin, eyes, lungs, and other tissues and organs.

The effect of viscoelastic properties of the matrix on the aging of organisms is confirmed by an investigation [176] of the disease called Hutchinson-Gilford Progeria Syndrome (Progeria). Progeria is a rare hereditary disease that manifests itself in the premature aging of the organism. In progeria, there is a protein called progerin (a mutant form of the protein lamin A) produced in the cell, which destroys the extracellular matrix, in turn leading to cell death. It turns out that in natural aging in the vascular system of the human body, an excessive quantity of progerin is also found.

Another study [177] has shown that mutant forms of lamin A violate of the construction of the polymerised lamin matrix, which is an important structural element of the cell nucleus and plays a key role in the expression of certain genes due to the influence on diffusional transport of signalling proteins to and from the cell nucleus.

In addition, a recent study [212] investigating the movement of messenger RNA (mRNA) through the nuclear pores by direct observation shows that the movement of mRNA through the pore for export from the nucleus happens like this: a molecule of mRNA upon coming close to the entrance of the nuclear pore, is "stalled" (moves extremely slowly) for about eighty milliseconds. Then it passes the pore surprisingly quickly—in five milliseconds—and after leaving the pore it is "stalled" again for eighty milliseconds before it goes into the bulk of cytoplasm. The explanation of such aberrant behaviour of mRNA molecules, in my view, lies in the mechanism of viscoelastic phase separation. First, the nuclear pore is a highly organized complex polymer matrix-type structure of many different proteins. In that structure the viscoelastic phase separation occurs between protein polymers and low molecular weight components. The explanation of the fast diffusion of mRNA through the pore is in the low viscosity in the pore volume and its ability for signifi-cant elastic deformation.

Second, the entrance to the nuclear pore is organized into the basket walls, which have a matrix structure with a fine mesh size that lets in only the low-molecular phase of liquid. Therefore, mRNA molecules work like a piston when moving through the pore, and only small molecules can be pumped through the pore by big molecules. Outside the nuclear membrane, actin filaments, present at the outlet of the pores, may play the role of a matrix-filter that also limits the intake of large molecules from the cell cytoplasm into the nucleus through a pore. Thus, within the pore space a low-viscosity liquid phase state is maintained.

Third, it is apparent that the flow of low-viscosity fluid, pumped into and out of the nucleus through both ends of the pores creates the "diffusion shock wave" of chromatin compaction during nuclear interphase as well as protein compaction in the cytoplasm. When exiting the pore, the stream of small molecules passes rela-tively freely through the matrix of inactive chromatin. This causes hydrodynamic pressure and compaction of the chromatin lattice in the nucleus and a similar process in the cytoplasm protein matrix. This compacted phase is responsible for the "stall-ing" of mRNA observed in [212].

In addition, the viscoelasticity of the nuclear pore is confirmed by that fact that its cross-sectional diameter can be deformed reversibly from 9 nm to 26 nm during the passage of large molecules. Any changes in the degree of elastic deformation due to changes in the properties of the nuclear pore protein matrices during aging strongly influence the adaptive capacity of the cell in relation to the range of varia-tion of mass transport from the nucleus to the cytoplasm and vice versa. It is an interesting fact that the action of nuclear pores in fact leads to a three-phase system of the state of the nucleoplasm, and the cytoplasm in the vicinity of the pore exits due to the viscoelasticity of its constituent elements. Indeed, we have a low-viscosity phase, a polymer phase, and the phase of compacted chromatin or protein in the "diffusion shock wave" near the nuclear pores. It should be noted that a change in the chromatin structure and, consequently, the viscoelasticity of the chromatin ma-trix, as shown in [226] varies greatly with age in the cells of animals and humans. The results of these studies suggest an important role that the viscoelasticity of the matrix or of elements of the organelles (e.g., pores) within the cells and surrounding structures plays in the function of cells.

Aging is known to be a risk factor for degenerative diseases, but is not the cause. The main problem with aging is that it reduces the adaptive capacity of cellular metabolism, violating cellular homeostasis, particularly protein balance. At least, regarding the transport of large signalling proteins or protein-containing clusters in the intercellular space and in its organelles, viscoelastic properties of various pro-tein matrices play an important role in the cell's adaptive capacity. The mechanical properties of the matrix determine fundamental metabolic processes in the cell, which in turn determine the rate of production and movement and the spectrum of proteins in organelles, intercellular signal exchange, etc. A reduction in the adaptive capacity of the organism with age does not allow cells to, for example, respond properly to external stresses or effectively remove damaged proteins. The appearance of defective molecules in cells, triggered by such factors as disanthropic nutrition or other disanthropic environmental factors, is changing the set of metabolic reaction pathways.

In the cells of the young body, the protein matrix is more capable of elastic deformation. This changes the size of the cells depending on the osmotic pressure in them. Young body cells have an adaptive capability to withstand stresses of various types. For example, in the cells of the human body through food metabolism and respiration, the basic energy molecule, ATP, is produced. The amount of ATP produced by our body's cells at rest per day is huge: from 65 to 75 kg. If we do some work, production of ATP is greatly increased. The cells of our body have an extraordinary ability to adapt to the changing needs of ATP by accelerating its production. The viscoelastic flexibility of the structural protein matrices in the cell at least partially provides this ability to adapt to changes. In addition, as mentioned in [175], a decrease in ATP production, which just happens in old age, reduces the flexibility of polymer matrices and, consequently, their ability for adaptive responses to most external stresses.

During the life of the cell, due to external chemicals or accidental causes, additional cross-linking of structural macromolecules of proteins occurs, altering the viscoelastic properties of the matrices. These processes occur mainly due to external factors such as background radiation, mineral and chemical substances that enter the cell, and free radicals. But we cannot rule out the spontaneous nature of such processes. In the natural course of events, these random cross-linking processes are relatively rare, causing natural aging. A smaller capability for elastic deformation (greater stiffness) decreases the adaptive capacity of the cells by reducing the range of the elastic deformation of the protein matrix lattice and therefore the ability to cope with stresses by regulating nano-diffusion. A good example of this is atherosclerosis (hardening of the arteries), an inflammatory disease in which the walls of the blood vessels thicken and become less elastic.

Cells oppose unnatural cross-linking of polymer matrices by dissociation and re-polymerization. For example, in the cell cytoplasm there are actomyosin, actin, and tubulin matrices that are continuously dissociating and re-polymerizing; these modulate, due to their viscoelastic properties, the mechanical movement of signalling proteins and parts of the cytoplasm and organelles. As emphasized in [144], the viscoelastic properties of lamellipodia—labile dynamic structures of actin protein matrices—define the fundamental characteristics of the cells' motility. Since the increase of the viscoelastic module—the stiffness of the lamellipodia significantly reduces the motility of cells.

In general, we can say that the mechanical coupling between the soft microtubules of tubulin and the elastic actin cortex provides cells with high mechanical stability despite the softness of the cytoplasm [180]. Study [285] demonstrated that in an extracellular matrix attached to cells, microtubule disruption activates the integrin-dependent signalling cascade(integrin is a type of protein), which leads to the assembly of matrix adhesions and the induction of DNA synthesis. The increase in cell contractility, solely dependent on cell viscoelasticity, is an indispensable intermediate step in this signalling process. According to the data of the study [286], integrin signalling is central to the control of cell adhesion, formation of the viscoelastic actin cytoskeleton, and activation of signalling cascades in viscoelastic intracellular media. Using a method developed to watch moment to moment as they move a molecule to precise sites inside live human cells, scientists in one study [305] found that the proteins at one location may signal division and growth of cells, while the same proteins at another location signal death. In general, many proteins inside of cells move to the periphery, a locale known as the plasma membrane, and from the periphery back to different organelles. Any single protein plays multiple roles in a cell by changing its location. Chemical signalling inside cells influences the movement of protein molecules through complex feedback loops determined as well by viscoelasticity of cellular protein matrix. Thus, through their effect on the rate of diffusion of proteins and the character of excitation and attenuation of chemical reactions with their participation, the mechanical properties of different polymer matrixes of lipids and proteins within the cells and their membranes have a significant impact on the vital processes of cells. These examples clearly illustrate the dependence of physicochemical reactions (the signalling processes) in a cell from the viscoelastic properties of the structural matrix in it, outside of it, and on the cell membrane. It should be remembered that the extracellular matrix plays an important role in the transport of molecules. Consequently, the utilization of the molecular composition of food depends on the viscoelasticity of the protein matrix, which is itself dependent on the molecules entering food metabolites.

In the work [110] it is also shown that the dynamics of the cytoplasm are determined by the elasticity of the matrix and the concentration of free calcium ions, which apparently creates in this case reversible cross-linking of the labile matrix. But if there is an irreversible chemical cross-linking of proteins in these structures, their viscoelastic control over molecular transfer decreases, with consequences to the functioning of the cell as a whole. For example, the use of dietary supplements of calcium, which allegedly have a widely publicized effect against osteoporosis and other diseases of the human skeleton, in fact, lead to a 30 percent increase in the frequency of heart attacks in people consuming them [111]. The loss of the viscoelastic properties of cardiac tissues, because of the cross-linking of macromolecular matrix in the cell and intercellular space, causes the cells to die. Reduced adaptation, which is closely connected with the viscoelastic properties of tissues, leads to tissue rupture when the heart is under stress. Also, the cross-linking in the protein matrices of the ion channels makes them stiffer and less elastic. Consequently they are less adaptive to changes in ion flow. The action of calcium in this case has a universal character, since it does not depend on age, sex, or, most importantly, the calcium consumed through dietary supplements. This is why the incidence of osteoporosis in the population is not decreasing.

In my opinion, one of the confirmations of the proposed approach is that we are also seeing in different degrees of autophagy, a catabolic process involving the degradation of a cell's own components [107]. Since this process usually involves the transfer of proteins inside the cell, the less deformable frame with a high modulus of elasticity of the cross-linked macromolecular matrix cannot allow the proteins to move freely, which contributes to a slower rate of autophagy in old cells. In contrast, in young cells due to the greater flexibility of matrices, nano-diffusion is easy enough for the movement of autophagy-related proteins. Interestingly, the degree of autophagy increases in older cells during starvation. This can be interpreted as meaning that the decrease in the concentration of various proteins involved in nano-diffusion reduces the nano-viscosity of the cytosol—the fluid in the matrix pores—and, consequently, the movement associated with the autophagy protein is accelerated in the less viscous medium.

Generally speaking, a moderate caloric restriction diet plays a significant role in the actual rejuvenation of the human body, but only on the strict condition that anthropic food is consumed. Such rejuvenation manifests itself, inter alia, as improving the viscoelastic properties of tissues and blood vessels. Data from a study [283] conducted by members of the Calorie Restriction Society, a group that practices self-imposed calorie restriction in the belief that such restriction will extend their life span, have recently been reported. The group consisted of lean (mean BMI, 19.6) adult men and women (mean age, 51 years; range, 35-82 years) who had been eating about 1,800 kcal/day for an average of 6.5 years, which was 30 percent fewer calories than age- and sex-matched individuals consuming a typical Western diet. Moreover, the composition of foods consumed by the calorie-restricted group differed from those consuming a Western diet. The calorie-restricted group ate a diet of anthropic nutrient-rich foods, such as vegetables, fruits, nuts, dairy products, egg whites, soya proteins, fish, seafood, and meat; the diet supplied more than 100 percent of the recommended daily intake for all essential nutrients. Processed foods, which are rich in refined carbohydrates and partially hydrogenated oils, were avoided. Compared with control individuals consuming a Western diet, the members of the Calorie Restriction Society showed many of the same alterations in metabolic and organ function previously reported in calorie-restricted rodents. This included a low percentage of body fat, low systolic and diastolic blood pressures, a markedly improved lipid profile, increased insulin sensitivity, low plasma concentrations of inflammatory markers, and low levels of circulating growth factors. The results also showed that ventricular diastolic function (i.e., parameters of viscoelasticity) in calorie-restricted individuals was similar to the function in those who were approximately 16 years younger.

It appears that dissociation and re-polymerization of protein molecules and matrices plays a key role in the aging process. These processes are critical for the cell, because the duration of "life" of various proteins is different, but none of them exist "forever." They are constantly broken down and recycled in the cell's organelles. But when some cross-links form in the protein matrices due to disanthropic chemicals, the debris from such molecules cannot be destroyed and recycled in the cell and actually "clutter" it. Or if all these molecules are involved in the polymerization, the resulting matrices of these structures do not possess the necessary qualities. Their viscoelastic properties are different from the matrices of healthy young cells. However, if in the natural aging process our cells are evolutionarily adapted to cope with this problem, on admission of chemical substances evolutionarily deemed "foreign" from disanthropic food, essentially the cells "do not know how to deal with it." Cellular health and survival thus depends on a controlled removal of defective proteins. But it is not always possible to remove such proteins—metabolites of disanthropic food—due to the fact that the cell does not have the evolutionary mechanisms to eliminate them, so they disrupt cell function. Thus, the aging of cells speeds up in the case of a disanthropic diet.

Autophagy of old cells leading to their death and rebirth does not dramatically improve the tissue's properties. This is because new cells inherit, to some degree, senile changes from previous generations when parts of cross-linked proteins from debris are used to create them. The addition of the defective protein debris can form cross-linked aggregates of "spoiled" macromolecules within cells and lead to changes in the nano-diffusion rates of proteins and other nutrients in the cytoplasm. This process is especially dangerous for nerve cells, where a long, narrow, hollow tube connects neurones with each other [122]. In the biochemistry of nerve cells, it is a well-known phenomenon that the accumulation of amyloids—cross-linked insoluble protein deposits in mammalian cells—is linked with diseases caused by aging. That, in my opinion, confirms the above concept. Keeping cells free of damaged protein deposits is especially critical for neurones because unlike many cells, they do not divide or replace themselves once created at birth. This picture of changes in the cells occurs within the framework of natural aging. But there is no development of degenerative diseases. However, if there are additional sources of intense external chemical, oxidative, or radiation stresses, the processes of change affect the genetic apparatus of cells through the classical or epigenetic mutation, which leads eventually to the emergence of degenerative diseases. The source of these kinds of stress can be both external environmental conditions and disanthropy of our diet.

The practical application proposed in this book on the viscoelastic hypothesis of aging is a theoretical possibility to assess the state of homeostasis of the organism by analysing viscoelastic properties at different hierarchical levels of the organism. For example, studies of viscoelastic properties on a cellular level can be extremely useful for an individual estimate of the exercise loads on athletes with age-appropriate changes, or an objective assessment of the true age of people, which often does not correspond to their date of birth. Homeostasis of the organism, as a state of its metabolism, is very sensitive to the flow of energy through the body. As the flow moves from low to high, the viscoelastic properties make a fundamental contribution to the adaptation of organisms to various external stresses.

One of the most important characteristics of our brain as a whole and its tissues and cells is viscoelasticity. Physiological aging of the brain is accompanied by the ubiquitous degeneration of neurones and oligodendrocytes (a type of brain cell, called neuroglia). An alteration of the cellular matrix of the brain impacts its macroscopic viscoelastic properties. Study [255] shows that liquefaction (decrease of the modulus of elasticity) happens in our brains as we age. The gender-related aspects of brain viscoelasticity are also important. Female brains are 9 percent stiffer than those of males. This difference can be translated into a scale of viscoelastic age, according to which female brains were on average thirteen years younger than male brains. This could partially explain why women live longer than men.

It must be noted that the viscoelastic properties of polymer molecules are beginning to be used in areas such as the delivery of DNA to the cell nucleus in gene therapy [104]. Also, to better our understanding of the movement of proteins through pores and narrow channels [105], the features of movement and self-organization of polymer molecule clusters in solutions depends on the degree of branching of macromolecules. Interesting results were observed in the "walking" of signalling molecules on the walls of channels and microtubules. For the correct interpretation and application of the results of these studies, we must take into account the viscoelastic properties of macromolecules, as well as the modern clinical settings of protein-engineered viscoelastic hydrogels used to support three-dimensional cell culture [255].

Summing up the section concerning viscoelasticity within cells, the basic strategy for therapy should be through preventing the consumption of disanthropic chemicals, active free radicals, and even natural components that could contribute to cross-linking among protein molecules in cells. Irreversible cross-linking of proteins in the human organism is responsible not only for cell disorders, but for many age-related changes, such as calcification of vascular walls, formation of fibrous structures in a tissue, destruction of heart cells and their replacement by an intermediate tissue, and many others as well. A disanthropic diet accelerates cross-linking of intracellular proteins, causing deterioration of adaptive capabilities and premature aging. Due to the higher rate of cell degradation relative to conditions associated with natural age-related changes with a completely anthropic diet, the emergence of degenerative diseases such as cancer, diabetes, and strokes is inevitable. In some cases, the prevention of the cross-linking of the polymer matrix helps to fight infections. So the main action of penicillin is that it prevents cross-links being made between peptidoglycan polymers in cell walls of bacteria.

Besides this, the viscoelastic properties of cells play a significant role in the immune system. So in recent studies [438,439] conducted at the University of Washington, several important discoveries have been made on the ways in which mammals exploit the biochemical properties of nitric oxide to defend themselves from germs. Nitric oxide is a key factor in the body's innate immune defences. The results emphasise that nitric oxide's antimicrobial actions are due to its interference with the metabolism, or energy production, of pathogens. The team of scientists [438] looked at the multi-pronged action of nitric oxide on Salmonella enterica serovar Typhimurium. This type of Salmonella can contaminate food and is similar to the bacteria that cause typhoid fever [439]. Researchers found that nitric oxide could interrupt the cellular respiration cycle, when a food molecule is broken down to release energy for cell growth and division. The biochemical and genetic mechanism of such an action of nitric oxide, in my opinion, could be explained on the basis of a recently published study [440] containing data that supports the idea that nitric oxide release may influence the development of degenerative joint diseases by inhibiting viscoelastic matrix formation via inhibition of the synthesis of macromolecules. The same processes of inhibition of viscoelastic protein matrices in Salmonella could lead to interruptions in the bacteria's respiratory cycle observed in the study [438]. Thus, we see that the fundamental characteristics of cell function and the immune system may depend on the state of the viscoelastic matrix of the pathogenic cell.

The viscoelastic properties of biological objects at all levels of the hierarchy and the related aging processes are a necessary part of the study of living systems which is a relatively underdeveloped branch of biological research. In the approaches adopted in systems biology, scientists are trying to understand how the emergence of living matter and its evolution comes from the interaction of individual chemicals and parts of cells. Systems biology is a promising science for studying life on Earth in all its manifestations, including the aging process. It could eventually help to create more effective drugs, animal feeds, biofuels etc. In order to corroborate and deepen our understanding of the processes of molecular signaling within and outside the cell and put this understanding into practice by making effective medicines for pathological health conditions, we have to use mathematics for correlative analyses using the viscoelastic hypothesis, as well as causal models empowered with sufficient data to make predictions. The main advantage of using mathematics to model biochemical processes in cellular structures is that it is very easy to introduce into the equation the influence of viscoelasticity by selecting the appropriate rheological equations of state. Crucial breakthroughs in the treatment of many common diseases such as diabetes, Parkinson's, and heart and coronary diseases could be achieved by harnessing a powerful approach that combines empirical and computational techniques to gain an understanding of complex biological and physiological phenomena in the cells and body tissues. Thousands of proteins can be involved in biochemical signalling processes that ensure the proper functioning of a cell. If such a signalling network is disturbed through the changes in local viscoelasticity, diseases such as cancer and diabetes may result.

The conventional approaches of biology do not have the capacity to unravel the elaborate webs of the interactions of many signaling molecules and many biopolymer matrices involved, which is why drug design often fails. Simply sending out one or a few target molecules into a cellular biochemical pathway is turning out to be the wrong strategy for drug design. Because cells have a much more complicated structure and at least partially these complications lie within the viscoelastic cell dynamics. The viscoelastic approach is now shedding light on the complex phenomena of living cells and tissues by revealing how the subcellular and intercellular networks of biopolymers interfere with signaling processes. That will make it possible for researchers to develop better therapeutic treatments on a molecular level. This theoretically could lead to significant advances in drug development and anti-aging medicine. It should also be noted that aging is a program of somatic cells in our body, but not in the sense that this process is specifically encoded in the genome of cells, although this cannot be completely ruled out. Most likely aging and death are inherent in the cells at the system level of their structure, at least in the aging and death that occurs long before reaching the theoretically possible limit of divisions, as it usually does. In fact, eukaryotes have developed as a result of the symbiosis of two or possibly even a larger number of cells. For example, mitochondria and chloroplast (the organelle responsible for photosynthesis) were once independent bacteria. Also, many parts of the DNA of our cells are derived from a variety of other ancient bacteria, which are at the dawn of eukaryotes and possibly later on the basis of horizontal gene transfer have got in the DNA. Unspecialized embryonic stem cells similar to the early eukaryotes have the ability to undergo an unlimited (or more likely a very large) number of divisions (proliferation). Starting from the differentiation of stem cells in the embryo, cells become highly specialized—i.e., the wave function of the ensemble of possible states of the genome of specialized cells is narrowed, reflecting the functional specialization of cells. If in the stem cells symbiotic elements of cells work synchronously (apparently, as a result of evo-lution), in the specialized cells, because of narrowing of the wave function of the genome, with time may occur some dissonance in the work of the nucleus with other parts of cells, such as mitochondria. In specialized cells, due to the narrow exposure of the genome, or in another words a narrowing of the wave function of the possible states of the nucleus genome, the conditions of its interaction with the wave func-tion of mitochondrial genome changes. In a process of cell division and division of mitochondria, these changes accumulate and affect the work of the cell similar to the processes of nonlinear damping, which we can see in the interaction of oscillation of nonlinear systems with two or more oscillators with slow changes of frequen-cies. Consequently, the change in the functioning of mitochondria causes a change of elastic properties of matrix and membrane ion channels, etc., and through this change in the transport of ions and proteins into the nucleus. This in turn changes the conditions of the functioning of the nucleus, which in turn again affects the mi-tochondria. After a certain number of divisions of the cell nucleus and mitochondria, the nonlinear damping reaches a level such that the cell cannot continue to exist, and it ends its life by apoptosis or cessation of division. But even before that, in the cells gradually accumulate senile changes, which occur for example in the shortening of telomeres.

This means that the aging of the nucleus' genome and the aging of the mito-chondria's genome run on different "clocks." And that difference between clocks varies from type of cells, stage of life cycle, tissue, or organ and has a tendency to grow with aging. Rate of accumulation of these differences with time actually can be interpreted through the speed of aging in a particular organ or tissue and main-ly depends on genetic and environmental factors, including diet. In other words, it could be said that the aging and apoptosis of an animal's cells happen due to homeostatic anarchy provoked in them, at a rate that at some point ceases to be compatible with their functionality. When we talk about the de-synchronization life clocks of various cell organelles, we mean a very long-term process. But there are also so-called circadian rhythms, the biological "clocks" found in many ani-mals. De-synchronization of work between the nucleus and the mitochondria over time leads to a violation of this relatively short rhythm. The circadian rhythm, in humans and animals, is a complex genetically regulated function tuned to the twen-ty-four-hour-day regular cycle. It influences a wide range of biological processes. In humans, researchers have found strong correlations between disrupted circadian rhythm and neurologic diseases such as Alzheimer's disease [601]. Aging is closely associated with this process as well. Researchers showed through both environmen-tal and genetic approaches that disrupting the circadian rhythms accelerated neuro-degeneration. They demonstrated as well that disruption of the rhythm came first and led the system to degeneration. The aging speed according to this approach is lowest in fetal stem cells, high in most specialized cells, and highest in the process of malignant metamorphosis in the cell. But the irony of the latter is that the cell becomes immortal as a result of the complete loss of "clocks synchronization."

The above idea is partly supported by the results of studies in vitro, which found in various cell cultures that the more specialized (differentiated) normal cells in the culture, the less their ability to divide. For a man it means that aging and death is imminent in the exhaustion of the number of divisions when the cell ends up by apoptosis or enters into a state of senescence—ending the division with the loss of functional properties. Programmed cell death, or apoptosis, is extremely important for body functioning. Programmed cell death goes awry in many types of cancers. For example, in the hematopoietic (blood forming) system, inhibition of apoptosis leads to leukemia. Apoptosis in many cases is the best yield for the body because it makes it possible to use substances from dead cells left to supply to the other cells or make room for another cell of the same type. In the case of excessive accumulation of senescent cells in tissues, the result is the degradation of the functions of organs, which can be more dangerous for the organism. An indirect confirmation of this possibility is the complete degradation of tissue infected by HIV during the AIDS stage. In the case of AIDS, usually the number of infected cells in the tissue does not exceed 10 percent (often less than 7–8 percent), and all the rest in the tissue appear completely healthy. This example illustrates that a relatively small number of non-functioning or wrongly-functioning cells may heavily damage or completely destroy an organ or tissue.

Researchers at the Mayo Clinic [590] have shown that eliminating senescent cells that accumulate with age could prevent or delay the onset of age-related degenerative disorders. Whether and how these cells cause age-related diseases and dysfunction has been a major open question in the field of aging. One reason the question has been so difficult to answer is that the numbers of senescent cells are quite limited and comprise at most only 10 percent or a bit more of cells in an elderly individual. But as is demonstrated in case of AIDS, even a small percentage of these "deadbeat" cells could destroy functionality of the tissue. Therapeutic interventions to get rid of senescent cells or block their effects is a very good road for already ill people, but the anthropic approach to everyday diet represents an avenue to make us feel more vital, allow us to stay healthier, and help us live a much longer life.

So centenarians die primarily from dystrophy (reduction in the number of cells) of the muscles, but not from degenerative diseases like most of the population. The presence of a relatively large proportion of senescent cells leads to degenerative diseases. Also, on the hypothesis set out here, the process of aging is given by no specific genetic program, but is a system property of eukaryotic cells—the presence of relatively independent functioning of the genetic systems in the cell. Theoretically, this can manifest itself as a synthesis of specific proteins observed in the cell biology researches that are associated with apoptosis or cessation of cell division.

From this point of view, even the stem cells may not be capable of unlimited proliferation, because during evolution their genome was composed of regions of the genome of different organisms. Over millions of years, pieces of viruses became parts of some regions of cells' genome and "trained" cells to the new features, such as immunity, by transferring genes, proteins, enzymes, etc. [605]. But the uneven work of these regions in genome through influence on the structures of the cell (cytoplasm, organelles, membrane viscoelastic matrix, etc.) for which they are responsible might lead to the imbalance in the cell after a large number of divisions. This can cease the work of the cell according to the scenario similar to the interaction of the nucleus and mitochondria genomes.

In addition, any single-celled organisms, even bacteria dividing by binary fission, are susceptible to aging in consequence of the existence of the genome transcriptional noise that leads to the accumulation of genetic damages, such as "wrong" proteins. Another major reason for aging is possible oxidization over time or external stresses. That is, even in the ideal case, the structure of somatic cell divisions ends in division and cell death. But during cell life, that process keeps accelerating because the cell accumulates a different error in nuclear DNA and mitochondrial DNA (mutations) and changes in its supramolecular structure, leading to epigenetic mutations. The source of these errors and changes are different stresses affecting the body. The most important of these stresses, along with the quality of the environment and physical activity, is disanthropic food. Changes under the influence of a disanthropic diet can cause the cells to stop dividing before they reach the state in which exists the possibility of apoptosis or senescence. That leads to early accumulation of senescent cells. Excessive accumulation in tissues resistant to apoptosis senescent cells is probably that invisible underwater part of the iceberg, which accumulates multiple injuries of tissue, that ultimately leads to degenerative diseases as a result of a disanthropic diet.

Consider the development of the human organism from conception until death. The highest point of human evolution is a fertilized egg (zygote), which contains the complete human genome, and all of its continued existence as a multicellular organism is dependent on the goal through interaction with the opposite sex genome of getting the next generations of the zygotes as their joint offspring. The intensive epigenetic changes occurring in the genome during the normal fetal development and childhood are completely predetermined by the genetic program. If we omit the extreme stress situations such as long famine, natural disasters, etc., then the development of an organism goes in the direction of getting maximum health to get a healthy egg fertilized by healthy sperm. At the same time, due to the fact that the total number of divisions a fertilized egg is not yet great in a young organism, all cells in them work synchronously with the evolutionary predetermined state of the wave function of the cell. However, after reaching the sexually mature state, it appears that "evolution is losing interest" in the human organism. This is evidenced by a decrease in the number of epigenetic changes (in the absence of extreme environmental stress) in cells from the ages of fifteen to eighteen years. A further surge of epigenetic changes occur at about fifty years old, and this surge is associated with epigenetic changes of accumulated non-uniformity of genomes of the nucleus and mitochondria, including irregularity of work of different regions in the nucleus genome. The interaction of the fragmented wave functions of different fragments of the nucleus genome leads to the inhibition of certain regions of that genome, which appears as reduction of the level of genes' co-expression in old age.

Genes showing decreases in correlation and expression aren't randomly located on the chromosome; they form special clusters. As shown in the study [578], some changes in expressions after fifty are mostly happening as reversals in clusters of fetal expression changes. The specific reversal of fetal expression trajectories seen in infancy is mirrored by changes during aging. What does this mean? Preserved during the entire life of the individual clusters of fetal expressions, changes apparently were responsible for many important functions of cells. Those genes clusters obviously are very "affluent" and determine the role of hundreds of others in performance of many different functions. Their reversal after fifty constitutes a substantial deterioration of the functionality of the cells during aging. But at the same time, some groups of genes showed higher correlation and expression in old age. And those groups or clusters are located much more randomly. Only few genes or gene clusters in an entire genome have importance of the clusters where fetal expression changes happened. Random or stochastic expressions of not-so-affluent genes or clusters statistically rarely lead to immediate collapse of cell functions. This is evidenced by a significant increase in cell-to-cell variation in gene expression in all tissues with aging and indicates the partial destruction of the wave function of the state of the cell. It would seem that the appearance of instability of the genome's wave function is a major cause of degenerative diseases and aging. In the genome there is a growth of transcriptional noise and transcriptional pulsing. Increase in transcriptional noise/pulsing is associated with an increase in point mutations, and it is observed in vivo in response to aging and in vitro by induced external food stress to the cell. The vast majority of genetic disorders have different effects in people with different lifestyle and food consumption habits. Moreover, an individual carrying certain mutations can develop a disease, whereas another one with the same mutations may not. Of course, a gene expression pattern due to transcriptional noise varies among individuals, even in the absence of noticeable genetic and food consumption variation. From a purely theoretical point of view, somebody can get, for example, cancer through random expression of "wrong" (mutated) genes even if he has a very good anthropic diet during all his life. But in reality the chances of that happening if he eats proper food are very slim because the evolutionarily determined gene expression pattern gives for cells a very low probability for such an event as expression of that "wrong" gene. And this point is confirmed by human history—no bone cancer has ever been registered by anthropologists in hunter-gatherer societies' fossils. One can say that the life span in such societies was very short for different reasons not linked with diseases and people did not live long enough to suffer from cancer. But that is not absolutely true. In some hunter-gatherer societies today, we can find many old people without any signs of the degenerative diseases that are characteristic for modern societies.

A slightly different mechanism may also be in effect: as shown in the research [579], the expression of a protein in an incorrect site may trigger up to a few hundred messenger RNA molecules, which transmit gene information for the synthesis of proteins, without these genes being mutated. This process leads to the random or semi-random expression of clusters of "normal" genes in unsuitable amounts and times. That reprogramming mechanism for the genes' expression is also sometimes responsible for turning a healthy cell into a cancerous one. Thus, the increase of transcriptional noise can occur as a result of natural causes during aging and because of environmental stresses during life. Increased noise in the genomes of mitochondria and the nucleus under the influence of external stress apparently leads to much earlier aging compared to the natural course of events in a cell, and early pathological aging manifests itself as degenerative disease. As mentioned above, one of the major stressors in life for most people is a long-term deviation from anthropic diet. At a time when "evolution forgets about us" after the period of puberty and early adulthood, a gradual intensification of transcriptional noise due to disanthropic nutrition leads to the wrong genes' expression and cessation of production of many substances in the cell vital to their proper functioning. So the genetic-epigenetic composition, transcriptional noise, and the environment (including diet) determine to a great extent whether or not a mutation or reprogramming will affect an individual. Therefore, the importance of proper anthropic nutrition during this period is to ensure the signs of aging come as late as possible and, more importantly, do not come as a result of degenerative changes in tissues and organs, which eventually lead to cancer, diabetes, Alzheimer's, and so on. Obviously, maintaining protein and cell quality in genetically, epigenetically, and "reprogrammingly" aging body cells through the compensatory mechanism of the anthropic diet is the most important component of somatic longevity. However, it is very difficult to predict what will happen to each person from their genome sequence alone. That's because the same gene can simultaneously protect against cancer and favor its growth [626], and scientists have already identified more than 500 genes that may cause or contribute to the development of pancreatic cancer [627]. The "predictive medicine" clinics using genome sequencing techniques flourishing today in many countries fail to provide credible guidance to people about their risk for cancer, diabetes, Alzheimer's, and other diseases. Researchers in one study [634] examined the data of thousands of identical twins from Sweden, Denmark, Finland, and Norway, and found that whole genomic tests will not be substitutes for degenerative disease prevention and are especially useless for cancer prediction in most people. Apart from degenerative diseases, genome sequencing cannot reliably predict any other medical problems that may be encountered during a lifetime by people who spend money on such tests. And rather than to put a great deal of money and effort into developing personalized and predictive genomic medicine, it could be much more beneficial for individuals and society to promote and maintain an anthropic lifestyle and diet.

## 3.6   Extracellular food effects

Intercellular space is filled with an extremely complex heterogeneous gel, which consists of macromolecular protein matrices with an intermediate fluid. If, with respect to cells, we are talking about nano-viscosity, it may be more appropriate to speak of the micro-viscosity in relation to the intercellular gel, since the characteristic size of the lattice of the macromolecular matrices is usually bigger. As diffusion through the gel is difficult, the passage of biochemical signals (proteins and other substances), is supposed to be very slow. However, outside as well as inside the cells, there is a characteristic size of molecules, when smaller molecules have an effective rate of diffusion several orders of magnitude higher than for molecules that are bigger than the characteristic size.

At the same time, the speed of diffusion of molecules is at its highest in small concentrations in the intracellular space. As the concentration of food metabolites in the extracellular environment increases, the average diffusion rate may fall significantly due to increased micro-viscosity. This may explain the observed disproportionate impact of small doses, compared to exposure to higher doses of a substance, as is the case with cytokines—peptide transporters of information to the cell. This issue will be discussed below in the section on the impact of a disanthropic diet and small doses of radiation. When exposed to a cell in high dose—i.e., high concentrations of molecules in the gel micro-phase—it is necessary to consider the influence of viscoelasticity of extracellular gel matrices on micro-diffusion.

During cross-linking of the extracellular matrix, a change in its viscoelasticity occurs mainly with the participation of sugar molecules chemically binding to proteins and DNA. This process, called glycation, exerts a strong influence on the properties of the tissues of the body when modifying fragments of sugar molecules that are converted into so-called Advanced Glycation End products (AGEs). At this stage, combined with cross-linking in extracellular matrix proteins, the presence of cross-linked sugars leads to a wide range of diseases from erectile dysfunction and arthritis to Alzheimer's disease and asthma, with dozens in between.

For example, the inflammatory process in asthma that runs in the bronchial tree, changes the state of the glycoproteins due to some physical and chemical transformations, and as a result they become cross-linked. This process is reminiscent of curing of the viscous and fluid gum in a tough and elastic rubber because the main role in cross-linking between molecules of glycoproteins belongs to the disulphide bonds—the "bridges" of sulphur molecules. In addition to the disulphide bridges, in the formation of the glycoprotein gel, molecular hydrogen and calcium links also participate. Through the cross-linking of the glycoproteins, the sputum (mucus) of asthmatic patients during periods of deterioration becomes elastic , or "rubbery," and tough, or "glassy."

As proven in studies conducted in vitro and in vivo, to deal with such processes, the main method is the consumption of foods rich in anthropic natural vitamins C, E, A, and many other compounds yet unknown to science. Synthetic crosslink breakers sometimes prescribed by doctors can in fact be quite dangerous for the organism. This is because, like almost all synthetic drugs, they have a rather low selectivity at the molecular level. This leads to the destruction of not only the "bad" cross-links, but often many more as well that are responsible for the molecular structure of proteins. This is the clearest example of reductionism, when pharmacists begin solving problems using primitive chemical reagents. It is also a good example of naked commerce, since the overall beneficial effects (on the prevailing side effects) of most of these "potions" are not confirmed by detailed research and extensive practice.

It should also be noted that the intercellular matrix is the provider of nutrients into the cell. This is done because the cell makes and releases into its outer environment specific enzymes called gelatinases, a class of enzymes that destroys extracellular matrices and have been implicated in a host of human diseases from cancer to cardiovascular conditions and in particular neurological conditions such as stroke, aneurysm, and traumatic brain injury. Researchers [594] have increasingly focused on developing potent gelatinase inhibitor drugs to treat those diseases. Regulation of the intercellular matrix is critical for organ function. Cells build up a tissue by communicating with their surroundings and other cells, thereby receiving instructions on whether to die, divide and change their adhesion, or move.

Here again, the presence of chemical cross-linking within the protein matrix leads to the emergence of molecular debris of the cross-linked proteins and polypeptides that cannot enter the cell. This debris changes the properties of the newly formed viscoelastic extracellular matrices, and some of it remains in the intercellular space in the form of toxins, or at best inert materials that do not actively participate in the metabolism, but still increase the micro-viscosity of the intercellular medium.

In any case, this leads to a change in the trajectories of metabolic reactions. The importance of these mechanisms is outlined in [113], which shows that this condition is characteristic for gastritis and sometimes is a precursor for stomach cancer.

Moreover, there is a well-known Asian paradox, which lies in the fact that, despite the lifestyle of the Asian population being by today's standards not very healthy—intensive smoking, pollution, etc.—the statistical level of cardio vascular system diseases and cancer is significantly lower than in Western countries. There are various explanations for this paradox, the main of which is a higher consumption of green tea in Asian countries. One study [114] examined the effect of the state of the mucosa of gastric and cascades of transformations in it in order to explain the statistical variation of gastric cancer in the presence of the bacteria Helicobacter pylori (H. pylori) in different countries.

The transformation of the mucosal environment under the influence of stomach bacteria was significantly different in patients from Japan and Indonesia. The environment of the stomach mucosa, which covers the epithelial cells, is a typical viscoelastic gel with the protein matrix on the properties of which may depend on the rate of diffusion of products emitted by H. pylori. Due to differences in diet and the genes of these populations, cascade reactions in the mucosa are different. In one case, this leads to a higher incidence of cancer. Although the specific mechanisms are still not exactly known, the influence of the viscoelastic properties of the mucous membrane of the cells is very likely.

It is known that inhibition of intercellular communication (gap junctional intercellular communication, or GJIC) is manifested in a variety of cancers. Perhaps in the case of H. pylori-induced gastric cancer, bacteria secrete enzymes that alter the properties of the extracellular environment, changing the viscoelastic properties of matrices. As a result, epithelial cells, turn up to the state of outside the normal control of the surrounding matrix and other cells. That situation alters the diffusional protein signaling processes from neighboring cells and, eventually, the cell enters the path of cancer evolution. Confirmation of the possible validity of this approach are the data of study [120], which demonstrated that the rate of movement of H. pylori in the gastric mucosa was abnormally high due to increased pH of the mucus in its presence. It should be noted that the viscoelastic properties of the gel coat of the stomach are highly dependent on pH. With increasing pH, the viscosity and the elasticity of the gel membrane decreased. This means that H. pylorus really release substances affecting the gel coat of the stomach.

Another confirmation of the importance of viscoelasticity of body fluids (mucus, saliva, etc.) for speed of a microbe's penetration is a biomechanical experiment conducted at the University Of Pennsylvania School Of Engineering and Applied Science, which has answered a long-standing theoretical question: Will microorganisms swim faster or slower in elastic fluids? The findings were published in the journal *Physical Review Letters* [459]. The researchers experimented on the nematode C. elegans. They filmed them through a microscope while the creatures swam the course in many different liquids with different elasticity but the same viscosity. Their result is that with increase of the elasticity in viscoelastic liquids, the speed of bacteria's goes down. Keeping an elasticity level up in different body fluids could be a potentially powerful prevention therapy for people with ulcer and other diseases linked to microbes.

Penetration of H. pylori in gastric epithelial cells [173] or, for example, the bacteria Shigella in the intestinal epithelium [172] also occurs through the mechanism of change in the viscoelastic properties of the actin cytoskeleton in cells of the contact area. Through signal transduction reactions in the contact zone  changes the actin cytoskeleton—under the influence of certain enzymes is happening a "depolymerization" of actin filaments and reduction of the elastic modulus of the actin matrix. A weak actin matrix entirely loses the ability to maintain the membrane strength of epithelial cells.The membrane softens and the invading cell breaks into. This is the start of stomach cancer and ulcers.

Another example of how important it is to take into account the viscoelasticity of mucus in the human body is a therapy for cystic fibrosis, a hereditary genetic disease of the respiratory tract. The main method of treatment for this disease is using the medicine that cleaves extracellular DNA in the mucus of cystic fibrosis patients, reducing the adhesiveness and viscoelasticity of the mucus, and facilitating expectoration of sputum, thereby improving pulmonary function and reducing the risk of serious infection of the respiratory tract [499]. Successful infection of human cells by bacteria requires the change in spatial distribution of many existing proteins and the delivery of many new proteins into the host cells that alter various functions to turn the naturally hostile environment into friendly space for bacterial replication. One part of the job, as shown in the case of H. Pylori or E. coli invasion in the cell bacteria, is done with the initiation of an existing protein matrix structure disruption, which leads to chaos in hundreds (maybe even thousands) of proteins' spatial distribution—in normal conditions regulated by viscoelastic properties of cell matrixes—and thereof to the same or even greater number of signaling pathways disruptions. Another part of the deadly job is done by the new proteins from invading bacteria.

Disrupted spatial distribution of proteins in the cell and new proteins from invading bacteria alter existing signaling processes within the cells in which an external signal, such as a hormones or enzymes, triggers cascades of different-from-normal reactions that eventually turn on a gene that changes the cell's behavior. Infectious bacteria are successful because they disorganize cell matrixes and signaling systems: they arrange wrong signals, over-amplifying some of them and inhibiting others, in order to evade the immune system and keep the cell from defending itself.

Understanding the key role of viscoelasticity in such processes, scientists could more specifically study how the proteins delivered by the bacteria accomplish invasion. With such understanding, the problem could be studied using computational modeling and experiments on model animals because a general mechanism of both bacterial infection and cell signaling events in higher organisms including humans, are the same. By the way, on the basis of the above mechanisms of the effect of viscoelasticity of the intercellular gel on development of gastric and intestinal ulcers, it seems quite reasonable to use traditional treatments of ulcers and precancerous conditions, specifically the use of raw quail egg. The gel of egg protein appears to be up for protein concentration in gel membranes of the stomach or intestines, which prevents the penetration of H. pylori in the epithelial cells. The difference in statistics of cancer in different population groups shows that this process may be influenced by diet-induced gene expression and may come out from the food and food metabolite substances. The critical role in cell survival, as shown in [115], is the interaction of intracellular matrices of cells with the extracellular matrices. Interruption of this interaction triggers apoptosis of cells. The mechanism for this is, perhaps, that the destruction of the matrices provides an unregulated, chaotic access to cell-signaling molecules, proteins of the intercellular space. An excess of some of them switches on a mechanism of programmed cell death.

In a fight with cancer or viral infections in our body, specialized cells in the immune system smuggle small molecules (granzymes) into cancerous cells and those body cells that have fallen prey to viruses and to the other cells surroundings of the infected cells. The molecules then trigger the diseased cells' built-in suicide program by altering their protein matrix. There are two possible ways in which the granzymes kill cells.

First, the granzymes do this with the help of a protein called Perforin, which works by punching holes in cells that have become cancerous or have been invaded by viruses. The holes let granzymes into the cells, which then destroy the cellular matrix and nucleus [269]. The second way, shown in a study [270] of the mechanisms of granzyme-induced smooth-muscle-cell death in the absence of perforin, granzymes induced cell apoptosis via the cleavage of several proteins of the extracellular matrix, including fibronectin. But in both ways, the important element is that granzymes destroy viscoelastic matrixes used in a cell function for the fine tuning of signaling pathways. The importance of taking into consideration the extracellular matrix viscoelasticity in oncology is demonstrated also in study [261]. Researchers have been working out the molecular chain of events inside and outside cells that leads to cancer. They found that the protein that loses control in many cancers, plasmin, encourages the breakdown of protein clusters that the part of extracellular matrix around cells. This allows cancer cells to spread to other parts of the body.

In the cases considered above, the strategy for the prevention of cancer, its development, and its spread in the body by the anthropic diet is to use natural substances of foods that prevent changes in the viscoelastic properties. In some cases it may be suppression of the enzyme, preventing the destruction of intracellular and intercellular matrix proteins. In other cases, it might be helping the destruction of the matrix inside the diseased cells. In yet other instances, it may be a matter of strengthening the enzyme that causes cross-linking of the matrix of the cells' mucous membranes. The importance of cross-linking chemicals in the cells is very clearly demonstrated in [167], where we studied the effect of DNA repair enzymes in resistance to anticancer drugs. Anticancer drugs act by cross-linking the DNA of macromolecules to prevent their proliferation and to accelerate the death of cancer cells. We found that the enzyme-repairing DNA prevents chemical cross-linking in DNA, and therefore kills the cancer cells. The same thing happens in a healthy cell, since the prevention of DNA chemical cross-linking is critical for a cell's activity. Although this example shows the negative effect of preventing the formation of chemical cross-links, it still illustrates the fundamental importance of this process in the life of any cell.

The condition of the intercellular matrix also plays a special role in the brain and in the development of its illnesses. The strong role manifests itself in those cases where it is necessary to take into account the interaction of cells of various types. In such cases we have to take into account the tissue microenvironment and metabolic interactions between different cell types.

In the brains of people with Alzheimer's disease, certain cells, such as glutamatergic and cholinergic neurons, tend to die in much larger proportions in moderate stages of Alzheimer's disease, while GABAergic neurons are relatively unaffected until later stages of the disease [297]. Each cell type has different biochemical pathways that support the normal state of extracellular and intercellular matrixes. Through the viscoelastic intercellular matrix, the metabolism of specific human cells affects the metabolism of other cell types. If for some biochemical reason (aging, diet, etc.) the mechanochemical state of the intracellular polymolecular network changes, it starts to distort communication between different types of cells, and therefore distorts signaling protein pathways. That leads to of the death of some specific cells—glutamatergic and cholinergic neurons. The misfolding of abnormal proteins in brain cells is also a key element in Parkinson's disease development. A recent study [317] suggests that the sick proteins slowly move between cells, eventually triggering the destruction of the new host cell. But their movements are dependent on the viscoelastic properties of intracellular space. This discovery could potentially lead to new therapeutic strategies for neurodegenerative diseases aimed at slowing the spread of protein misfolding throughout the brain by increasing the viscoelasticity modulus of the extracellular matrix.

It appears that the viscoelastic properties of cells play a crucial role as well in the development of the neural systems of organisms. Recently published results of research [325,326] demonstrate that mechanical stress is instrumental in several key phenomena in neuronal development. Once a neuron has developed, the authors explain, it is attracted to and then attaches to another neuron, which pulls it to the appropriate place within the neurosystem. In the human brain, mechanical stress—the amount of pressure applied to a particular area—is regulated by viscoelastic properties of neurons. Mechanical forces keep neurons together and functioning as a system within the body, and proper nerve function is dependent on this tension. Researchers at Tel Aviv University say that mechanical stress plays an even more important role than medical science previously believed. Research [325,326] has the potential to tell us more than ever before about the form and function of neuronal systems: as the neurosystem develops, some cells are eliminated, while others are stabilized and preserved; cells that successfully connect with one another maintain this connection through mechanical stress. This tension draws cells to their destined locations throughout the neurosystem. As neurons develop, they migrate to the appropriate location in the body, and it is mechanical stress that draws them there. For designing a proper model of such behavior of neurons, to take into consideration viscoelastic properties of cells and tissue is an absolute necessity.

Another interesting example of the importance of considering the viscoelastic properties of cells in the body is considered in [170]: the influence of the state of the actin cytoskeleton in cell processes of cell infection with pathogenic bacteria. Contrary to the popular belief that the bacteria quickly try to get inside a cell in order to use most of its contents as their own food, this study found that infecting bacteria sometimes prefer, having attached to the cell-victim, to wait some time to prepare for the final attack through the synthesis of certain proteins. To prevent the attacking pathogen being swallowed up and destroyed by defenses of attacked cells in this period, the attacking bacteria induce a sequence of signals that cause an increase in the modulus (stiffness) matrix of actin within the attacked cell directly below the attachment of pathogenic bacteria. After thatthe pathogen finally readies itself to attack, it modifies the actin matrix in the attacked cell, which is then locally destroyed or reorganized by pathogen signals (softening) that provide penetration of pathogenic bacteria into the cells of the victim. As we see, directed regulation of the viscoelastic properties of the matrix cells of the pathogen in this case plays a significant role in the mechanism of infection. A very similar mechanism of introduction of pathogens into the cell during infection acts with Salmonella as well [265]. The bacteria attach to cells of the intestinal wall and induce their own ingestion by cells of the intestinal epithelium. The bacteria inject a protein cocktail using a "molecular syringe" into host cells, leading to dramatic rearrangements of the cytoskeleton below the cell membrane that apparently facilitate their invasion. Thus, in all cases the penetration of bacteria into the epithelium in the event of cancer [171,172] or the case of infection [170,265], there is one and the same general mechanism for changes in the local (in the zone of invasion) viscoelastic properties of cells. The change in these properties does not occur as a side consequence of certain biochemical processes. That change in the elastic moduli of the viscoelastic component allows, using military terms, to conduct the operation in cell invasion of alien bacteria.

Another good example of the influence of viscoelasticity is the action of macrophages, white blood cells that specialize in the human body to heal wounds based on the creation of a viscoelastic intercellular matrix. These macrophages are converted into arginine ornithine, which is a precursor of high molecular polyamine and collagen in the intercellular matrix structures, which, in turn, due to its viscoelastic properties, provide mechanical protection of the wound surfaces and simultaneously regulate the transport processes of intercellular signaling proteins in the communication between cells. It is a well-known fact that arginine is essential for the function of macrophages. But until the study [277], which is published in the August 2010 edition of the journal *Science Signaling*, no one realized that arginine has a much bigger role. Researchers from University of Alberta have illustrated that an arginine is required to let the body "know" that it's being attacked by an infection. They uncovered a role for the extracellular nutrient arginine in the activation of macrophages and found that arginine is critical for two aspects of the innate immune response in macrophages: It is the precursor used in the generation of the antimicrobial mediator nitric oxide, and it facilitates the production of cytokines (small protein molecules that are secreted by numerous cells of the immune system and used as chemical messenger molecules to communicate). This finding has implications for the millions of people who do not get proper (anthropic)) food with enough supply of arginine or it precursors in their everyday diet. It shows that the level of amino acids is a very critical aspect of nutrition for our wellbeing. But we must remember, as will be shown in the Chapter 7 of this book, that only arginine naturally coming from the anthropic food plays a positive role in the body. Therefore, it is important to distinguish between saturated synthesized or extracted arginine in sports nutrition products that are  presented on the market, and the natural arginine sources of arginine-rich foods. It should be noted that the same applies to all other essential amino acids.

The mechanical properties of surrounding of cells in the body also significantly influence the processes of their reproduction. For example, the influence of viscoelastic properties of the surrounding stem cells in their growth and reproduction is shown in [224,225]. One of the major challenges in stem cell transplants is how to obtain sufficient numbers of healthy cells to put into patients. On study [224] demonstrated that mechanical forces created by elasticity inside of body play a key role in blood-forming stem cell growth. Researchers could generate up to three times more stem cells if they mimic the environment of cells inside our body than using current methods (rigid surface) alone. If to consider results of another study [272] (see below), the method could be helpful for eliminating the tumor-risk factor in processes of growth and utilizing human stem cells. Researchers at the University of Pennsylvania have shown [225] that the elasticity of a stem cell's environment is a major determinant of what type of tissue the stem cell becomes. In laboratory tests they grew stem cells derived from bone marrow in viscoelastic polymer hydrogels with low, medium, or high modulus of elasticity. Cells grown in low modulus media such as brain tissue tended to produce nerve-like cells; those grown in environments with medium elasticity modulus, similar to muscle, produced muscle-like cells; the cells grown in high modulus medias, like bones, produced bone-like cells. Both studies demonstrate how fundamental the influence of viscoelasticity in body environment is on cells' development. According the study [368], cultured so-called mesenchymal stem cells can "feel" at least several microns below the surface of an artificial microfilm matrix, gauging the viscoelasticity of the extracellular bedding that is a crucial variable in determining their fate. The higher elasticity modulus of the surface, the shallower the cells could feel; the lower modulus of the surface, the deeper they could feel. Because in a body the surrounding matrices varied, the scientists found significant differences between stem cells grown on varying viscoelasticity thick surfaces that represent human tissue microenvironments—such as brain tissue, which has a lower elasticity modulus than muscle, which has a lower elasticity than cartilage, which has a lower elasticity than pre-calcified bone.

It should also be noted that these data support my strongly advocated thesis that the viscoelastic medium in the structures of the body is in effect the essential property of cells. The use of a synthetic polymer hydrogel in [225] simulates the intercellular environment, showed that only the mechanical properties played role in this case, because its chemical composition had no relationship to living matter. It also confirms the importance of studying the properties of the cell cultures in vitro in the three-dimensional (3D) artificial structures that mimic the cells environment in a real body. The similar result obtained in research [557] by a team working at the National Institute of Standards and Technology (NIST) reinforces the idea that stem cells can be induced to develop into specific types of cells solely by controlling their shape. The experiments examined the effect of architecture alone on bone marrow cells without adding any biochemical supplements other than cell growth medium. The scaffolds, made of a biocompatible polymer, are meant to provide a temporary implant that gives cells a firm structure on which to grow and ultimately rebuild tissue. The results [557] show that the stem cells will differentiate quite efficiently in a bone cell on the nano-fiber scaffolds, even without any hormone additives.

The importance of a three-dimensional structure of cell cultures in order to accurately reflect the processes of development of tumors in the body is shown in [288]. Researchers at the Stanford University School of Medicine have successfully transformed normal human tissue into three-dimensional cancers in a tissue culture dish for the first time. Researchers added pre-cancerous epithelial cells to a tissue culture dish containing other components of human skin. Epithelial cells normally sit on a thin partition called the basement membrane that separates them from a lower layer of skin called the stroma. They found that at first the cells nestled down on the basement membrane and formed what looked like a normal, three-dimensional cross-section of skin. But within about six days, the cells started to behave more ominously, punching through the membrane and invading the stromal tissue below. In usual conditions of the human body, cells go from a pre-malignant state to invasive cancers over the course of many months or even years [289]. But in this 3D cultured human-tissue model, that process occurs much more quickly. The 3D culture system used in this research also indicated that the stromal cells themselves somehow encourage the invasion of the altered epithelial cells, and that the cells don't need to be dividing wildly in order to be able to invade.

Why is a three-dimensional culture of cancer much faster than the human body? It is obvious that the main difference between the two is the absence of the normal metabolism of the human body. That is, these studies demonstrate that the copying with a sufficient measure of the accuracy of the three-dimensional structure of the body's condition in the model can repeat the process of tumor development at an accelerated rate. But the acceleration in turn demonstrates that the absence of a natural anthropic metabolism speeds up the development of tumors. From this it follows that malnutrition, as the main cause of distortion of the trajectories of cascades of reactions of metabolism (the chemical and the epigenetic level) and violations of the mechanical properties of three-dimensional structures of tissues and intercellular spaces, may be a source of accelerating the development of malignant tumors. Also, according to some studies, the three-dimensional cellular environment and the homeostatic pressure in the organs of the body play an important role in the spread of cancer throughout the body—the development of cancer metastasis. Combining the laws of mechanics and the biological state of homeostasis, the authors of study [272] propose that every biological tissue regulates to a preferred pressure called homeostatic pressure, and that an increased homeostatic pressure is a generic trait of neoplastic tissues. This property can drive tumor growth at the expense of the host tissue. Eventually the process is highly dependent on mechanical viscoelastic properties of the organ tissue. It potentially explains the observed preferential growth of metastases on tissue surfaces and membranes such as the pleural and peritoneal layers and the failure of cancerous cells to grow inside invaded organs. Hence, the efficiency of the metastatic process depends on 3D mechanical interactions via homeostatic pressure between the invading cancer cells and the viscoelasticity of local organ tissues.

One study [274] has shown that competition between surface and bulk effects leads to the existence of a critical size that must be overcome by metastases to reach macroscopic sizes. The homeostatic pressure as introduced in that work combined with viscoelastic properties of organs could constitute a quantitative, experimentally accessible measure for the metastatic potential of early malignant growths in the organs. Thus, maintaining a high modulus of elasticity inherent in the tissues and organs at a young age is a good prevention of the emergence and growth of the metastasis in internal organs, and the viscoelastic properties of soft biological tissues are essential to their physiological function.

And it is a well-known fact that tumor cells are extremely mobile in a process of penetrating into healthy tissue to form metastases. They very effectively adapt to the consistency of the respective tissue by changing their shapes and altering the viscoelastic property of the membrane and internal protein matrix constantly and attach flexibly to surrounding tissues during movement with the help of special membrane surface receptors [299]. That cell that moves more rapidly is a more aggressive form of cancer cell. It penetrates surrounding tissue extremely fast in a process—aggressive spreading of the tumorof the formation of metastases.  Results of study [306] indicate that all known patho-mechanisms of malignancy require changes in the active and passive biomechanics of the tumor cell and its stroma. Changes in the viscoelasticity of the cytoskeleton can therefore be a general prerequisite for malignancy independent of the peculiar molecular manifestation in individual cancers. The cytoskeleton is one of the most important parts of a cell because it stabilizes and organizes the cell and the cell motility and mechanotransduction. If the cytoskeletal alterations in a tumor are necessary, they have to trigger biomechanical changes that impact cellular function. From a medical perspective, insights into the changes in viscoelasticity of cells that occur during tumor progression may lead to novel selective treatments by altering tumor cells, viscoelastic properties. Such drugs would probably not cure by killing cancer cells, but may effectively hinder the propagation of the neoplasm. These possible treatments would cause only mild side effects and may be an option for older and frail patients who can no longer tolerate radical surgery and cytostatic drugs [308]. Thus, the results presented in the above studies show that the viscoelastic properties of most cancer cells and surrounding tissues play a significant role in the development of tumors.

We have in results of study [268] another independent confirmation of the need to put greater attention on the viscoelastic properties of the extracellular matrix. Mad Cow disease and its human variant Creutzfeldt (Jakob disease), which are incurable and fatal, have been on a welcome hiatus from the news for years, but because mammals remain as vulnerable as ever to infectious diseases caused by enigmatic proteins called prions, scientists have taken no respite of their own. In the October 2010 edition of the journal *Science*, researchers at Brown University report a key new insight into how prion proteins—the infectious agents—become transmissible: In yeast at least, it is the size of prion complexes, not their number, which determines their efficiency in spreading.

"The dogma in the field was that the misfolding of the protein is sufficient to cause disease, and the clinical course of the infection depended on the amplification of the misfolded protein," said Tricia Serio, associate professor of molecular biology, cell biology, and biochemistry [268]. "But over the years in mammals it has become clear that the abundance of misfolded protein is not a good predictor of disease progression. The question is what else has to happen for you to get the clinical pathology?" Cells make prion proteins naturally, although biologists do not understand what their normal role is in mammals. When those proteins misfold in cells, they assemble into aggregates, but other proteins, known as chaperones, attempt to break down the aggregates. The rates at which this assembly and disassembly occurs are determined by the shape or conformation that the prion protein has adopted. Different conformations of the same prion protein can dramatically alter the spread of pathology and the incubation time of prion diseases. By combining experiments in yeast cells with mathematical models, the Brown University team found that what affects a prion's ability to transmit from cell to cell is the size of the structures into which they assemble. If the aggregates become too large, they lose their transmissibility among cells. Prion aggregates that remain small are transmitted with greater efficiency. In this study, researchers changed the transmissibility just by shifting the size. Researchers monitored differently sized prion aggregates as they moved among cells under the microscope and could see that smaller ones fared better than larger ones. The computer simulation that best replicated experimental observations was the model in which aggregate size, rather than abundance, was the key factor.

Previously it was not clear why you would have those outcomes. But due to viscoelastic phase separation and diffusional asymmetry in the extracellular protein matrix, it is easy to explain this, as in some diapason of sizes bigger aggregates have abruptly less diffusional rates with exponential relation between size and rate of diffusion. Ultimately the findings could influence future strategies for developing a treatment for prion infection. If researchers unaware of the importance of aggregate size and viscoelastic properties of matrix developed a therapy, they might inadvertently make things worse by producing smaller aggregates and/or decreasing elastic modulus of the viscoelastic extracellular matrix. A more effective strategy might be to control the size of the aggregates and the extracellular viscoelastic environment rather than the number of aggregates alone. The findings theoretically may also relate to other neurodegenerative diseases that depend on misfolding proteins, such as Alzheimer's or Parkinson's diseases.

The viscoelastic properties play an extremely important role at the tissue level of the organism. The viscoelastic properties of tissues arise from the viscoelastic properties of cells through desmosomes, which are one of the types of intercellular contacts, providing a strong connection of cells (usually epithelial or muscle tissue). The function of desmosomes is primarily to provide a mechanical connection between the cells. Desmosomes are formed mainly between the epithelial cells. To the desmosomes attached intermediate filaments that form a matrix that has high tensile strength and, together with the cytoplasm, form a viscoelastic gel. Through desmosomes, intermediate filaments of adjacent cells combine together into a continuous viscoelastic network covering the entire fabric of tissue. Therefore, all tissues have strong viscoelastic properties that will undoubtedly affect the processes of the transfer of signals between cells. The need to consider the mechanical properties of the tissues of the epithelium is caused by the fact that 90 percent of all human cancers originate in epithelial tissue. Epithelial tissues are constantly regenerating, creating ample opportunities for errors to occur during DNA replication that can promote tumor growth. Also, these tissues are exposed to the environment—the skin to different types of radiation, the digestive system to dietary disanthropic substances and radiation from radionuclides, the lungs to inhaled toxins and radionuclides, and so forth.

In the research [253], scientists at the Stanford University School of Medicine have implicated the lack of a protein important in hooking human skin cells together in the most common variety of skin cancer. Depletion of this protein, called Perp, could be an early indicator of skin cancer development, and could be useful for staging and establishing prognoses. Perp, a desmosome component, weaves in and out of a cell's surface like a thread through fabric. The study [253] shows not only that desmosomes are crucial to maintaining epithelial tissues' integrity, but that the loss of Perp, which is crucial to desmosomes' function, promotes cancer. Disrupted function of another kind of adhesion junction has been implicated in late-stage cancers. But desmosome disturbances may occur earlier on, during tumors' initial development. Desmosome disturbances also influence the viscoelastic property of tissue, which is why it could be used to detect cancer early as the tumor-progression marker. It means that for proper understanding of the molecular and cellular basis of regulation for all biochemical functions of living tissues, we have to take into account the importance of viscoelasticity as main structural features of extracellular matrixes which determine tissues' property, formation, and repair. But it is important not only for normal body tissue, but also for body liquids. Two studies [469, 470] are devoted to an important aspect of the influence of the viscoelastic properties of fluids and tissues of the body to the vital processes. The loss of a protein that coats sperm prevents it from proper mechanical interaction with viscoelastic cervical mucus and may explain a significant proportion of infertility in men worldwide, according to one of the studies [469].One of the mysteries of human fertility is that sperm quality and quantity seem to have little do with whether or not a man is fertile, said T. Tollner, a member of the international team that carried out the study. Sperm from men with some defective genes look normal under a microscope and swim around like normal sperm. But they are far less able to swim through a viscoelastic gel made to resemble the viscoelasticity of the human cervical mucus. Tollner noted that compared to sperm from monkeys and other mammals, human sperm are typically poor quality, slow-swimming, and with a high rate of defective cells. It's possible that because humans, unlike most mammals, breed in long-term monogamous relationships, sperm quality just does not matter very much [470]. Starting from the Neolithic revolution, mankind began to move from dominant polygamous relations or a brief period of monogamy up to two to three years age of the child, to long monogamous relationships. This transition to monogamy was also important because, with the rise of the disanthropy of the farmer's diet compared to the hunter-gatherer diet, the deterioration of semen quality started and reduced the likelihood of conception down to its current state. But monogamy helped to compensate the degradation of men's health (to be precise, the semen health) by increasing the frequency of sexual intercourse with one partner. Thus, the viscoelastic properties of cervical mucus may affect the likelihood of fertilization. Moreover, the interaction of sperm and viscoelasticity of mucus affect not only the biological processes in the body, but also to the fundamental social processes in human populations due to the effect on fertility.

Infertility is a huge problem in developed countries and affects 10 to 15 percent of the male population. In 70 percent of afflicted men, you can't explain their infertility on the basis of sperm count and quality [470]. This means that it is possible for more than two-thirds of all cases of male infertility to be caused by disanthropy of lifestyle, the most crucial part of which is disanthropy of food. Besides, I personally suspect that epigenetic reasons may be the underlying cause for male infertility through the aforementioned loss of a protein that coats sperm. A team of researchers [471] studied semen samples from male members of couples attending an infertility clinic. Using highly specialized molecular biology techniques, the researchers studied the epigenetic state of DNA from each man's sperm. They found that sperm DNA from men with low sperm counts or abnormal sperm had high levels of methylation, which is one of the ways the body regulates gene expression. However, DNA from normal sperm samples showed no abnormalities of methylation [471]. The results of the study suggest that the underlying mechanism for these epigenetic changes may be improper erasure of DNA methylation during epigenetic reprogramming of the male germ line that has resulted from fetal origins. But as mentioned repeatedly in this book, epigenetic changes, at least in part, have their origin in the food that your parents ate, and perhaps more distant ancestors, and from your diet starting from your mother's womb.

By the way, some interesting examples of the influence of the viscoelastic properties of the structures of the human body on its functioning come from studies of astronauts in space flight. All parts and cells of our body are under the influence of gravity. Therefore, the viscoelastic properties of macromolecular gel structures of the human body adapted during evolution to compensate for the effect of gravity and homeostatic pressure. Weightlessness in space flight alters the homeostatic level of pressure and influence on the state of the matrices of biopolymers in cells, mainly cytoskeleton matrix [266]. Structural elements of the cytoskeleton—actin filaments—which normally uniformly fill the volume of the cells are shifted to the edges. This changes the functioning of receptors and ion channels and the mesh size of the matrix. It also changes the properties of the extracellular matrix. All of this obviously affects the nonlinear transport processes of the cell-signaling proteins and beyond, and will activate some processes of transport and suppresses others. Due to the reorganization of the actin matrix, there is an additional spatial heterogeneity of the viscoelastic phase separation. This explains why, in my opinion, astronauts have increased susceptibility to infections during flight and immediately post-flight—they seem more vulnerable to cold and flu viruses and urinary tract infections, and viruses like Epstein-Barr, which infect most people and then remain dormant, but can reactivate under the cellular stress of space flight [267]. As reported in the publication [268], data also shows that during weightlessness (microgravity), the processes of intercellular interaction are not broken, but the activity of cells, such as the immune system, changes.

The viscoelasticity of tissues of embryos, as shown in [282], completely determines the pattern and development of the vascular network in our bodies for the rest of our life. A multidisciplinary team made up of physicists and biologists has discovered how, in the embryo, arteries and veins develop in parallel pairs. Using physical measurements, theoretical models and numerical simulations, the researchers showed how the growth of the arteries directly controls that of the veins through a process that depends solely on the mechanical viscoelastic properties presented in an embryo. An amazingly complex vascular network, made up of arteries, capillaries and veins, runs through our bodies and carries the necessary oxygen and nutrients to each cell and is used to remove the metabolites produced. The network has such a huge number of branches that the positions of each vessel cannot possibly be coded for genetically. But our genes are in control of the mechanical viscoelastic property of our tissues, and through them they are de facto in a control of the growth of arteries, veins, and all capillaries. A detailed study of the spatial and temporal development of the arteries and veins at the embryonic stage shows that a transformation of vascular branching takes place spontaneously during growth. By using images of the vascular network and measurements of local mechanical parameters carried out in situ, the researchers [284] showed that this transformation is triggered by the growth of the arteries. In their vicinity a viscoelastic response of the living tissue is observed, causing swelling. This response leads in turn to an increase in the permeability of the capillary bed which is highly localized in areas that are perfectly parallel to the previously formed arteries. The same we can see in developing plant leaf veins, branches, and roots were viscoelasticity play crucial role [366].

The universal importance of viscoelastic effects for all living systems and, in particular, for the life of plants is also demonstrated in [377]. This study demonstrated that turgor (a hydrostatic pressure in a cell) serves as a necessary condition rather than as a direct driving force for growth cells in the plant. The driving force of growth of plant cells is water uptake. That is provided by wall stress viscoelastic relaxation in turgid cells. This mechanical hydraulic process produces irreversible cell enlargement (elongation) in a growing plant. Attenuating growth by increasing the viscoelasticity modulus of the cell wall seems to be usual in most plants. For example, treatment with the some stress hormone elicits an increase in cell wall viscoelasticity modulus (stiffness) of shoot organs, leading to a reduction in elongation rate at the same turgor. On the other hand, drought inhibits the growth of shoot organs even at high turgor (osmotic adaptation) by increasing the cell wall viscoelasticity modulus. But roots respond the opposite way under these conditions by decreased wall viscoelasticity modulus, allowing the maintenance of growth even at low moisture level in the soil. The adaptive value of viscoelasticity in these changes in mechanic of cell wall is obvious. These findings confirm the notion that changes in cell wall extensibility, produced by changes in cell wall viscoelasticity, are the basic and universal mediators controlling plant cell growth.

The results of the above studies show that by taking into account the viscoelastic properties of various biopolymer matrices, we can formulate more adequate models of transport processes in cells and tissues of the intercellular space and functioning of organs and body fluids. Modern biochemistry seeks to describe physiological processes in terms of gene functions and specific physicochemical molecular mechanisms. Dietology and medicine add the practical goals of understanding aging and disease progression and developing treatments. The molecular pathway identification phase of modern biochemistry is approaching completion, and the sheer size of the organs' "parts list" highlights the importance of understanding function, not only at the level of every single gene or every single metabolic pathway, but rather at all higher levels of hierarchy, including cells and organs, involving their mechanical properties and particularly the viscoelasticity of different types of biopolymer matrix.

But from the published results of many studies, I am under the impression that the conception of modeling cells' and organs' functions is at the breaking point—it is cognitively  impossible even with the use of modern computing to juggle large numbers of metabolic pathways involving an even bigger number of molecular components. The problem is that the amount of data in molecular biology is growing exponentially. Although the computing power by Moore's law is also growing exponentially, the rate of exponential growth in computer performance is less than the rate of exponential growth in the volume of genomic and other molecular data. Therefore, cognitive analysis, despite all the clever tricks in the programming, begins to choke. And even more difficulties arise from situations in which the Ergodicity tenet, which is generally recognized as a law of nature, would be not adhered to molecular transport and signalling pathways in living cells [314]. (The Ergodicity theorem predicts that statistically, the result of throwing ten dice once would have the same average distribution as throwing one die ten times [315]). The Ergodicity theorem is expected to apply for molecular transfer processes in biological systems. Therefore, it means that if you observe the movement of some molecules in many cells at once, you can expect to get the same result as by looking a single cell repeatedly over a long period of time.

The researchers [314] studied some molecules that are naturally occurring in cells. Using a special state-of-the-art instrument, an optical tweezers, they were able to hold onto the small molecules inside living yeast cells using an extremely focused laser light. By measuring the movement of the specific molecules over several hours, they could observe that they were not behaving with the distinct pattern expected according to the Ergodocity theorem. But conflict with ergodicity in this study [314] had happened, in my opinion, only because the researchers failed to take into account the viscoelastic state of cells and cells' interior. In reality they were watching the molecules in different viscoelastic surroundings, which was why their behavior looked abnormal. Thus, the results [314] show that neglecting the viscoelastic component of mechanotransduction molecules makes any model of transport processes in the cell and outside it irrelevant. That is, in principle such an approach cannot be used for modeling molecular processes of metabolism of food, drugs, toxins, etc. Current understanding among most biologists of regulation processes in the cell is based on concepts of pure chemistry. But mechanical states of biopolymer matrix in the cell and in the surroundings, processes of mechanotransduction in viscoelastic media that play a very important role in biological regulation, escape their attention. Hidden from the attention of scientists is the role of the viscoelastic matrix in the study [314], like the famous Cheshire Cat [472], which is known for its disappearance into the air, leaving for observers his smile—very viperous in this case—an apparent violation of ergodicity.

The most difficult problem of biology at the present time is the problem of biological complexity. It would seem that the amount of biological and medical knowledge is incredible: the human genome is decoded; genetic and molecular manipulation techniques have achieved an incredible sophistication; but a complete victory over disease still seems very far away.

The work of the simplest living cell is much more complicated than it seemed before, and the most modern methods of research have revealed their weakness. Only in the last decade, it has been finally realized that the deep knowledge of the general pattern of the cell's functioning is absolutely necessary for the understanding of disease: cancer, aeing, degenerative and metabolic disorders, etc. As a real biological system, the cell is a network of hundreds or thousands of different enzymes, proteins, and metabolites, which have millions of ways to influence each other. Scientists may know almost all the components of living cells and the parameters of most reactions, but they still cannot say how the system works. As a result, even "armed with a computer mind," they are cognitively powerless before the creation of the simplest reliable description of metabolism, or cellular signalling process. In practice, this means that they cannot univocally manage the cell's work with medications or diet.

Because of the molecular transport of proteins, the concentrations of million proteins in the cell have their spatial and temporal heterogeneity, which are determined by several relatively simple physical and chemical mechanisms. First, proteins that must be delivered to a specific place in the cells (organelles, membranes, etc.) acquire a signal molecule that determines the "destination" of a given protein. It is based on some simple chemical process. For example, a protein that contains a part of the amino acid cysteine, on the surface of the Golgi apparatus (cells organelle) gain as a signal molecule the so-called lipid anchor, and therefore are automatically transported to the cell membrane or the membranes of organelles. This process, known as palmitoylation, equips the membrane proteins with a kind of address label and ships them off to the cell membrane. The cell uses this directed transportation from the Golgi apparatus to the cell membrane as a means of countering the permanent "leakage" into other membranes that occurs. This is important because, besides the cell membrane, the cell is filled with membranes from organelles connected to one another via vesicles. Consequently, palmitoylated membrane proteins, originally intended for the cell membrane only, also reach other locations. With time, these proteins would then be distributed throughout the cell [340]. The pattern of distribution depends on interaction between the cell biopolymer matrix and palmitoylated membrane proteins.

Second, proteins can also attach themselves to the molecular motors that help them move in the desired direction. Molecular motors are a low-molecular-weight compound, joining the protein molecule, while maintaining a degree of freedom of their terminal groups, allowing them, for example, to rotate and act as a "propeller" to move the ship of a protein molecule.

Third, due to the dynamic asymmetry of diffusion in viscoelastic phase, separation in the cell speed of different-sized protein complexes is different. Therefore, the viscoelastic matrices cells begin to do the work as the parametric filter and set the

spatial pattern of concentrations and control them by means of mechanochemical changes of properties and stresses in the matrices. That is, seemingly complicated processes of regulation of spatial concentration, selection, and transport of
proteins in the cell are controlled by relatively simple physical and chemical mechanisms "lapped" to each other during evolution. This fine-tuned "lapping" explains the exceptional difficulty of regulation of such processes in cells through the
chemical action of drugs. The action of foreign chemicals in a separate stage of a
process confuses "tuned" mechanisms of other processes. From it also follows the
critical importance for body cells of the anthropy of the set of chemical elements and
compounds entering the body through food.

But viscoelasticity manifests itself not only at the cellular or other body level but
also at other levels of the organization biocoenosis. Moreover the viscoelastic properties of living organisms can influence the general conditions of life on earth. For
example, one recent study [350] revealed that the physical properties of Arctic sea ice
determine its habitability. Whether ice-dwelling organisms can change those properties has not been systematically addressed before. Following the discovery that sea ice
contains an abundance of gelatinous viscoelastic extracellular polymeric substances,
researchers found that their existence affects ice and pore microstructure and sea ice
habitability, survivability, and potential for increased primary productivity, even as
they may alter the persistence and biogeochemical imprint of sea ice on the surface
ocean in a warming climate.

In concluding this part, I can draw an unambiguous conclusion that the viscoelastic properties, which manifest themselves in any and all hierarchical levels of the
structure of organisms, are the true genius loci (lat.) of living systems. They affect
all aspects of life: from molecular responses to mechanical properties of organisms as
a whole and, of course, the processes of transformation of food and its metabolites at
the cellular level of organisms.

## 3.7   Food and genes interaction in viscoelastic beings

Full knowledge of all processes in the cell and the organism is a task of unimaginable
complexity which is comparable with the problem of artificial reproduction of the
organism. However, regardless of who created all living things—God or evolution
of nature—our modern knowledge is not comparable in scale with either of these
possible driving forces of the universe—with the first, by definition, and with the
second, because of the scantiness of our knowledge at this point.

The set of biochemical reactions determining the metabolism of a certain
organism may differ from other similar organisms. For example, one study [28]
showed small deviations within the structure of human metabolism as individual

as fingerprints. This of course does not alter significantly the region of attraction of trajectories of the reactions in the phase space, just as pattern prints of our fingers do not determine the function of the hand, but it is, nevertheless, clearly an important evolutionary feature of our body. Individuality of metabolism was used by our ancestors and is now used by many animals (usually males) in the wilderness by using their feces or urine to indicate the boundaries of the territory. Through subtle variations in individual metabolism, animals can easily distinguish their own smell from the smell of a competitor, another animal of the same species. Results of the study [213] provide an example of ways in which normal people differ in their metabolism. Most people produce a specific distinct odor in their urine shortly after eating asparagus. But a significant minority—approximately 8 percent of the people tested—did not produce in the body metabolism the odorous chemicals related to asparagus. Although this fact looks like just a curiosity, the individual differences in metabolism could be important in other realms. For example, metabolic changes due to obesity in children, because of genetic conditions or diets, significantly alter the trajectory of drug metabolism [29] and determines subsequent adult disease [30].

Also, as shown in study [281], all humans have a gene responsible for making enzymes, but different populations of humans vary in which version of the gene they carry, and thus, which versions of enzymes they produce. "Such variations lead to differences in the way people metabolize and respond to particular drugs or food. With a drug such as caffeine, for example, one population of people might be fast metabolizers, while another might metabolize the drug more slowly," M. Green, lead author of the study, explained. Because the risk of caffeine-induced heart attack may be higher in slow metabolizers, the ability to actually take a snapshot of the phase changes of the specific enzymes could help to understand better how certain chemicals can affect people in vastly different ways.

Also, according to the studies [58, 398], nutrition during the first weeks and months of life may have long-term consequences on health, potentially via a phenomenon known as metabolic programming. This is the concept that differences in nutritional experiences at critical periods early in life can program a person's metabolism and health for the future. That demonstrates the stability of individual metabolic changes during the entire life of the organism and is particularly sensitivity of the child's body to disanthropic influences. It should also be noted that despite the presumption that most childhood obesity results from little movement and lack of sports, these larger studies [63] show that the situation is closer to the reverse—obesity is caused by poor diet, which leads to a sedentary life.

As registered in many studies, the presence of subtle differences in metabolism among similar animals and humans suggests that the metabolism is a highly sensitive system with regard to the expression of the organism's genes and to the diet.

Furthermore, the diet-related genes also appear to have evolved faster than other genes—protein and promoter sequences of these genes changed faster than expected, possibly because of adaptation to new diets [361]. In studies [34], cultures of yeast cells showed that feeding conditions alter the work of specific genes of yeast cells that lead to a change in metabolism and adaptation to the new food source—i.e., to a set of completely different trajectories of metabolic reactions. At the same time, when there are two sources of food for the cells, they prefer to use the one they like more and which is better adapted to them. Conversely, when the better one is disappearing they turn to the worse one for their food source.

The same thing happens with iron deficiency in the cells, since iron is used in the mitochondria of cells for the processes of energy production in the form of ATP. Iron deficiency alters the trajectory of cell metabolism, and finally, when the iron concentration becomes very small, the alternative mechanism of energy production—the direct conversion of glucose—is activated in cells [108]. This is a very inefficient process compared to the process in the mitochondria, but it allows the cell to survive for a time. In general, the most important factor for the proper metabolism is the balance of iron in the body cells because elevated levels of iron have a toxic effect. We can say that our body has a love-hate relationship with iron. Just the right amount is needed for proper cell function, yet too much is associated, for example, with brain diseases like Alzheimer's and Parkinson's, according the study [589]. Iron accumulates in our bodies as we age, and does so more quickly if we are using iron-containing medication or dietary suplements (DS). Higher brain iron levels in men may be part of the explanation for why men develop these age-related neurodegenerative diseases at a younger age than women.

A study conducted in the United States reveals the existence of at least two independent mechanisms for iron absorption from non-meat sources and a potential treatment for iron deficiency, the most common nutrient deficiency worldwide [560]. Researchers demonstrated that there is an alternative mechanism for the absorption of plant ferritin—a large, protein-coated iron mineral abundant in legumes—in addition to the more well-known mechanism for iron absorption of small iron complexes like those found in artificial dietary supplements. The study shows that during digestion, ferritin is not converted from its large, mineral complex, which contains a thousand iron atoms, to individual iron atoms like those found in many iron supplements. When the intestine takes in a single molecule of ferritin, however, it gets a thousand atoms inside that one ferritin molecule, making iron absorption that much more efficient. Natural ferritin iron is absorbed in its protein-coated, iron mineral form by a different, independent mechanism; iron absorbed as ferritin leaves the intestine more slowly, but may provide greater safety to the intestines than iron rich DS. Absorption of iron in a natural form as ferritin also causes less irritation to the intestine. Plus, this different mechanism of iron absorption from plant ferritin is thousands of times more efficient and gives the intestinal cells more control. Also, iron supplements frequently cause uncomfortable and potentially dangerous side effects linked to iron accumulation in tissue, but absorption from plant ferritin causes no side effects An alternative and highly efficient mechanism for iron absorption from natural anthropic plant food, therefore, could provide the key to helping solve worldwide iron deficiency by providing a readily available and affordable source of iron.

The interaction with oxygen is a major source of proper iron metabolism, but at the same time, it is also the source of many diseases. Also, the interaction with proteins for transportation into the cell is critical in maintaining the concentration of iron in the cell [109]. The complexity of metabolic processes involving iron means that the correction of imbalances in the long run is not possible with medication and dietary supplements. They will inevitably have side effects in long-term use. This is because the regulation of such delicate metabolic trajectories is only possible via a natural way to obtain iron from food in the anthropic state that has evolutionarily good metabolic qualities, and unused balances of such iron can be easily excreted by the body. The findings bear out what geneticists long have said: there is nothing that works for every phenotype and genotype, because an organism is specific and each has a unique set of genes whose expressions pattern changes constantly. The individuality of body metabolism could have different sources, but as a result, all people process foods and utilize their metabolic contents differently.

1. First, the individuality of metabolism is determined by our ancestral genetic heritage: populations in all parts of the world lived on their regional food supply for thousand years and so fulfilled their metabolic needs from different food sources. Those living in a northern or other harsh climate tend to burn food quickly. They require high energy foods to sustain them. Eskimos, for example, can easily assimilate in their metabolism large quantities of fat and protein. In southern climates zones, however, people live on lighter, vegetarian food. Because we all have mixed genetic inheritance, we are utilising food slightly differently.

2. Second, the individuality of our metabolism is also determined by the epigenetic status of gene expressions inherited or accrued in a lifetime.

3. Third, the state of our metabolism is determined by the quality of food in our everyday diet.

4. And fourth, the state of our metabolism highly depends on our health state and how organs and systems of the body are functioning.

And of course, the anthropic diet must take account of these four sources of individual metabolic state of each person. Genetic features of metabolism often occur as the absence of tolerance to a particular food. Good examples are the lack of tolerance of alcohol or milk in some parts of the world's population. From the other side is well known fact about food addiction. Scientists [474] have found that some disanthropic substances may have hijacked the genetic neural programs of metabolism that serve as pathways for anthropic elements of food. When the genetic program is operating with disanthropic elements of food, experiences that are part of the execution of the program become deeply embodied as addiction in the overall patterns of an individual's behavior. This explains how the different food addictions (for example, excessive fat, salty chips, and soda and sugary drinks consumption) arise. The researchers found that the genes expressed by stimulating an instinctive behavior, like a salt appetite, were the same groups of genes regulated by heroin addiction. Deeply embedded pathways of an ancient instinct may explain why treatment of food- and drug-addicted individuals with the chief objective of abstinence is so difficult. We should expect very similar difficulties when people switch from a modern diet to an anthropic one. Though wrong-food addictions are using existing genetic neural metabolic programs, they may be substantially, but slowly, changed by training the organism to accept the right anthropic diet, through education and cognition.

Epigenetic changes are present in all populations as a mechanism for adaptation to local food rapid change. It is clear that the state of cell metabolism, which is due to the expression of certain genes activated by the steady cascade of biochemical transformations, is determined mainly by the signals received from food. This is, so to speak, a biochemical interpretation of the nonlinear processes mentioned above, such as the activation of food particles, formation of the cumulative jets, and occurrence of coherent interactions in the phase space of a strange attractor metabolism. It should be noted that this is a wonderful confirmation that the impact of food on the cells of the body quickly starts the epigenetic excitation of various parts of the genome. It was assumed before that the effect of chemical compounds of food may occur in the genome only as a result of a relatively slow process of formation of mutant genes in the cells.

Food stress has become one of the major disease sources in our everyday life. Stress appear at the cellular level after exposure to such disanthropic factors as chemical pollution, the wrong food, and bacterial toxins. Stressed cells at whole and organelles in order to survive and maintain their normal function have to react to disturbances. A recent study [213] found that stress factors from food can control genes by turning on certain genes that were supposed to be activated. This also confirms that without changing our genetic code, food stress factors can control the activity of our genes and our genetic appearance on individual and population levels.

When it comes to how they epigenetically respond to the environment, humans may not be that different from other living beings. Many studies show that even genetically identical human twins could have a different chance of getting a disease.

This is because each twin has unique personal lifestyle—mainly diet—throughout his or her life. It turns out that the same is likely true for forest trees as well, according to new research {441}. These authors looked at the theory that trees and other plants, even when they were genetically identical, grew differently and responded to stress differently depending on the nursery that the plants were obtained from {442}. The researchers {441} found that there is a specific "molecular memory" in trees where a tree's previous individual "lifestyle" influences how it responds to the environment. In the study, they used genetically identical poplar trees that had been grown in two different regions of Canada. These stem cuttings were then used to re-grow the trees under identical climate-controlled conditions in Toronto. The authors subjected half of the trees to drought conditions while the remaining trees were well watered. Because the trees were re-grown under identical conditions, the research team initially predicted that all the specimens would respond to drought in the same manner, regardless of where they had come from. Remarkably, genetically identical specimens of two poplar varieties responded differently to the drought treatment depending on their place of origin. The researchers also showed that this difference occurred at the fundamental level, the level of gene expression. Even though the specimens were all genetically identical, trees that had been obtained from one region used a different set of genes expressed to respond to drought than the ones that had been obtained from another region.

The findings of this study are relevant for all organisms. It means that universal rules applied for epigenetic interaction with the environment for all living beings, and thus the anthropic approach to human dieting, should be exceptionally productive in practice. The fact that epigenetic "mutations" under the influence of the various components of food are not exotic fantasies is illustrated in another recent study [295]. A good example is the influence of folate (folic acid) during pregnancy, which is found in many natural sources of our food, on the epigenetic modification of the genome of infants. This study showed that the levels of a critical metabolite of folic acid, homocysteine, in the blood of newborn babies is linked to modifications of their DNA (DNA methylation) in key genes, and that such modifications might be used to predict birth weight. One author of [295], Professor Farrell said, "It has been known for many years that dietary folic acid is essential for women during pregnancy to reduce the risk of neural tube defects and low birth weight delivery. However, we had little idea as to how this worked. This study is the first to suggest that methylation of particular genes in the baby›s DNA may be the key to unlocking the secret of the action of folic acid" [296].

Incidentally, it should be noted that the processes of methylation of genes play a crucial role in many biological processes related to people. From a general point of view, a methylation is a process by which enzymes add a small molecular tag to a particular location on a nucleotide, a molecule that is the structural unit of

RNA and DNA. Proteins created by bacteria genes had been found to play a key role in the bacterium's mechanisms of antibiotic resistance. In bacteria that are not drug-resistant, when the antibiotic molecules are tagged to the specific nucleotide, it facilitates the proper functioning of the bacterial ribosome, a macromolecular biopolymer structure that is responsible for making proteins that bacteria need to survive. One study [397] demonstrated the existence of a new chemical mechanism for methylation in which the specific protein molecular tag binds at some location on the nucleotide. The addition of this new tag blocks the binding of antibiotic molecules to the ribosome but without disrupting its function. Such a methylation process is involved in the process of evolving very dangerous antibiotic resistant "superbugs" that often are found in hospitals.

It should be noted that the processes of methylation occur in the viscoelastic biopolymer environment, but to isolate the role of viscoelasticity in the macromolecular matrix of the ribosome is not possible due to lack of information about the molecular mechanisms of its work. Recent years have seen growing interest in the phenomenon of epigenetic inheritance—the idea that our genome, through epigenetic tags and other structural encoding, transmits more information than the sequence of letters encoded in its DNA base pairs alone. Stresses of various kinds have been shown to induce such epigenetic change.

Epigenetics will not replace classical genetics. It is, however, providing for us the biochemical explanation for non-DNA, heritable changes that tell cells what their parents have been and what they themselves are going to be. Many studies show that adult stem or progenitor cells, like all of us, come with a "family history" that affects their functional destiny. The processes in chromatin structures on behalf of epigenetic memory are of great importance to human diseases. But we have to keep in mind that the mechanisms of epigenetic memory and of dynamic chromatin changes leading to differential expression of genes throughout our life are different processes. Reversible chromatin structure changes under the influence of environmental factors are an element of flexible adaptation of cell metabolism. Theoretically it could lead to diseases as well. But generally, it is minor changes within the boundaries of normal. The external factors that lead to a heritable epigenetic state of metabolism are called epigenetic memory.

But the existence of inheritable epigenetic changes is a very controversial matter in modern science. This is because, on the one hand, there is experimental evidence of the existence of transgenerational persistence of DNA methylation-dependent phenotypes. But on the other hand, during sexual reproduction epigenetic reprogramming takes place, which consists of the practical removal of all methylation from the cell [476]. Thus the main question: how the DNA methylation patterns can be inherited under a context of complete de-methylation. Personally, I think the answer to this question may lie in a partial inheritance, unique to each partner conformational 3D chromatin structure in the sex cells, which in the fetus leads to methylation pattern, and partially manifested in traits of both parents. From above it follows that parental diet stress during their life can leave an imprint on their children's genes - an imprint that lasts into later life and affects how the genes are expressed in the individual. Due to the nutritional stress of disanthropic food in the parent's diet, the genes have consistent changes in methylation levels at multiple sites on the DNA, including many of those crucial in keeping the healthy state of cells, as for example, in the production of insulin and other hormones. Recently, many studies focus on the work of the miRNA (microRNA) in cells as a regulator of synthesis and transport of proteins in the cell. The role of RNA is crucial in epigenetic processes of gene expressions. But it turns out that miRNAs are present in human food, such as in breast milk, and therefore have a significant effect on metabolism [55]. In the artificial feeding of infants, no microRNAs presented. This confirms disanthropy of artificial mixtures and the inability to replace mother's milk. From this follows the idea strongly advocated in this book that all levels of the body's metabolism depend on the chemical composition of incoming food, which has an influence on the expression of certain genes. Moreover, according to the study [396], microRNA mediates gene-diet interaction related to obesity and other food-related gene expressions. In the study [572], scientists report the surprising finding that exogenous plant miRNAs are present in the tissues of various animals and that these exogenous plant miRNAs are primarily acquired orally, through food intake. I am sure the same should be true for people.

Research [35, 56] also confirms the fact that aging, diabetes, and life expectancy are influenced by genes and diet together. From the above relationships in the expression of certain genes and the composition of food, it becomes clear that the mere interaction of the genome and the food is extremely complex. Many medical studies have shown that the onset of manifestations of inherited diseases in the body such as cancer, diabetes, heart and blood vessel changes, and osteoporosis to a large extent depend on lifestyle and, basically, on the type of food eaten. That is, implementation of a hereditary predisposition to the disease occurs due to improper metabolism of disanthropic diet, to which people with impaired (aberrant) heredity are usually the most sensitive. For a long time, it has been known that there is a link between red meat consumption and prostate cancer. In one study [74], genes whose expression occurs in the digestion of red meat also have a nine-fold activity in prostate cancer. Consequently, excessive red meat consumption and the associated constant expression of the corresponding gene can lead to cancer.

Another study [209] demonstrated that Alzheimer's disease and type 2 diabetes are both controlled (at least partially) by the expression of single gene. The data show that a gene for a certain protein, which can cause type 2 diabetes, impacts the accumulation of amyloid-beta (A-beta) peptides in the brain. A-beta is the main constituent of amyloid plaques in the brains of Alzheimer's disease patients. But it follows from the results of other studies in this book that the link between type 2 diabetes and diet is very important. That is, quite the same relationship may exist between Alzheimer's and diet. Incidentally, this is corroborated by the fact that both Alzheimer's and type 2 diabetes are reaching pandemic levels, afflicting millions of Americans. One study [208] offers the best evidence that the connection between these two diseases originates on a molecular level where food (and food metabolites) and genes interaction are most crucial. In light of the study [353], the link between diabetes and Alzheimer's disease is not surprising. Results from this study completely altered scientists' ideas about Alzheimer's disease—pointing to the liver instead of the brain as the source of the amyloids that deposits as brain plaques associated with this devastating condition. The findings could offer a relatively simple dieting approach for Alzheimer's prevention and treatment, because the functioning of the liver as the organ with a very high level of metabolism could be significantly altered via diet. This finding also suggested that significant concentrations of amyloids might originate in the liver, circulate in the blood, and enter the brain. It means that blocking production of amyloids in the liver should protect the brain. And it is much easier to block production in the liver than in the brain because usually it is very difficult for medicine and food metabolites to penetrate to the brain tissue through the blood-brain barrier.

In light of the foregoing information, it is important to take steps to correct the organization of the anthropic nutrition to prevent childhood obesity. Medical experts fear that the childhood obesity epidemic could lead to large numbers of adults developing initially diabetes and later Alzheimer's, causing serious health complications for decades in the next one to two generations of Americans. Even a relatively short and small changes in diet as shown in [478, 479] may significantly change the condition of the people regarding the development of degenerative diseases such as Alzheimer's or obesity. For example, following a low-saturated fat and low-glycemic index diet appears to modulate the risk of developing dementia that proceeds to Alzheimer's disease, and making a switch to this dietary pattern may provide some benefit to those who are experiencing cognitive difficulty [477]. As the study [478] also shows, changes in specific dietary factors may have a big impact on long-term weight gain. These changes include very simple recommendations, eating less liquid sugars (e.g., soda) and other sweets, as well as fewer starches (e.g., potatoes) and refined grains (e.g., white bread, white rice, breakfast cereals low in fiber, other refined carbohydrates) and eating more minimally processed foods (e.g., fruits, vegetables, whole grains, nuts, yogurt) and fewer highly processed foods (e.g., white breads, processed meats, sugary beverages). However, even if these relatively small changes in the diet of modern man lead to such a visible result, it is easy to imagine that even a partial transition to the anthropic diet for prolonged enough exposure may lead to dramatic positive changes in the individual's health status in terms of the risk for degenerative diseases. Of course, any relatively healthy dietary pattern may influence the development of dementia or long-term weight gain in many ways, including, for example, through biologic effects, such as changing hunger or insulin levels. But specifically only the anthropic diet could influence the development of metabolic degenerative diseases through epigenetic effects leading to long-term health benefits and improving our phenotype on cell and subcell levels. It has paramount importance for a lot of people in the developed world because researchers [479] forecast a looming global epidemic of Alzheimer's, with the disease quadrupling worldwide by 2050. The main reason for the rising cases of Alzheimer's is not the rise in number of people and aging population, but the wrong diets of most of the population—the disanthropy of modern food. In a recent report [548], an international team of researchers analyzed and compared changes in gene expression associated with aging and disease in a region of the brain known to be affected in Alzheimer's. Comparing samples from healthy individuals ranging from 16 to 102 years old with samples from diseased individuals, the investigation found striking similarity in the changes in gene expression patterns associated with aging and Alzheimer's disease. The same epigenetic expression changes that take place in healthy individuals at an advanced age may happen in a potentially diseased person at a much younger age (up to 20 years). But as we know, epigenetic changes at least partially could be successfully altered by environmental factors. This gives us hope to manage in some degree those genes' expressions via diet containing anthropic substances.

The power of even a partially anthropic diet compared to ordinary dietary intervention is demonstrated in research [535]. Patients with high cholesterol who received counseling regarding a diet that combined cholesterol-lowering foods such as soya protein, nuts, and plants containing sterolsover for six months experienced a greater reduction (13 percent) in their low-density lipoprotein cholesterol (LDL-C) levels than individuals who followed the ordinary advice of conventional dietary therapy on a low-saturated fat diet (only 3-percent reduction in range of experimental error). A meaningful LDL-C reduction obtained via the anthropic intervention distinctively show the great advantage of the anthropic approach to dieting in comparison to the useless and incorrect paradigm of conventional dieting. Researchers from the Boston University School of Medicine (BUSM) have reported similar results [536]: African American women who consume more vegetables are less likely to develop estrogen receptor-negative breast cancer than women with low-vegetable intake.

Even such small changes in diet as presented in review [540] could prevent some types of cancer. The risk of breast cancer dropped significantly in animals when their regular diet included a modest amount of walnut. According to the review, results of the genetic analysis showed that the walnut-containing diet changed the activity of multiple genes that are relevant to breast cancer in both animals and humans. Other testing showed that increases in omega 3 fatty acids did not fully account for the anti-cancer effect, and found that tumor growth decreased when natural dietary vitamin E increased. In addition, adding healthy fat and other components from walnut meant that unhealthy fat was reduced to keep total dietary fat balanced. During the study period, the group whose diet included walnut at both stages developed breast cancer at less than half the rate of the group with the typical diet. The number of tumors and their sizes were significantly smaller. These reductions are particularly important when you consider that the mice were genetically programmed to develop cancer at a high rate. Walnuts added to the diet were able to reduce the risk for cancer even in the artificially predisposed genetic conditions.

Also interesting is the relationship of many degenerative diseases simultaneously with a genetic predisposition and diet. One study [167] analyzed several diseases with a genetic predisposition, including diabetes, coronary insufficiency, and rheumatoid arthritis. These diseases are associated with mutations in single genes, where the structures of nucleotides vary from person to person. Variations of this type are called single-nucleotide polymorphisms (SNPs)—DNA sequence differences in the size of a single nucleotide (A, T, G, or C) in the genome. Some of these SNPs are associated with the risk of a disease or, conversely, with protecting the body against certain diseases. To calculate the genetic risk of a disease, it is necessary to calculate the net effect of addition SNP-risks.

But it turned out that some variations reduced the risk of one disease while at the same time increasing the risk of another. For example, the genetic variation of the gene responsible for enhanced antiviral resistance to enteroviruses, also increases the risk of Type 1 diabetes. The same there is, for example, between TB and rheumatoid arthritis [166]. It is possible that in populations in which there are no external triggers for the disease, most people only have a positive effect from the relevant SNPs. However, if the external conditions—chief among which is the diet—change, then the negative qualities of SNPs are beginning to be realized in the form of statistically relevant manifestations of the disease in the population.

In Africa, there are populations with a genetic predisposition to a certain type of anemia. But this same gene protects against malaria. The positive effect of protection is much greater than the risk of disease. However, if the living conditions and nutrition of individuals vary, for example, when moving to Western countries, then there is no risk of malaria, and the risk of anemia is manifested and amplified by the disanthropy of diet. The very survival of this evolutionary disease in the population is connected, apparently, with the presence of a positive effect on health for the SNPs associated with disease. In the natural anthropic environment, at least some of the genes responsible for disease were epigenetically repressed, and thus manifestations of the disease are almost nonexistent, and the positive effects of protection against malaria in a population prevail.

As if the recent prediction that half of all Americans will have diabetes or prediabetes by the year 2020 isn't alarming enough, a new genetic discovery published online in the *FASEB Journal* provides a disturbing explanation as to why: we took an evolutionary "wrong turn." In the research report [352], scientists show that human evolution leading to the loss of function in a gene called "CMAH" may make humans more prone to obesity and diabetes than other mammals.

But the propensity to diabetes does not make it inevitable. A well-known illustration is the Indians in North America. Most of them did not hear about diabetes at least for a few thousand years. But in the last century or so, many of them have adopted a main feature of American lifestyle: unhealthy food. Almost overnight in evolutionary scale, a very high percentage (up to 50 percent in some areas) of Indians ended up with type 2 diabetes. Their genes cannot change so quickly. The genetic differences for increased diabetes risk were always there. But with their old style of eating, it didn't matter. In other words, the gene's variation itself wasn't enough to cause the diabetes. Their food sources and eating habits had changed before they developed the disease, and that led to the epigenetic changes—wrong genes' expression. This suggests that the best thing you can do to prevent getting type 2 diabetes is to eat the right food.

Similarly, people of South Asian ancestry are up to four times more likely than Europeans to develop type 2 diabetes, which is a major risk factor for heart disease and stroke. In the recent study [524] that is the first to focus on genes underlying diabetes amongst people originating from South Asia, the researchers examined the DNA of 18,731 people with type 2 diabetes and 39,856 healthy controls. The genomes of the participants were analyzed to look for locations where variations were more common in those with diabetes. The results identified six positions where differences of a single letter in the genetic code were associated with type 2 diabetes, suggesting that nearby genes have a role in the disease. It means that evolutionary changes in ancestors of South Asians that brought some benefits to the population made them more susceptible to diabetes. And so a high number of genes' loci involved in diabetes show that evolutionary changes happened in a relatively later stage of evolution, probably within the last 1–4 million years when the complication of human ancestors' genome turned them into the direction to became Sapiens. But it was not any problem until the last century, when changes in diet released hidden forces in the genome, making South Asians vulnerable to diabetes.

So obesity and diabetes are not purely a downside of human evolution, as some researchers suggest [523]; they are rather the downside of our diet's evolution since the beginning of the Neolithic revolution and the interaction of the disanthropy of our food with our genes. The evolutionary changes help us to win some traits, which keep our species ahead of the extinction curve. As long we are living in the evolutionarily inherent environment with evolutionarily inherent (anthropic) diet, everything is OK. But evolution cannot cope with much faster, dramatic and unexpected changes in our diet that lead to diabetes and many other diseases. Changing their normal diet to the unhealthy Western-style diet for people from different part of the world almost inevitably put them at higher risk of metabolic diseases. So South Asian adults in the UK have approximately three times the risk of acquiring type 2 diabetes compared with the white European UK population, while people of African-Caribbean origin in the UK have roughly a two-fold greater risk[444]. Despite strong evidence that type 2 diabetes is genetically predisposed, it can be prevented or at least delayed by a combination of lifestyle changes and anthropic dietary intervention. Many studies indicate that dietary advice alone could play an important role [537]. One study randomly assigned people to either a control group or a dietary group. After six years, 67.7 percent of people in the control group had diabetes, compared with only 43.8 percent in the dietary group. This was a 33-percent reduction. In another study, 12 months of dietary intervention led to significant reductions in many diabetes-related factors, such as insulin resistance, fasting C-peptide, fasting pro-insulin, fasting blood glucose, fasting triglycerides, and fasting cholesterol.

At the present time, which many call the era of DNA, many studies of various diseases have focused on genetic causes. But after extensive research over the last ten to fifteen years, we can reliably say that most of the data linking certain genetic variants and disease in the majority (70 percent) of cases are incorrect. For example, known commercial tests for the prediction of cardiovascular disease, Alzheimer's disease, and others were totally inadequate to the declared objectives. In fact, even knowing the variant gene that leads to a particular disease, doctors cannot reliably predict the risk of developing the disease. First, most diseases are not determined by a single gene, but by their combination in the whole genome, which sometimes reaches several tens and perhaps hundreds of genes. Second, the vast influence of epigenetic factors that determine the pattern of gene expression depend on the effect of diet, environment, and lifestyle of the individual. In particular, genome-wide studies cannot be certain that the genetic variation identified directly increases the risk of diseases. The DNA change could be a flag or marker that the important gene lies nearby, or the DNA change could lie within a control element that regulates a different gene some distance away. I am not arguing to ignore DNA, but for most common diseases, genes alone only tell us part of the story. That's because the food and food metabolites interact with DNA in ways that are difficult to predict, even in simple organisms like single-celled yeast. Thus the effects of a person's genes—and their risk of disease—are greatly influenced by their diet. To understand gene-environment interactions at the most basic level—at the individual

DNA letters that make up the genetic code—the researchers [293] turned to a well-known good model for the human organism, the yeast Saccharomyces cerevisiae, culled from North American oak trees and vineyards, where it grows naturally. They asked a simple question: would growing the yeast in different available food environments influence the rate at which the yeast produce spores? In their study, researchers grew the two yeast strains with all sixteen combinations of four SNPs in different simple sugars: glucose, fructose, sucrose, maltose, raffinose (a combination of sucrose, glucose, and fructose), grape juice, and galactose. These were all mono- or disaccharides, so the foods are not radically different from one another.

Surprisingly, the scientists found that the effects of the four SNPs on spore production were dramatically different in the different food environments. The effects of different combinations of SNPs in one type of food were not an accurate predictor of the effects of those same SNPs in other types. In relation to humans, this means that any food or medication is likely to leave a measurable molecular signature in epigenetic factors and in the chemistry of metabolism simultaneously. But exact results are unpredictable even for organisms as simple as the yeast. Only simple conclusions in well-known cases could be derived. For example, eating a lot of fatty foods raises triglycerides; smoking raises nicotine levels; and eating high-fat, high-sugar foods raises blood sugar levels and increases the risk of diabetes. Having a particular genetic variant may not have much of an effect, but combined with a person's everyday wrong diet, it may have a huge effect. The key is to figure out what are good metabolic readouts of the food and factor those into statistical models that assess genetic susceptibility and epigenetic changes to disease and response to food or medication. As scientists conduct ever-larger studies to identify rare and common genetic variants underlying diseases such as cancer and diabetes, they hope to uncover variants that have larger effects on disease.

But I think we have to have the humility to accept that our current understanding of nutrition and how the food could influence and correct the body's metabolism is meager at best and possibly wrong in many parts. For the last– twenty to thirty years, modern nutriology and its part - nutrigenomics, has fallen for the widespread delusion that the glimmerings of insights emerging from genomic research, cell biology, and other areas would provide a Golden Road to new functional food and herald an era of personalized nutrition and medications. Although "personalized" dieting is a term used in science and dietology that holds significant promise of improved treatment and prevention of many diseases, it may set up unrealistic expectations in people. Despite confirmed success, patients would be foolhardy to expect anything more than a small number of insignificant additional tailored interventions in their diet. We should not expect from "personalized" dieting the cures for all common diseases, and should understand that a simple, targeted solution is unrealistic because of the complex interplay between genes, their expression patterns, cell structures, metabolism, and diet influences. Only a complex approach through maintaining an anthropic diet as a lifestyle, could give results such as preventing diseases or postponing their development. In my hard opinion, it is not feasible to spend huge funds on projects like Food4Me, the new, EU-funded project investigating the potential of this personalized nutrition.

At present, personalized dieting and medicine centers are being opened in which scammers, under the guise of an individual approach to the patient, instead universally facilitate their wallets. The wrong paradigms of modern dietology as a whole and thousands of charlatans in the field of dietology prevent the return of people to the diets evolutionarily inherent to the human body and show in this case that a little knowledge is worse than no knowledge at all. Nevertheless, however, a person's diet, environment, and lifestyle will be very important and in many cases will have prevailing significance. Genome-wide association studies have done a good job narrowing down the areas in the genome responsible for diseases. They are providing signposts of where to look, but these are only first steps that then enable detailed research and practice to pin down the mechanisms of interaction between food, genes, and viscoelastic matrixes.

Food affects the physicochemical processes and the state of the viscoelastic matrix of the body at all hierarchical levels of structure and the epigenetic manifestation of the genome. All this, together with numerous feedback mechanisms, determines the state of metabolism of the organism as a whole. If the anthropic nutrition is to be used for therapeutic purposes in order to eliminate certain defects of metabolism, then the personal metabolic characteristics of different people must be taken into account. This applies particularly when active substances of food are used as drugs.

The idea of biochemical individuality had its beginnings nearly 100 years ago with researches into the autonomic nervous system and what role it plays in regulating the body's metabolism. What this research revealed is that the metabolism of the body is dependent on a dominance of one or two of the three parts of the autonomic nervous system, which are the sympathetic, the parasympathetic, and "nonadrenergic and non-cholinergic" neuron. Some of the conditions that are found to be associated with sympathetic nervous system dominance include symptoms such as heartburn, insomnia, hypertension, high blood pressure, low appetite, irritability, and hyperactivity. Parasympathetic dominance demonstrates tendencies to allergies, low blood sugar, chronic fatigue, and excessive appetite. Our metabolic individuality is dynamic and can be influenced by food, stress, and environmental conditions.

Even the lack of certain minerals in the body can result not only from a lack them in a food, but also a malfunctioning metabolism of the individual at the genetic level. For example, a magnesium deficiency, with symptoms ranging from fatigue and muscle weakness to severe seizures and heart rhythm disturbances, may also be associated with diabetes and high blood pressure. Up until now, it has been mostly explained by dietary insufficiencies. But a recent study [383] shows that changes in a gene (or in a gene's epigenetic expression) that is involved in the regulation of magnesium processes might be the cause of magnesium deficiencies. This research opens the way to understanding and possible future dietological treatment of genetically caused magnesium deficiencies via possible alteration of relevant gene expression through the diet.

The same point relates to a lot of other nutrition. That is, treating such states must not engage in the forehead—just saturating food-related minerals—but must try to change the status of genes in the genome of the organism through the use of either anthropic diet or medication. These two alternative ways of treatment have their advantages and disadvantages.

Adjustment of metabolism with respect to anthropic diet is a long and difficult process. But this process also gives a chance to rebuild many other damaged metabolic pathways. Drugs in most cases offer a faster result. However, due to the unselective action of the drugs, they may break many other cascades of metabolic pathways. In addition, it should be noted that drug therapy for the specific case described above of the genetic cause of magnesium deficiencies does not exist yet. So the only hope for the correction of these conditions is anthropic food.

Yet if we speak about individuality of metabolism, we have to admit that people from all regions of the world and in any state of health and age have much more common in the metabolism than the individual. Humans were evolutionary created to be healthy by consuming the wide varieties of anthropic foods for use in the evolutionarily designed metabolism, which nicely regulate and regenerate our healthy state.

An even more complicated picture of the interaction of food and the body appears from the recent studies of one type of cell response to external signals coming from the food, its metabolites, and the environment. In studying the effects of various signaling proteins in single cells revealed [54] that there is a wide range of responses of identical cells to the same signals. Until recently, researchers collected information on the effects of various treatments by studying the cell population. The paper [54] examined the response of single cells to stimuli and showed that earlier studies of these issues in cultured cell lines  often incorrectly interpreted results. The study of cell population did not reveal surprisingly intricate results of transfer of information from cell to cell and from the environment to the cell. Getting the signal to the cell and its reply to the cell community controls the activity of cells in an organ or tissue and coordinates the work of all cells and organs in the body. Only the superposition of different reactions of identical cells to the same stimulus in the organ leads to the correct reaction of the organ. This is the basis of recovery and growth of tissues, the functioning of the immune system, and the functioning of other systems of metabolism. In general, the metabolism of the organs, which is largely determined by the metabolism of cells, is much more complex than previously imagined. This again confirms the appropriateness of viewing the body and its organs as a strange attractor of the metabolism in interaction with the strange attractor of the food.

The search for a cure within the paradigm of reductionism in the form of medicines, nutritional supplements, vitamins, and other things at the level of partial understanding of the trajectories of individual reactions is virtually useless. But only the non-linear approaches for studying the effects of nutritional signals in the phase space of metabolism may lead to understanding the common functioning of living organisms. And this is especially true for humans.

Based on these principles and the scientific evidence underlying them, a new scientific field has formed: nutrigenomics. Nutrigenomics is a rapidly developing area combining molecular biology, genetics, and nutrition and studying the regulation of gene expression through the influence of food, its components, and its metabolites. As was shown in the examples in the above studies, the food in general and its active components may influence the gene expression, mainly through the transcription factor, which joins with DNA or enhances transcription of the gene, and thus its expression, or, conversely, depresses it. Nutrigenomics, although a new direction, is in my opinion not entirely free from the chronic disease of all modern dietology: reductionism. Experts hope that a role of specific components of the food items or medicines, such as various RNA and genetics of a particular person will allow us to find a diet for disease prevention and cure. However, in my opinion successes are possible only through the joint efforts of nutrigenomics and the use of the anthropic food.

In the absence of disanthropic elements in the diet there is achieved a minimum of entropy in the body—the maximum production of Schrödinger's negative entropy or, in other words, the absolute (true) state of health of the body. However, in reality, the picture is actually more complicated. Sometimes the anthropic elements of food or their metabolites can fix or replace the missing chemical compounds of natural metabolism. It is known, for example, that with age in humans, amino acid proteins (hormones, enzymes, etc.) are less essential because of the interruption of some trajectories of chemical reactions in the whole body or in part of it. The sensitivity of general metabolism to such cracks in its phase space is extremely high.

It turns out that the anthropic compounds of food and their metabolites, called precursors, can replace the endogenous substances on the individual pathways of chemical reactions, restoring the natural result of proper metabolism. An excellent illustration of this is the results of a study [62] that compared the signal function of cells in the body's endogenous compounds (the hormones) and substances that come from food.

Normally, combining with specific molecules such as hormones, receptors receive chemical signals, which get into the cell and cause certain types of reactions. In the absence of necessary proteins, in this case, inhibition of production of certain hormones, in the cell may begin other reactions that may not be necessary at all for cell functioning. Study [62], incidentally showed for the first time that the amino acids coming from food can replace the missing hormones and enzymes in the transport processes inside the cell, restoring the normal course of the cell metabolism.

Recent article by Marcin Imielinski and Calin Belta [87] devoted to the consideration of a non-linear approach to the metabolism of the human body by analyzing the destruction of trajectories reactions on a set of parallel pathways of metabolism. The study suggests that the sets of reactions frequently interact with other cascades of reactions, which leads to suppression of mutations, or in other words, to a strong epistasis—the interaction of genes in which the expression of certain genes is under the influence of other genes between which seems to be no obvious relationship in functions. By the way, already cited work [62] essentially confirms the faithfulness of a theoretical approach [87], when a cascade of reactions of metabolism some substances are beginning to replace the other and ensure the implementation necessary functions in the cells by the replaced compounds, including, obviously, and the activation of genes. But it is vice versa. An imbalance in the expression of certain genes can lead to the domination of certain pathways of metabolism that differ from the norm.

For example, the research paper published in the *FASEB Journal* [220] examines the roles in response to cellular stresses of the p53 tumor-suppressor gene and the oncogene NF-kappaB. The p53 gene restricts the consequences of stress by initiating cell apoptosis. The oncogene NF-kappaB in stress situation promotes cell division via the synthesis of regulatory proteins for cell division. In other words, one gene controls cancer, the other immunity. If they get out of balance in a cellular stress situation, the parallel metabolic cascade reactions originated from the work of each gene could go out of balance too, and suppress one other, with the consequence being either cancer or autoimmunity.

P53 is considered a gene that prevents genomic abnormalities. The studies [516,517] showed that a chronic stress of any origin leads to prolonged lowering of p53 levels. Research [516] has shown that a variety of bioactive food components, including both essential and nonessential nutrients, may influence cancer risk and tumor cell behavior by modulating the p53 tumor-suppressor gene pathway. The impact the dietary components have on this pathway depends on a host of genetic events that determine whether cells die or proliferate. Results of in vitro studies show that a strong link exists between a number of nutrients derived from fiber, red grapes, soyaa, cacao, tea leafs, etc. and an increase in apoptosis or a decrease in proliferation of tumor cells. Stress from chronic nutrient deficiency in a diet may be the reason for the chromosomal irregularities in body cells. Food or food metabolites and p53 gene interaction probably is the mechanism that may help to explain the stress response in terms of DNA damage from a disanthropic diet. Here the term "damage" refers as well to the chromatin 3D structure damage and similar changes under the interaction of diet and genes.

Regarding the replacement of substances in cascades of metabolic reactions, it can be noted that work [81] demonstrated that Sulforaphane, a natural chemical compound abundantly contained in broccoli and other plants, interacting with cells lacking the expression of a gene responsible for the anticancer activity of the cell, effectively performs the function of this gene and reduces the risk of malignant degeneration, characteristic for the cells in the case of prostate cancer. This explains the long-recognized anti-cancer properties of broccoli and some other plants.

The amazing fact that different substances can quite easy reprogram the genetic apparatus of cells and run them in new organized cascades of chemical reactions is shown in [292]. In this study the researchers created 80,000 double-mutant cell lines in which each line carried mutations in a different pair of the 400 genes. When double-mutant cells grow much more slowly or quickly than expected, these mutant genes are said to interact. To create the differential map, interactions were identified both before and after exposure to a DNA-damaging compound similar to drugs used in chemotherapy. These two networks were then subtracted, one from the other, to reveal differences. Remarkably, researchers found that most of the interactions identified with the drug were not present without it, and vice versa. In other words, the genetic network was completely reprogrammed by DNA damage. In extending this approach to mammalian systems and ultimately to human cells, of course, new challenges will arise—the ability to selectively control the genetic makeup of cells, the redundancy in genes, transcription factors, and properties of inter- and extracellular protein matrixes that make more advanced systems steadier and stronger, but also more complicated to study [297,298].

Of course, the medicinal chemicals used for these model experiments were much more potent agents than the substances contained in your normal diet. But the fact that various substances, including those contained in food or in the air, etc., may epigenetically reprogram a genetic system (network) is not in doubt. For example, lower oxygen levels turn on certain genes, and those genes may change the way heart muscles function. But oxygen is not unique in this sense. The same applies to any natural chemical compounds from food. A lack of anthropic compounds in our diet or a lot of disanthropic unnatural chemicals turn on and off certain genes and take the human body's metabolism away from its normal evolutionarily determined state. Another good example of this is the results [311] that show that the reproductive success of men and women is influenced by the food they receive at an early stage in life.

What is evident from the results of these studies? They show that proper an-thropic nutrition is not only the basis for conservation of the proper metabolism, but also allows you to "repair" (re-program), some regions of the phase space of metabo-lism by correcting to the normal state the set of reactions arising from violations of hormone production or gene expression after a long-term disanthropic diet. And vice versa: Even small shifts in the balance of substances in the body, leading to a shift in the balance of the genes' expression, could shift the balance of cell metabo-lism and lead to diseased state.

However, such changes may be useful for the organism if the vector of changes aimed at restoring the patient metabolism. Restoration of disturbed phase space of metabolism via anthropic dieting may also be due to precursors—the intermediates for endogenous production of hormones, enzymes, etc. on the trajectories of chemical reactions—that are contained in a natural anthropic food. The end of the endogenous production of certain precursors is due to a lack of raw materials for their production in the food or caused by disanthropic food that may violate the pattern of the genes' expression responsible for the production of precursors.

An independent external supply of precursors directly from the food or its me-tabolites partially restores the trajectory of the reactions and provides endogenous-ly certain compounds required for healthy metabolism. A study on extra virgin olive oil, which was part of a larger investigation of the effects on different aspects of bodi-ly health of natural polyphenols in the human diet [64], led to a truly remarkable result. It turned out that the impact of oil polyphenols that appear on gene ex-pression of peripheral blood mononuclear cells (Peripheral Blood Mononuclear Cell - PBMC) is responsible for atherosclerosis and heart disease. This means that the function of genes can be modulated by certain anthropic dietary interventions. The results of this study also provide a new key to the understanding of the anthrop-ic compounds' action on human metabolism. It is hard to say what specific phe-nolic component of the diet in this case supports the epigenetic modulation of the genome. Most likely this is a synergistic effect of all the polyphenols and phy-tosterols and what surrounds them in the oil compounds.

The fact that the genetic effect of phenols oil are very noticeable is supported by the data of the study [68], which found that the polyphenol components of the oils can alter the expression of up to ninety-eight genes, and this can influence vari-ous inflammatory reactions in the body. The presentation of research [75] for the study of vegetable oil phytosterols showed that they can significantly inhibit tumor growth through the diet.

It should be noted that known a wide range of polyphenols which is nec-essary for man and contained, except olive oil, in a variety of wildlife prod-ucts: grapes, cocoa, nuts of various kinds, etc. The effectiveness of olive oil in these studies is due to the fact that in vivo studies used it in live form: fresh oil from first cold pressing. Such processing of olives—with no heat—allows the oil to retain all the life-giving properties of anthropic oil. Polyphenols from living products have in the human body positive effects on lipid and DNA, metabolism of insulin, and the development of tumors. Without any exaggeration, the effect of these compounds suggests that nutritional treatment of diseases with the use of anthropic food is very useful.

In studying obesity [181] researchers have found that genes interacting with diet, rather than diet alone, are the main cause of variation in metabolic traits, such as body weight. This helps explain why some diets work better than others, and suggests that future diets should be more tailored from an anthropic point of view. Understanding genetic and food factors, and nonlinear epigenetic interactions between them in gene expression and its contribution to the human metabolic state, we should stop looking for a panacea and start accepting that this is a complex problem that may have a solution different from the mainstream of modern dietology. Thus, the effect of anthropic elements of food manifest in the processes of metabolism, not only because they are best suited to our body chemistry. Rather, the metabolism of food in the body also includes much deeper levels of interaction, up to the modulation of the genetic apparatus of cells epigenetically.

DNA, the genetic apparatus of cells, is wound around proteins known as histones, becoming 50,000 times shorter as a result. Other proteins then aggregate on it to form chromatin and, finally, the chromosomes. This not only allows the long DNA (up to 1.5 meters) chain to fit into a cell but also plays a role in gene expression or silencing because the way the genes are wound affects which ones are exposed and which are not. Histones have small chemical attachments, such as acetyl, methyl, and phosphate groups, in different places. For instance, when histones are acetylated, genes are transcribed; when they are de-acetylated, genes are turned off. It means that the histone modifications are in the control of gene activity and complement the genetic code. When a cell divides, this chemical attachment pattern of the histones is inherited by the daughter cells. This so-called epigenetic inheritance is controlled by a cell-specific "histone code." Food has a significant effect on the epigenetic state of histones. For example, resveratrol, found in grapes, de-acetylates histones, causing tighter packing of the chromatin and a lower level of transcription of DNA. In general, other flavonoids from raw cacao and other products have similar effects. This mechanism allows us to understand how a wrong metabolism is remembered by the body. That is, for an irreversible or nearly irreversible changes in the metabolism of the body it is enough for long time to eat disanthropic food. Or not so long, if you are an embryo and your mother eats such food. All of this is compounded even more if your mother ate such foods for a long time before she met your father. During this time, the histones code of most cells in various organs has changed under the influence of substances from food. At the same time, the inherited genetic code of the body is left mostly unaffected. A return to normal metabolism requires reverse epigenetic changes at the cellular level, which is not always possible or takes a long time. That is why the correction of improper metabolic states is not an easy task. This problem is generally not subject to modern medicine and, of course, there are many charlatans promising miracles, such as the pill Resveratrol or something similar. The only hope, however—but also without a 100-percent guarantee—is to attempt to reconstruct the body metabolism by eating an anthropic diet.

Epigenetic effects of chemical pollution in the food affect the body's metabolic status in early childhood, in the state of the fetus and even earlier. That the impact of chemicals on disanthropic metabolism of the mother during pregnancy and before it leads to childhood obesity is shown in the study [233]. Babies whose mothers had high levels of the endocrine disruptor DDE in their blood were more likely to both grow rapidly during their first six months and to have a high BMI by fourteen months. DDE is a by-product of the famous pesticide DDT and is also called "Fattening Pollutants." The bodies of virtually all (99 percent) of pregnant women in the United States carry multiple chemicals, including some banned since the 1970s and others used in common products such as non-stick cookware, processed foods, and personal care products, according to a new study from UCSF [310].

Moreover, a recent study [294] shows that a pregnant mother's diet not only sensitizes the fetus to those smells and flavors, but physically alters the fetus' metabolism and the brain, directly impacting what the infant and later adult eats and drinks in the future. An expectant mother's diet has long-term effects—for better or worse—on her child's future metabolism and food preferences, most likely through epigenetic changes in the fetus. This research highlights the importance of eating an anthropic diet and refraining from disanthropic food for woman at least during pregnancy and nursing.

During fetal development, the properties of the food that the future mother consumes and the epigenetic modulation of her genome associated with them determine the state of the phase space of a fetus' metabolism in later adult life. With a disanthropic mother's diet, there may be sustained change of metabolism in the child from the evolutionarily determined state.

Fathers, by the way, can also contribute to the epigenetic changes in offspring. This happens through the mechanism of genetic imprinting, in which the expression of certain genes depends upon which parent provides certain alleles to the child. According to two studies [323, 324], we aren't just what we eat; we are what our parents ate too. That's an emerging idea that is bolstered by a new study showing that mice sired by fathers fed on a low-protein diet show distinct and reproducible changes in the activity of key metabolic genes in their livers. Those changes occurred despite the fact that the fathers never saw their offspring and spent

minimal time with their mothers, the researchers say, suggesting that the nutritional information is passed on to the next generation via the sperm, not through some
sort of social influence. According to the authors [324], "the observations provide an
inbred mammalian model for transgenerational reprogramming of metabolic phenotype that will enable dissection of the exposure history necessary for reprogramming and genetic analysis of the machinery involved in reprogramming."

Epigenetic changes due to the diet can be long term and can spread in the cell communities. A good example of this interesting phenomenon is the bystander effect. The
essence of this phenomenon lies in the fact that after an impact, such as exposure to
radiation in a cell culture, certain changes occur. But after removing all of the irradiated cells from the medium and placing in it the other, non-irradiated cells, those
other cells develop the same changes as in the irradiated cells. This means that epigenetic changes are transmitted through a medium surrounding the cell via some signaling mechanisms. That is, processes of epigenetic changes caused by the diet can
be transmitted by inheritance during cell division, and by another way: through the
signals of other cells in the community.

Epigenetic changes most strongly manifest themselves in fetal development
but may also occur in the adult organism under high-intensity impact of external
stress factors, such as the presence of disanthropic compounds, poisons, and toxins in
food. Epigenetic changes usually play a role in a many cell generations within the
life of an organism, if disanthropic stresses continue. But there is evidence from
animal experiments that these changes can be inherited within three to four generations of animals.

It is hard to overestimate the role of epigenetic mutations from the diet in influencing the human metabolic pathways, as well as on overall health, the development of degenerative diseases, and on correction of metabolic processes by the
anthropic nutrition. Thus, the state metabolism is defined by an existing genetic
program that specifies the possible range of epigenetic manifestation of the genome, and by environmental factors, the main one of which is the diet, which modulates the manifestation of genome. An anthropic diet, along with providing the
body with necessary metabolic compounds and energy, modulates the work of the
genome to achieve the status of ontogenesis (development from conception to death)
of metabolism, ensuring maximum life span. This is what in the different sections of
this book I have called the genetically or evolutionarily determined state of metabolism. The essence of the correction of the metabolism by the anthropic diet is epigenetic effects in the form of repression of the genome regions, expression of
which was made possible by disanthropic nutrition; and expression of the genome,
which was in turn repressed by disanthropic nutrition.

The theory of anthropic diet proposes that, by affecting the epigenetic process,
it is possible to change the metabolism in the desired direction and thereby prevent

the development of epigenetically created illnesses, such as metabolic syndrome, cancer, etc. The number of different combinations of epigenetic expressions of the more than 200 types of human cells is colossal. To decode these might require more work than decoding the human genome. The cell has a very stable genome, but the epigenome of all cell types may vary, so the emphasis on epigenetic changes under the influence of food and chemical pollution are made in this book consciously, because they usually are the most difficult to reverse.

It is necessary to offer the caveat that the presence of disanthropic food substances could lead to metabolic disorders even without epigenetic changes in the body. However, exclusion of these substances usually makes it quite easy to restore metabolism.

In the examples of many living organisms such as ants and bees, we can see how much the epigenetic factors of diet early in life affect life expectancy. In a community of bees, a longer supply of royal jelly to the queen bee uterus in early life provides a life expectancy ten times more than for ordinary bees. This is despite the fact that the queen bee is genetically identical to all other bees in the hive. The same thing happens in ants and in some other insects.

Equally, analysis of the epigenetic state of identical twins (monozygotic twins) shows that their epigenetic characteristics are virtually identical. However, with age, especially if the twins have different lifestyles from living independently of each other, between them appear significant epigenetic differences. These differences are greater when the distance between the habitats of twins is greater—that is, when their living conditions and nutrition are more different. Being genetically identical, identical twins, "epigenetic" with time differ from each other more and more that, ultimately, affect their metabolism and the probability of incidences of certain degenerative diseases.

A study [246] published in the *Journal of Women's Health* shows a rapid increase in the number of hospitalizations due to diabetes for young adults, particularly young women. This pattern of hospitalizations echoes the dramatic increase in rates of obesity across the United States in the last thirty years. And according to the study, rates of diabetes continue to increase. Diabetes hospitalizations were up by 66 percent for all ages and sexes, but the number of diabetes hospitalizations among younger adults, age thirty to thirty-nine, more than doubled from 1993 to 2006. What's the point? Why is this age group is most heavily influenced by some factor causing diabetes? Let's look back fifty to sixty years. In the postwar years, the United States began advertising  breast milk substitutes for infant feeding. Advertising in those years, in its extreme manifestations, argued that artificial feeding is even better than natural feeding because the formula uses special blends of manufacturing, based on the latest "scientific" data. Artificial breastfeeding during the baby boom propagated like an epidemic of bubonic plague in the Middle Ages. Born after the war, the children entered the child-bearing age around the 1960s. Artificial feeding apparently increased the frequency of epigenetic changes in the metabolism of postwar babies that they then passed down by inheritance to their children. These children display a more-than-twofold increase of diabetes in the period 1993–2006 among a group of thirty to thirty-nine year olds.

For an unknown reason, women in these ages were 1.3 times more likely to be diabetic than young men. Starting in 1993, more women than men were hospitalized with diabetes even after the exclusion of hospitalizations associated with pregnancy. An estimated 24 million Americans have diabetes. There is no cure, but those from that generation can prevent the disease with a healthy diet and exercise. Diabetes is the sixth leading cause of death in the United States. The epidemic of epigenetic changes today could affect the future health of Americans. As concern about children's health grows, many experts fear that the childhood obesity epidemic could lead to large numbers of younger adults developing type 2 diabetes, causing serious and lasting health complications. Correction of the epigenetic roots of metabolic disorders in this age group and in following them in the generations, especially among children, is absolutely essential. Because there are no credible nonspecific drugs for such correction, the use of anthropic diets is the only way to prevent the manifestation of epigenetic mutations, and possibly also a way to end their future inheritance. Similar data for the "epidemic" of diabetes exist for Canada.

The last thirty years registered the rapid increase of type 2 diabetes in all groups of population, with a particularly strong trend among First Nations people and other indigenous and developing populations has been precipitated by eating habits leading to epigenetic changes rather than pure genetic factors. A study published in *Science Daily* [247] indicates that in the UK, girls whose mothers are classified as clinically obese are significantly more likely to struggle with weight problems in childhood, with a similar relationship existing between obese fathers and their sons. The findings showed that the same trend does not exist between mothers and their sons, and fathers and their daughters—meaning that diet and other behavioral factors rather than genetic factors could be the key to unraveling the causes of the current obesity epidemic affecting children. A genetic link between obese parents and their children would be indiscriminate of gender. The clearly defined gender-dependent pattern that research has uncovered is an exciting one because it points towards dieting and behavioral factors at work in childhood obesity. But this type of inheritance is more consistent with epigenetic mutations. This, incidentally, seems to indicate a higher percentage of cases among young women with diabetes, noted in [246], who obtained it from their mothers.

In the study [248], the researcher found that the mechanism of epigenetic inheritance of metabolic programming occurs when an insult during a critical period of development, either in the womb or soon after birth, triggers permanent changes in metabolism. The scientists looked at the effects of a diet high in saturated fat on mice and their offspring. As expected, they found that a high-fat diet induced type 2 diabetes in the adult mice and that this effect was reversed by stopping the diet. However, if female mice continued a high-fat diet during pregnancy and/or suckling, their offspring also had a greater frequency of diabetes development, even though the offspring were given a moderate-fat diet. These mice were then mated with healthy mice, and the next generation offspring (grandchildren of the original high-fat fed generation) could develop diabetes as well. In effect, exposing a fetal mouse to high levels of saturated fats can cause it and its offspring to acquire diabetes epigenetically, even if the mouse goes off the high-fat diet and its young are never directly exposed.

In the study [373], researchers observed that female mice born of mice with a deficiency of the hormone ghrelin had diminished fertility and produced smaller litters than mice born of mice with normal ghrelin levels. Mice exposed to a ghrelin deficiency in-utero demonstrated alterations in uterine gene expression which led to low fertility. But the mice in this regard are not unique. The same thing happens to people when they are obese.

Findings published in the paper [358] reveal a novel mechanism by which maternal diet and an offspring's aging interact through epigenetic processes to determine the risk of age-associated diseases. The research demonstrated that children born to mothers who consumed an unhealthy diet during pregnancy have a significantly increased risk of type 2 diabetes, heart diseases, and cancer in adult life. Researchers concluded that the epigenetic changes resulting from maternal diet and aging lead to the reduced expression of some genes, decreasing the function of the pancreas and therefore its ability to make insulin (and thereby increasing the risk of diabetes). In vitro study of the DNA from rats' insulin-secreting cells show that expression of this important gene was controlled in the same way as in humans. Scientists in an international study [392], led by University of Southampton (UK) researchers and including teams from New Zealand and Singapore, have shown that during pregnancy, a mother's diet can epigenetically alter the function of her child's DNA and lead to the obesity of her child many years later.

Providing further understanding of the link between low birth weights and obesity later in life, researchers [359] found that nutritionally deprived newborns are "programmed" to eat more because they develop fewer neurons in the region of the brain that controls food intake. This means that it almost does not matter whether a woman eats enough or too little food during pregnancy, but if that food is low in certain nutrients—i.e., with a distorted anthropic range—the result is the same: underdevelopment of the brain, leading to obesity.

Some people would take the results of this study as simple advice that pregnant women should try to eat a healthy, balanced diet. But the real significance of this study is that maternal diet has a big influence on the health of her offspring in later life. A "healthy, balanced" diet in modern understanding is not enough to have really healthy offspring. Only an anthropic diet with the right spectral content of all necessary nutrition in natural form could provide the right development and health state for the offspring during their life.

Under the influence of diet, there is a change in the "shape" of the phase space of the strange attractor of metabolism, or to put it another way, the transition of ontogenesis from one trajectory to another. Evolutionary basis for epigenetic mechanisms is in the need for organisms to adapt to changing environmental conditions, including changes in available food. Epigenetics is a common field for the interaction of both genes and the diet. Today, we already know that repression and expression of genes carried by DNA methylation occurs due to histone acetylation.

Research [241] shows that an imbalance of methylation-acetylation in the cells occurs even before the appearance of their malignant phenotype in epigenetic mechanisms of cancer emergences. The reversibility of the epigenetic process of DNA methylation, in contrast to genetic changes, gives grounds to consider a new correction strategy based on the anthropic diet to prevent the occurrence of precancerous states. The cited paper noted that there are opportunities for drug modulation and regulation of the level of DNA methylation of key genes. However, it also emphasized that the use of medications for de-methylation can have serious side effects and even cause the expression of genes associated with malignant transformation of cells. This is due to the fact that de-methylation by chemical compounds does not selectively act on all genes. Adding to the diet artificial versions of B vitamins (B2, B4, and B12), folic acid, and the amino acid methionine also has non-selective and unpredictable effects on the genes and carries the risk of side effects. However, it is also known that the supply of these substances from the natural food in the anthropic state does not cause any negative effects. Therefore, in the anthropic diet for cancer prevention should be the use of superfoods rich in these compounds beneficial to man, in the daily diet. For example, it is well known from numerous publications in the media that certain components of green tea, due to de-methylation of DNA, cause the expression of genes that suppress tumor growth. However, for reasons which will follow below, just to drink brewed green tea in the form in which it is sold in stores will not be sufficient for effective prevention of tumors.

Obviously, the effect of external stresses on epigenetic changes in the cell cannot easily occur through protein coding RNA since the work of the coding RNA is determined by the DNA code, and if it is not broken, the RNA itself has a relatively high resistance to external influences. It is therefore possible that in epigenetic changes a major role is played by the interaction of external factors with the so-called noncoding RNAs.

The epigenetic mechanisms of gene silencing by methyl groups is accomplished by specialized enzymes called methyltransferases, which attach methyl labels to specific parts of a gene whereby access to the whole gene is blocked. One study [249] demonstrated that noncoding RNAs also play a crucial role in gene silencing. Usually RNA is best known as a working copy of the DNA sequence of genes. In this role, it is a carrier of the genes' instructions to the cell, which manufactures proteins according to information in the RNA molecule. These noncoding RNAs do not contain recipes for proteins, but they nevertheless are important in the cell for their involvement in epigenetic regulation of chromatin structure and gene activity. The authors [249] reasonably speculate, "It is very well possible that there are exactly matching noncoding RNA molecules for all genes that are temporarily silenced. This would explain how such a large number of genes can be selectively turned on and off."

Additional research [278] demonstrated that long (bigger than 200 nucleotides) noncoding RNAs have a wide variety of functions. A few examples are modifying chromosomes, regulating genes, influencing cell structure, and serving as precursors for small RNAs and microRNAs, which are involved in interactions with nutrients and viruses. The library of RNA transcripts inside of a cell is called its transcriptome, and is a reflection of gene activity. A host of different RNAs can be transcribed from a single gene. That is why a transcriptome contains much more complex and flexible instructions than seems possible from the DNA code. Unlike the genome, the transcriptome varies in different types of cells in the body and in accordance with the ever-changing environment inside and outside the cell depending on cell functions, nutritional supply, or virus attack. The study [278] demonstrated as well that virus infection alters the expression of numerous long noncoding RNAs. These findings suggest that noncoding RNAs may be a new class of regulatory molecules that play a role in determining the outcome of infection. The long noncoding RNAs may be helping to manage the infected body's response to the virus, including the basic, first-line defense against infection—the body's innate, or inborn, immunity. But an array of alterations of these RNA expressions could be achieved as well via interactions with dietary nutritionals rather than with viruses alone.

Many dietary factors ranging from alcohol to zinc have been reported to influence the supply of methyl groups for the DNA methylation process [249]. In addition, dietary modulation was recently shown [250] to influence microRNA expression, which suggests subsequent functional changes, including modification of noncoding RNA. Using human cells and tissue from adult animals, Johns Hopkins scientists [563] have uncovered the process involving external proteins that make normally stable DNA undergo the crucial epigenetic chemical changes implicated in cancers and other degenerative diseases. While DNA, as we already know, is the stable building block of all of an individual's genome, the presence or absence of a methyl group at specific locations chemically alters DNA and changes the expression of the genes. Both methylation and de-methylation have long been linked to genetic alterations and a wide range of diseases. Using human cells in a dish, the Hopkins team focused its investigation on the actions of a chemical base of DNA known as cytosine. The team added different proteins to the cell. As a result, some of the cytosine became methylated, but some reverted to plain cytosine, indicating de-methylation. This tells us that the proteins are promoters of this process of DNA changing status from the methylated to the de-methylated state. This knowledge gives us an entry point in understanding the fundamentally important process of dependence of our epigenetic status on food that we are eating. While under natural conditions of the healthy human cells it is very infrequent process of de-methylation, the researchers found that they could enhance the de-methylation process by adding at least two types of proteins. This finding also tells us that changes in the epigenetic status of genes could happen under the influence of external signals coming from such disanthropic food factors as a lack of or distorted balance of nutrients, chemical pollutants, and steroids and enzymes from modern meat.

From the data presented in [248–252, 279, 563], we can conclude that the diet has a significant effect on all cellular components involved in epigenetic regulation. The study [250] also shows that genomic imprinting may be modifiable not only in utero but also after birth. This study found that altering the diet in later life affects the expression of the certain locus in an imprinted gene. This result allows us to talk about the reversibility of epigenetic processes linked to the metabolic disorders and the potential effectiveness of the anthropic diet to correct such conditions.

And of course at the end of this part, it must be emphasized that the interaction of food components and genes occurs through the transport of molecules and clusters of food metabolites into the cell and the nucleus. But in the process of sorting and transporting proteins in the cells, an essential role, as mentioned earlier, is played by the viscoelastic properties. Sorting proteins is fundamental to the gene expression of every organism—from bacteria to humans. Particularly important during biosynthesis is sorting secretory and the cell's or organelles' membrane proteins, which have to find the way to their final signaling destination through viscoelastic matrixes inside or outside the cell or organelles.

One study [235] raised the interesting idea of parametric resonance as a possible mechanism of action of signaling molecules at the cellular level. According to the approach [236] at the coincidence of timing parameters triggered signaling molecules in the cell processes in general or in its organelles and characteristic times of transport of signaling molecules through the cell and extracellular matrix structures may occur, so-called parametric resonance. Transfer times in such systems are defined by the viscoelastic properties of matrices. This means that certain characteristic relaxation times of the system are influencing parametric amplification of transmission for certain signals. However, when characteristic relaxation time differs significantly enough from characteristic times of signaling protein translation, then the inhibition of translation is possible. Changing the transmission factor leads to increased expression or repression of certain genes. Similarly, when the viscoelastic properties change due to aging, for example, the characteristics of parametric resonance and, therefore, the conditions of signal transduction at the cellular level change also. The same approach, I think, is quite applicable in considering processes at the subcellular level, such as translation signals into the cell nucleus and back. In principle, the emergence of the so-called higher harmonic parametric resonance in the signals translation in such systems is possible. However, consideration of this complicates the problem excessively even at the descriptive level adopted in this book.

In two papers [257, 258], the interaction of the outer surround waves with the viscoelastic medium can result in a formation in the ordered structure, in some ways comparable to a phased array. In wave theory, a phased array is a group of antennas in which the relative phases of the respective signals feeding the antennas are varied in such a way that the effective radiation pattern of the array is reinforced in a desired direction and suppressed in an undesired direction [256]. In our application, an analogue of the phase array will be a range of states of the mesh in the polymer matrix through which go the transport of signaling molecules. The authors of [257, 258] introduced and justify the notion of the dissipative resonance and have concluded that in this case there is an important new class of physical phenomena that play an important role in the structural organization of the physicochemical systems and biological objects. One of the characteristics of the dissipative resonance is that the system in this case, the polymer matrix, has the ability to "tune in" to any external signal, and the time of rising-damping of the oscillations (the speed of signal translation protein) is determined not only by the period of oscillation (the diffusion characteristics proteins), but also by the spectrum of the system setup time (the spectrum of relaxation times of the polymer matrix).

The phenomenon of dissipative resonance is one of the possible mechanisms by which the viscoelastic properties of polymer matrices influence the diffusion rate of proteins. In this case can be the cooperative amplification–inhibition of the signal translation due to the interaction of signaling molecules and matrix. In principle, the phenomena of dissipative and parametric resonance are just different methods of describing the same nonlinear processes of interaction of viscoelastic polymer matrices and of moving protein molecules through them. However, there are other possible approaches. For example, for polymer gels there exists the well-known phenomenon of the susceptibility to small changes of external parameters in the state on the verge of "collapse." In this state, small changes in the gel chemical or physical parameters can dramatically change the properties of the gel, and even cause it to "collapse" [238]. The state of matrixes "on the verge of collapse" is quite typical for living structures. But a significant change in the characteristics of the matrix, in turn, strongly affects the signal transferring processes and the related gene expressions ensemble.

Confirmation that the gene expression may be determined by the mechanical properties of cells and external stresses are given in [351]. This study used a new method to observe and track individual cell behaviors, characterizing for the first time what happens when human endothelial cells move from an initial dispersed state to the formation of capillary structures. This is one of the first explicit studies to look at the variations between cells during 3D tissue formation, and it overturns the assumption that genetically identical cells behave in generally similar ways. Using a systematic approach to quantifying the changes in cell shape and movement for every single endothelial cell over time, the researchers found unexpected patterns in behavior. First was the discovery that most cells behave differently from the average. Secondly, the team also observed that groups of cells behaved in similar fashions, and that some of these clusters of behavior resulted in distinct structural roles in the final capillary network. The origins of the variations in behavior lie down in viscoelastic effects. As I have mentioned above, the researchers [282] showed that these types of transformations are triggered by the changes of viscoelastic mechanical parameters around individual cells. It means that viscoelastic properties have an important role in producing biologically significant variations in gene expression. But according to the study [366], not only the structure of the blood capillaries of organisms of animals but also elastic stresses may play a crucial role in determining a leaf's venation pattern. It appears that plants and animals use the viscoelasticity of the cells and surrounding tissue as a mechanism to regulate vein formation in the bodies and the leaves and branch formation on the main trunk and on the main root.

This study once again underscores the critical importance of viscoelasticity of tissues for growth and regulation of cells and tissues at the epigenetic level and functioning of all organisms on Earth. That is, viscoelastic properties of the matrix protein are involved in fine-tuning the biochemical and epigenetic processes of the functioning of the cell and its organelles. The molecular mechanism of these changes may lie in the peculiarities of the molecular interaction of proteins and ligands—molecules involved in the transport of proteins and in their interaction with the target of a signaling process-specific receptor or enzyme. Conformational change of the ligand-protein complex can change its effective size and diffusion rate through the mesh of a viscoelastic matrix. This can occur both because of changes in the "geometry" of the combined ligand-protein complex, and because of its exposure to changes in the individual parts, such as terminal groups of atoms interacting with the macromolecular matrix. Last, fairly obviously, since the molecules of proteins and ligands, collectively and individually, and their parts, each have a unique affinity, and activity profile in different conformation state, they are together defines the signaling output.

Both the changes in the interaction with the target receptor and changes in the diffusion rate in the matrix alter the conditions of excitation of parametric resonance and, therefore, the work of the cells and organelles and the epigenetic state of histones. But it should be noted that when we are talking here about the biopolymer matrices, this does not necessarily mean that they have a network lattice structure. For instance, the role of the matrix could be performed by microtubules— long thread-shaped structures about 25 nanometers in diameter and several micrometers in length, which extend through the whole of the cell. Viscoelastic phase separation provides a stable two-phase state around microtubules in a certain range of nanoscale. Viscosity of the cytosol is minimal at certain distances, in my estimation of the order of 25-35 nm from microtubule. This allows us to consider a cross-section of the low viscosity zone as a mesh of the matrix. In this zone, as in tunnels, there is rapid transport of the signaling ligand-protein complexes to intracellular targets. Typically, ligands are still working as so-called motor proteins, and indeed perform many other functions.

Life implies movement. Most forms of movement in the cells are powered by tiny protein machines known as molecular motors. Among the best known are motors that use sophisticated intramolecular amplification mechanisms to take nanometer steps along protein (microtubule) tracks in the cytoplasm. These motors transport a wide variety of cargo, power cell locomotion, drive cell division, and, when combined in large ensembles, allow organisms to move. Motor defects or change in cell environment such as nano-viscosity of cytoplasm can lead to severe diseases. Basic principles of motor design and mechanism of movement have now been derived, and an understanding of their complex cellular and extracellular roles is emerging. The cell functioning is controlled by signaling pathways which deliver molecules derived from the extracellular space tissue to the cell organelles and back from intracellular compartments to different parts and outside of cell. For both efficient and tightly regulated signaling, as I said above, cells organize their proteins into distinct cellular structures—microtubules [538]. Flexible adaptive regulation of cell and tissue properties is critical for organ function. The cells in population organize themselves to a tissue by communicating with other cells, thereby receiving information on whether to proliferate, divide, or migrate.

In study [237], scientists from the Max Planck Institute for Molecular Cell Biology and Genetics (MPI-CBG) in Dresden, along with a colleague from the University of Florida in the United States, have been carrying out research into how motor proteins and their cargo can move in cells without bumping into or sticking to anything. They found that motor proteins, of which kinesin-1 is a prime example, keep distance between a cargo and microtubules filament of around 17 nm. It would appear that this is how refined motor proteins succeed in getting cargo to its destination without any significant viscous resistance due to viscoelastic phase separation. In my opinion, this is a good example of a role that viscoelastic properties of cell proteins possibly play in cell functioning. Moreover, these viscoelastic micro-tubules filaments keep extracellular space and tissue well organized [538].

The change in the conditions of translation of biochemical signals, from food, toxins, and drugs, leads to changes in the work of cells, including the processes of gene expression. This means that the interaction of food, toxins, drugs, and the genome cannot be looked at separately from the mechanical state of cellular struc-tures. Also, of course, these interactions are extremely complex. Therefore, it is naive to expect that adding to your diet some chemical compound, of artificial or natu-ral origin, will dramatically resolve any problem of the body. Only the combination of compounds of wildlife in its natural form, properly selected for resolving a specific health issue, can help your health.

Recent studies uncovered another interesting aspect of the structure of DNA, which may be connected with epigenetic processes in the genome. Researchers [362] recently found in DNA huge "superstructures," millions of nucleotides (the elementary "building blocks" of DNA) in length. These are separated from the surrounding genetic "background" by the characteristic content of certain nucleo-tides. As you know, DNA contains the genetic information of an organism and is composed of four types of nucleotides: A, C, G, and T. This long range order presented in the genome as the probability of a given nucleotide appearing is not independent of the nucleotides that came before it. Authors of study [362] found that each human chromosome segmented into a few huge segments tens of millions of nucleotides long—bigger than any previously known structure in the genome. To find out what these structures are biologically, researchers used a separate database that assigns to each gene a series of descriptive terms characterizing its biological function, such as "membrane," "metabolic," and "signaling." Usually any two genes randomly selected in DNA are likely to share six of these descriptive terms. They found that, remarkably, any two genes in the same superstructure are likely to share roughly eighteen terms.

I think that superstructures do have a biological function, which is to facilitate the epigenetic "mutations" under the influence of external factors on these super-structures. Since each gene is shared with others in the superstructure of a set of functions, the changes in the expression of a gene do not cause catastrophic chang-es (features in the superstructure supported by other genes), but at the same time pro-vide the ability to change enough for adaptation to external influences. Among these external factors, the change in the diet is one of the most significant. It is also interesting to imagine how such a superstructure could arise in the genome. The existence of the spectrum of mechanical relaxation times in the matrix led to the emergence of different characteristic scale structures. In particular, we can assume that one of the scales that is appropriate to the superstructures observed due to its functionality (unfortunately we do not know exactly what it is) was preserved in the course of evolution. The key idea in this hypothesis is the idea that the viscoelastic properties and the special set of nonlinear feedbacks that exist because of them in these systems may lead to the emergence of the long scale structures due to the existence of big relaxation times of the DNA matrix.

In addition, it should be noted that the polymerization of the DNA molecule is a typical example of co-polymerization of four nucleotides. It is well known from modern research in the field of polymerization processes, that the sequence of monomers units and their spatial grouping in the macromolecule of the co-polymer change the mechanical properties of the co-polymer, and vice versa. Similarly, the appearance of superstructures in the DNA chains of the polymer matrix is due to the evolution of the matrix through the emergence of new mechanical properties of new macrostructures and back feeding them into the co-polymer chain structure. In addition to the above-described superstructures in the cell nuclei, there are present three-dimensional spatial organization of chromosomes. It was found that coding in the genome is existing not only at the level of DNA double helixes "wound" on the histones and chromatin-forming filaments form the chromosome, but also at the level of spatial organization of chromosomes themselves. A DNA molecule has a length in an expanded form of a few meters wrapped (coiled) up in a ball in a spherical volume of 10–15 micrometers in diameter. This coil has a pronounced structure, which varies between different cell types. As has been shown in numerous studies of these structures, in some way they may affect gene expression. In the nodes of the interlocking coils of chromosomes is an interaction of genes located at great distances by the linear length of DNA, which causes their expression and possibly the expression of noncoding regions of DNA. Many researchers suspect that epigenetic expression of the so-called noncoding regions of DNA might affect the genome and can cause diseases such as cancer. In general, it can be argued that the spatial organization of chromatin in the chromatin matrix defines a certain level of hierarchy in the spatial regulation in epigenesis of the genome. The nodes of the matrix are sites of interaction between distant genes (gene sets), which in the absence of such a structure would not have the chance to interact with each other.

However, if the normal epigenetic regulation (switching genes on and off) is achieved by chemical modification of DNA and histones, the dynamics of chromosomes in the three-dimensional matrices of chromatin are unpredictable and poorly understood. Under normal conditions of cells in the absence of external stress, apparently, the dynamics of physical interactions between chromatin sites is random and reversible. At the same time in the different cells of the same organ, epigenetic "mu-tations" of various types are manifested. However, generally the set of mutations is statistically defined by the physicochemical properties of cells and their nuclei, chro-matin, and environments. In some very unfortunate cases, it may be that these muta-tions can lead to disease. However, the main purpose of this mechanism, apparently, is the primary subtle epigenetic modifications of the work of the nucleus under the influence of external stress. On arrival of some of the chemical signals, which come from the food metabolites, the conformation of chromatin structure changes and thus changes the structure of the chromatin matrix—i.e., changes the position of the matrix nods. This in turn leads to the expression of other genes and to the synthesis of "new" proteins and to changes in the work of the nucleus and cell.

After some time in the set of cells, the organ or tissue develops a new statisti-cal distribution of mutations, which partially defines its work. It is obvious that viscoelastic properties of the chromatin matrix should play some role in this pro-cess. Through changes in the viscoelasticity of the matrix comes a change in the diffusion asymmetry, which, along with the "new" proteins synthesized and new epimutations, alters the overall picture of the proteins' distribution in the nucleus and the cell and, consequently, cell metabolism. Through this process, bacterial cells, for example, may gain the ability to resist antibiotics by a mechanism dis-covered in work [397]. Cells of organs and tissue can adapt relatively quickly to changes in the set of metabolites from disanthropic food, for example. When you restore a normal supply of the anthropic nutrition, this process can be reversed. This fact gives hope that the relatively small external stresses on the nutrition can be cor-rected relatively easily. But if the net effect of external nutritional stresses is great (in duration or intensity), the epigenetic changes lead to chemical modifications of DNA and histones under the direct influence of metabolites that are alien to the cells. Handling such processes that affect the deeper levels of the organization of the genome is a more complicated problem.

But the epigenetic state of chromatin structures obviously influences the muta-tion frequency of involved and linked-with-involved genes. It has recently been dis-cussed in article [403], that evolution may take advantage of differential mutation rates associated with epigenetic features to control the mutation rates of particular genes or groups of genes. This means that the partially genetic mutations lead-ing to a specific phenotype may be the result of the epigenetic changes in the 3D chromatin matrix under the influence of external factors (environment, food, and metabolites). Because changes in the 3D matrix involve changes on many genes' expression, this can lead to changes in the mutation rates of these genes and to so-called simultaneous (or cooperative) multiple mutations. One study [404] already demonstrated such a possibility: if mutations can happen cooperatively in twos, threes, or even more points, cells could make large evolutionary leaps and reach a different "fitness state" by acquiring multiple mutations simultaneously. Frequent mutations of chromatin remodeling genes in transitional cell carcinoma were recently discovered in work [498]. This means that aberrations of the chromatin structure remodeling genes may directly lead to the wrong regulation of multiple genes, consequently promoting cancer or other degenerative diseases. It also means that when 3D chromatin structure changes under the influence of disanthropic food metabolites, if those changes are involving expression of chromatin remodeling genes, the risk of cancer could rise significantly. So according to researchers at the Stanford University School of Medicine [552], about 15 percent of cases of an aggressive, difficult-to-detect form of ovarian cancer contain a unique fusion between two neighboring, normally separate genes. Furthermore, the "hybrid" protein arising from the fused genes that are certainly foreign for cell metabolism may initiate or contribute to cancer progression. This kind of genetic lesion may be happening due to a 3D chromatin rearrangement involving pairs of genes that are normally located at some distance turn out to be near one another in the nodal point of the chromatin matrix. And it is not possible to detect before the unique chromatin rearrangements lead to cancerous fusion, and it is possibly too late after.

The importance of keeping 3D chromatin structures intact from the influence of disanthropic food factors and the consequences of unwanted epigenetic changes is affirmed as well by study [510]. According to publication [511] researchers [510] have discovered a mechanism causing cancer susceptibility, clearly showing that spontaneous or induced small epigenetic changes in some anti-cancer genes can act as "magnets" to attract modifying biochemical tags, effectively silencing them and predisposing people to an increased risk of the cancer.

But apart from 3D structure of chromatin, a new structure of it was recently discovered as well. So according to the textbooks, chromatin, the natural state of DNA in the cell, is made up of nucleosomes. And nucleosomes are the basic repeating unit of chromatin [513]. But researchers [512] recently discovered a new type chromatin particle halfway between DNA and a nucleosome. This new particle is very similar to a nucleosome, but in fact it is a distinct particle of its own. According to the authors of paper [512], this new particle is a precursor to a nucleosome. They call it a "pre-nucleosome" and suggest that it is necessary to reconsider what chromatin is. The pre-nucleosome might be an important player in how the genetic materials interact with food metabolites. Therefore, the discovery of pre-nucleosomes suggests that a chromatin is a mixture of nucleosomes and pre-nucleosomes. The packaging of DNA with histone proteins to form chromatin and the 3D structure of chromatin as we have shown above plays an important role in regulating gene expression. The existence of a novel intermediate DNA-histone complex theoretically could help to explain so active a role of food metabolites in processes of epigenetic expressions.

For example, if we will look to the health of skin, we see that it has the epidermis, which is the barrier than separates the body from the outer world. According to [553] the deepest of the epidermis' five sub-layers consists of progenitor cells (cells that have to differentiate into a specific type of cell) that form keratinocytes, the most common type of skin cells. These gradually migrate upward, differentiating into cells with distinct properties in the upper layers of the epidermis, and eventually flake off—a continuous process that in human skin takes two weeks or more. At each new level, the keratinocytes accumulate more and more tough, connective keratin proteins. The keratins and other proteins form filaments that begin to interlink. The cells stiffen, lose their nuclei and other internal structures, and finally program themselves to apoptosis. The top layers of skin form a tough barrier, made of dead cells that fit together like tiles. At every stage of the process, the appropriate genes must be turned on and off to regulate cell signaling, cell adhesion, metabolism, and gene transcription. But for many of these genes, to be clustered together in chromosomal regions specific to the expression of keratin and related proteins is essential to the skin barrier's ability to maintain proper (healthy) 3D chromatin structure. These gene assemblies could be destroyed by chromatin remodeling under the influence of disanthropic substances coming mainly from the food metabolites from inside the body. That is why it is almost useless to influence the health of the skin from the outside using a variety of cosmetic means. Health to the skin only comes from the inside through the right anthropic diet.

The results of studies [554, 555] give us two pieces of good news and two bad as well on epigenetic changes in cells. We will start with the good news first. The comprehensive inventory of epigenetic changes over several generations shows that these often do not last and therefore the genome expression ensemble probably could be turned back to the normal state, according to study [554]. The scientists found that many epigenetic mutations are apparently not stable and return to their original state after a few generations if an external stress does not persist. Experiments show that methylation changes are often reversible. This means that after the removal of disanthropic diet factors, the epigenome of cells may return relatively quickly (in very few generations) to the evolutionarily normal state.

But the first bad news is that externally induced epigenetic loss of methylation has a disproportionately greater effect on the activity of transposons than of regular genes. Transposons were previously thought to be non-functional parts of the genome and were even labeled as "junk DNA." But accumulating evidence indicates that these elements play an important role in genome regulation. Researcher [555] found that the majority of DNA damage and associated chromatin remodeling that occurred with cell aging was due to transposons. The bad thing in this finding is that disanthropy of our diet leads to the changes in genome of the state of transposons that are very similar to changes in a senescent cell. It means that if we do not eliminate hazards in the diet, the cells and body begin to age much faster than in the evolutionarily normal state.

Although the body is constantly replacing cells and cell constituents, damage and imperfections accumulate over time. From damaged proteins may arise such serious problems as age-related diseases like Parkinson's, Alzheimer's, diabetes, and cancer. The second good news is that in some ways the body is at least partially preventing inheritance of epigenetic damages. A few days after conception, the cells in the embryo all look the same—they are unspecified stem cells that can develop into any bodily cell type. Researchers [556] found that the level of protein damage was relatively high in the embryo's unspecified cells, but then it decreased dramatically. A few days after the onset of cell differentiation, the protein damage level had gone down by 80-90 percent. In the past, researchers have believed that the body keeps cells involved in reproduction isolated and protected from damage. But it has been shown in [556] that these types of cells go through a rejuvenation process that rids them from the inherited epigenetic damages. In my opinion, though, it is possible that one of the reasons for the elimination of epigenetic mutations in the embryo may be the existence of viscoelasticity, which is responsible for the emergence and frequency of synchronized pulsations in the embryo during the first hours of its existence (see Section 3.4).

Mechanical pulsation forces from the extra-cellular matrix in the embryo that are exerted on the surface of cells are also channeled along an intracellular matrix (cytoskeletal filaments) and concentrated at distant sites of cells in the nucleus. Acting at a distance, they induce mechanochemical conversion in the nucleus, leading to specific chromatin modifications [600], which are chemical tags that promote or hinder gene expression. In the initial state of the embryo development, the mechanical pulsations at a certain frequency provide the same pattern of gene expression in all stem cells and thereby provide the reconstruction of 3D chromatin structure to an evolutionarily given state ensured by the intact genetic code. However, as the embryo grows and its interaction with the border of the surrounding tissues of the uterus, having other viscoelastic properties, is changing, the additional mode of pulsation emerges. This additional mode appears because the viscoelastic properties of the embryo-border system at this stage of development are described by at least two mechanical relaxation times. The emergence of a new relaxation time has the result that pulsation mechanical forces on the cell surface are different at the outer edge of the embryo and inside it. The difference in action on the cells' forces leads to a difference in gene expression in cell layers at the surface and inside the embryo and causes the emergence of primary specialization of stem cells. Further, continued division of cells in the embryo and changes in the mechanical conditions of their existence complicate more and more the wave function of the pulsations and thus the time-spatial differentiation of mechanical stresses in the embryo. The variation of mechanical stress amplitude leads to variations in genes expression and marks the beginning of the cells' specialization, which will in later life form muscles, neurons, and other bodily parts. Thus, the viscoelastic properties and their changes are, in my opinion, the only governing factor, at least in the initial cells' differentiation process in the embryo.

But the second bad news is that some epigenetic mutations are still susceptible to inheritance. In the cells, the epigenetic changes, due to changes in diet, in the chromatin matrix of the nucleus could theoretically act also as a kind of precursor to genetic changes in cells of the body, the so-called de nova mutations. Moreover, due to the process described above in Section 3.3 of this book, the possibility of renormalization of probabilities of states of the ensemble of genes involved region of the genome, the evolutionary process can gain speed and direction determined by the phenotype arising as a result of the initial epigenetic change. Also important to point out is the fact that simultaneous multiple mutations, according the study [404], could be much more beneficial for cells than any individual mutation from the same set. Through individual mutations, the cell would not be able to reach the improved "fitness state," as the less-fit intermediate states would be eliminated by natural selection. Cases like this are referred to as "fitness valleys" [404]. The most interesting result of the existence of the 3D chromatin matrix structure in the nucleus is that it raises the possibility that a cell could leap across "fitness valley" and reach a "better fitness state" by acquiring multiple mutations simultaneously.

But "better fitness state" in this sense does not always mean better results for the organism as a whole. For example, in a lot of senses, a cancerous cell is a "better fit" for survival than a normal cell. In the case of cancer, crossing the fitness valley happens in the opposite direction (degradation of cell structure) than in the case of evolutionary development. That could be illustrated by results of the UT Southwestern research [405] that contradicts the previous thinking that only a few de nova mutated genes are important in the development of cancerous mutations involved in the formation of colon cancer—far more than scientists previously thought. I think so many genes involved in the formation of cancer could only mean that the cancerous state of cells in this particular type of cancer might be achieved through the simultaneous multiple mutations leap mechanism [404]. Based on the study [406], researchers are suggesting a new approach to colon cancer treatments targeting multiple genes and pathways simultaneously. Current cancer treatments target just one or two known cancer-driver genes, believing this will be beneficial to patients. While patients may get transient tumor burden reduction, almost universally the tumor growth returns.

"Fitness valley"(FV) in the case of tumor development is resistance of the cell to turn to the cancerous state under too few mutations. In such a situation, the natural defense system is working in a cell to eliminate the consequences of "bad" mutations via some repair mechanism or apoptosis. But in the case of multiple mutations, damage to the cell could be so big that natural defense forces in the cell will lose the fight. Induced via diet, de nova mutations in the genome (not inherited mutations) could lead to the widest spectrum of diseases from cancer to schizophrenia [502, 503]. The study's results help to explain the rise of global incidence of schizophrenia, despite large environmental variations. What is common in various environments in our world? The fast spreading of modern, utterly disanthropic food. It has been well established that food-induced mutagenesis is activated in response to adverse conditions, such as starvation, lack of nutritional elements (vitamins, etc.) in food, and toxins. The development of mutations depends on the intensity and timing of these factors. Generally, mutations accelerate adaptation to environmental changes, such as the acquisition of resistance to antibiotics, for example. But this is an evolutionary adaptation. For the individual it can result in a very poor outcome.

How does the cell overcome this through the mechanism of the multiple "simultaneous" mutations the FV? First, it should be noted that the mutations apparently were not quite simultaneous, but rather very fast. Consider the mechanism for the acceleration and orientation process: Once the cell is faced with a situation of strong external stress (radiation, antibiotics, disanthropic compounds, toxins, starvation, etc.), then it creates stress-induced changes in the 3D structure of chromatin. That is, the matrix structure of chromatin adopts a new configuration, but rather in different cells bit different configurations. Set or spectrum of these configurations determines, but only probabilistically spectrum of the ensemble of possible states epigenome, which provides primary epigenetic response of the cell to external stress. Such spectrum of possible ensembles of states of epigenome provides to some cells a small probability of survival and the possibility of division, and thus of all cells out of the bands dying or loosing the ability to divide.

But the situation is still far from ideal. In front of the cell, in an evolutionary sense, lies the FV in which all mutations taken separately do not ensure its long-term survival, or even the contrary—they may lead to its inevitable death. That is, the cell cannot cross this valley of "death" only due to epigenetic changes, in order to be able to thrive in a cell under the existing stress.

Suppose that a set of mutations exists that can provide new cells with an existence in a stressful environment, so much so that these conditions have ceased to be stressful for the cell and have become the new norm. But the probability of the simultaneous random occurrence of such a set of mutations in the cell population is extremely low. However, suppose that the de nova mutation "A" that occurred in the cell slightly increased the probability of survival (including ability for division) to the probability of $Pa > Po$, where $Po$ is the initial survival rate under stress. Assume also that there was a second mutation "B," which in itself may not be beneficial to the cells, except that in conjunction with mutation "A," a cell increases the survival rate to the level of Pa > P . Incidentally, that genes interact with each other and that this affects the survival rate of organisms is a well-known fact. What happens with the spectrum of possible states of ensembles of chromatin matrix in the cells' population in the process? This spectrum narrows, since there is a greater number of cells with mutations type "AB." That is, genes A and B must be expressed, but that usually happens in the vicinity of the chromatin matrix nodes. And that means that at least the nodes in charge of these genes are constantly expressed, or rather, are expressed more often than others. So that cells with such nodes survive and divide more often. But such double fixation of nodes narrows even further the spectrum of an ensemble of possible states of other nodes of the chromatin matrix. That in turn reduces the ensemble of possible states of the epigenome of the cell. As a result, the frequency of mutations induced by external stress in expressed genes increases in the course of mutation relative to the initial state, because in the cell population, they expressed more often due to narrowing of the ensemble of possible states, relative to the initial epigenome at the beginning of stress.

Next, suppose that among these mutations there randomly appears a third mutation "C" with Pa c > a. As a result, the frequency of mutations in expressed genes increases even more. And so on. That is, the process in the following steps is accelerating even more, and as a result, quickly gets full adaptation, which looks like a "jump" through the FV. That is, these mutations occurred, though not, strictly speaking, at the same time or simultaneously, but fast enough because of the increasing number of mutations in a relatively small number of expressed genes. The ratio of the total number of mutations to the total number of genes remains constant and determined by the intensity of external stress influence. It should be noted that in conditions of external stress, this ratio is higher than in the absence of stress. But in addition, this mechanism can work if some mutations necessary for the "jump" already exist in the genome in a repressed state. Then the change in chromatin structure by external signals and feedback loops causing the expression of new regions of the genome may help to find and fix these mutations in the 3D structure of chromatin. It should be noted that the proposed type of process complies with the processes of renormalizations of probabilities of ensembles of states of the particles in living systems by the final state of the system, proposed in the aforementioned work [222] (earlier in Section 3.3) in the context of the evolution of molecules in processes with a viscoelastic feedback in the primordial soup. That is, the renormalization principle in living systems, apparently, is universal. It is interesting to emphasize here that the trajectory of the evolution of the cell is determined by a final state of the system, which the cell during the process, generally speaking, does not know.

If the cells are under the influence of a so-called carcinogenic chemical compound under which the cell cannot virtually live, function, and divide, the jump across the FV leads the cell to the acquisition of immortality, which is actually the cancerous state. However, this process goes a bit differently than described above. A strong external impact initially brings disruption chaos to the work of the cells' genetic apparatus due to the high intensity mutations, but it increases the number of degrees of freedom of the chromatin matrix—i.e., expands the range of possible states of chromatin matrix. This increases the pool of possible mutations and increases the probability of occurrence (or expression if already exists) of the first beneficial mutations with which the cell begins its path through the FV.

This approach could help to explain the phenomenon of genetic instability, a common feature of cancer in which cells mutate at an abnormally fast rate. These mutations can cause cancer cells to grow, or they can cause the cells to die. According to many observations, a cancerous cells grows best when they are very unstable in early stages of tumor growth; the pattern of genetic changes in this stage is very similar to jumping through a fitness valley, which I have discussed above. Therefore, it is possible that the phenomenon of genetic instability is an innate property of the cancerous cells.

An external stress of the strong carcinogen can be replaced by the weak but acting in long term. In this case, it is possible that the initial degradation of the genome requires more time. However, once this is happens, the cell starts down the path of evolution to a cancerous state. The process can go fast enough on the mechanism described above. Indeed, a cell in a precancerous condition is in a very hostile environment among the healthy cells. Its signal communication with surrounding cells is breached, and hence less necessary nutrition comes to it. In a condition of such external stress, the cell can jump in the direction of becoming cancerous. Turning a single cell into a cancer is sufficient to begin the process of unstoppable division, leading to illness and death of the organism. Theoretically, a cell can jump into a different direction of evolution and "recover," or become still different from the surrounding cells, but not cancerous. An increase in the population of such cells is so-called benign tumors. An interesting aspect of this approach is that mutations occurring in the process also change the 3D structure of chromatin matrix. Thus there is a joint cooperative action of gene mutation and changes in chromatin structure of the matrix leading to the best fitness state of cell. We can therefore assume that for all types of cells there exists its own fairly narrow spectrum of ensemble probabilities of 3D chromatin matrix states in the nucleus, with its spectrum of ensemble viscoelastic states of chromatin. Consequently, the conditions of diffusional transportation of macromolecules in the chromatin matrix cell nuclei of different types are different.

However, there is also, perhaps, another mechanism for the effect the state of the chromatin matrix has on gene expression. In fact the nucleoplasm and chromatin matrix together are the co-polymer gel. On the "filaments" of DNA and histones in this matrix are constantly taking place processes of self-oscillating complementary replication of RNA molecules or binding-unbinding of protein molecules. These processes lead to changes in the local state of the gel—the volume occupied by them—similar to the process of binding-unbinding ions, for example, in the heart muscle. Thus, the chromatin matrix is in the processes of many simultaneous BZ reactions. That is, the chromatin gel is in a constant oscillation with a complex set of frequencies organized in the wave packet. This wave packet determines the dynamic state of nodes of the 3D chromatin structure and thus affects (statistically) the expressed genes network. Each local change of the gel state is changing, in some degree, the characteristics of the whole wave packet oscillations of chromatin and, consequently, is changing the spectrum of genome expression and the spectrum of RNA and proteome. On the other hand, the chromatin matrix very likely also has a certain buffering ability: disrupting one locus can affect its immediate neighbors, but the problems may not propagate throughout the whole matrix if there are no "resonant" parts for this particular locus in the genome.

The interaction of genes and how they alter each other's work was studied by researchers of the University of California [642]. To create a map of gene interaction, the researchers focused on 400 genes that manage the signaling pathways in a yeast cell. They then created 80,000 double-mutant lines in which each cell line carried damages in a different pair of these 400 genes. When double-mutant cells grow slower or quicker than expected, it does mean that these mutant genes are interacting. To create the gene interaction networks map, the interactions were identified both before and after exposure to DNA-damaging chemicals. These two gene sets were then subtracted one from the other to reveal differences. The researchers found that most of the interactions linked to the damaging chemicals were not present without them, and vice versa. In other words, the genetic network was re-normalized by DNA damage to the new state. This result is fully consistent with the above approach: if there is DNA damage in the genome, randomization and the re-normalization (reprogramming) toward a new steady state of the entire genome network is happening.

This is a possible mechanism of adaptation of the genome to the changing environment. Viscoelasticity of chromatin is the secret to fine-tuning a cell's genome. This approach may also be used to explain remote genes' interactions. If pulsations from the work of a pair of distant genes are in a state of some sort of "resonance," then they may influence the work of each other—suppress, amplify, or in another way alter each other's functions. But different genes' activity may have different effects on other genes. So, for example, some genes—those that are in the most "important nodes" of network from the standpoint of impact on the state of the chromatin gel, called the regulatory genes—can increase or decrease the expression of hundreds of downstream genes. But the most interesting fact is that in different stages of cell life, the different downstream genes can be affected. This means that in different stages there exist different structure of chromatin. That leads to the "resonance" of different genes under the molecular replication work of the same controlling gene.

Everything from sun tanning to environmental factors and food and metabolic processes inside the cell damages the DNA of that cell every day. This in turn can lead to production of faulty RNAs and proteins that, if not repaired, could go on to become the driver of cancer or other diseases' development. To prevent these devastating effects, such bifurcation as damaged DNA triggers an elaborate alarm system, which sets off a chain reaction in the cell, to slow some processes, terminate others, and wait while "legions" of molecules go to work on the damaged DNA [625]. When the DNA of cells is damaged using radiation or chemical drugs, thousands of protein chemical modifications are produced in the whole genome in the process of repair. The mechanism of initiation of these proteins' production may be the process described above in the gel structure of chromatin. The appearance of the DNA damage alters the wave function of chromatin oscillations. This change leads to the production of a large number of new chemical signals in the various parts of the genome. Since the state of the chromatin before damage appears can be assumed to be locally stable for relatively small changes in the parameters of the system, the interaction of various molecules from the "legions" with the damaged part of the matrix changes the state of chromatin in either direction. But since we're talking about local stability of the previous 3D chromatin state in a thermodynamic sense, then there should be a positive selection of molecules partially returning chromatin to its original unperturbed state. The change of the wave function of chromatin in a process of returning to the stable state generates new cascades of chemical modifications of proteins, among which are proteins that work to bring total wave function even closer to a stable state. That is, there is a process of renormalization of the ensemble probabilities of states of particles in phase space of the genome of the cells, which I quoted above for the description of the primary gel-like structures in the origin of life. The role of chromatin matrix viscoelasticity is in the occurrence of temporal fractal structures in the wave function of the chromatin oscillations. These fractal structures govern the processes of production of the proteins needed to repair the damaged section of DNA. Of course, we must bear in mind that this process leaves room for error in the repair of DNA damage that is the source of all diseases, but also of human evolution.

I do not think that the above approach will actually help to control the processes of DNA repair in the present state of biochemistry and molecular biology. However, it can be useful when you want to kill cancer cells, for example. We know that radiotherapy and chemotherapy mostly kills cancer cells by damaging their DNA. In order to prevent the repair of DNA in cancer cells, according to the proposed approach, the therapy must, before or in conjunction with radiotherapy, chemically or otherwise affect the 3D structure of chromatin in order to reduce the degrees of freedom of the system (the ability to reformat), and therefore its wave function. This will significantly reduce or modify the spectrum of the production of initial molecular signals in response to the DNA damage induced by the applied therapy. Or maybe this spectrum may contain a set of proteins suitable for DNA repair. Or the development of the process does not allow re-normalization of the ensemble of probabilities in required for cell survival way. That is, in this case, the therapy should be supplemented by the action, which may decrease the adaptive capacity of the genome at the level of chromatin matrix. With appropriate research and ingenuity of scientists, this problem can be solved at the present level of molecular biology theory and practice—for example, via introduction of cross-linking chemical agents in the chromatin matrix.

Also, the microstructure of the chromatin itself is a fractal structure. The emergence in biological objects as a set of fractal structures is obviously the result of long process of evolution, which at each point of bifurcation (mutation) is followed by a dynamic chaos, leading to the emergence of new structures and their transformation. A good example of a dynamic chaos in the case of DNA damage is how "legions" of molecules go to work on the damaged DNA. Fractal structure arises naturally in any viscoelastic system. That is, in one way or another, the fractal properties of the viscoelastic system have a similarity with the coastline map, in which the degree of irregularity depends on the scale of the map, and it is higher in more detailed image, as it was considered in the first works on the fractal geometry in nature done by B. B. Mandelbrot. The theoretical study [431] of the viscoelastic properties of heterogeneous media shows that upon application of external forces to a medium with a fractal structure, there are peaks ("resonances") in a wide range of concentrations of fractal phase. Resonance parameters of micro heterogeneous (temporal and spatial) chromatin structure are parameters in which these peaks appear—that is, the situation in which the parameters of the applied external force is in resonance with parameters (its own frequency) of individual fractals. In such an approach to the work of the genome and other body systems (cells, tissues, and organisms), there is a theoretical possibility for using the dependence of the macroscopic wave functions of biological objects from the changing of their fractal characteristics (fractal micro wave functions) for the "wave diagnostics" of a state of these objects. When the external stress signals, such as from food, reach the DNA in the nucleus of the cell, they cause some randomization of the genome in the new environment. This changes the wave function of the chromatin structure, which, however, rushes to the new steady state. This *is called reformatting the 3D structure of chromatin. In this new steady* state, randomization of the genome comes to the "normal" level of a working cell, which begins to synthesize a slightly different ensemble of RNA molecules and proteome that *is optimal for the new stress conditions.*

In the case of extremely severe stress, *finding a new steady state may require more time, since much*-changed conditions may require finding already existing appropriate mutation in the genome, which probably happened long before, necessary to adapt to new conditions. In this case, the randomization of the genome of cells involves a large number of sites and, accordingly, there is a large number (tens of thousands) of new molecules. Reformatting the chromatin happens in a much larger set of gene expression for finding the required number of regulatory genes that control many other genes, leading to steady-state cells under conditions of severe stress. The process of finding these mutations is similar to the above-described jump through a fitness valley (FV).

If stress leads to mutations, particularly of regulatory genes, the process of reformatting and randomization leads to finding a new steady state. Thus, any external stress factors lead to the expansion of the function of an ensemble of states *of the genome. The randomization of genome work* results in a situation as it goes into the evolutionary antecedent state. And out of this relatively chaotic state due to reformatting of a viscoelastic matrix, the *chromatin system finds a way to a new steady state, unless of course* such a state exists in nature. Randomization and reformatting actually comprise an elementary act of what is usually called evolution. Viscoelasticity plays a key role in this act, since only the viscoelastic matrix has the ability to create a stable fractal structure of the wave function of the mechanical pulsations.

I must also point out that in the case of cancerous damage to its DNA, the cell turns to the state of relatively higher randomization and destruction of the functional genomic structures of the later stage of evolution. Because of that genome randomization, cancerous cells are able to survive and repair their DNA much better than normal cells. Also, the cell population in a tumor should be much more genetically/epigenetically diverse than normal cells in tissue. This is why they often retain under chemotherapy the self-renewal property that makes cancer so deadly. I have found results [636] that confirm this idea, published in the journal *nature online.* This study is the largest genetic analysis of triple negative breast cancer. The scientists expected to see similar gene profiles when they mapped the genomes of 100 tumors. But to their amazement, they did not find any similar genomes. At a genetic-molecular level, scientists have observed a continuum of nearly 100 different types of cancer.

Researchers at the European Molecular Biology Laboratory's European Bioinformatics Institute (EMBL-EBI) examined 120,000 tiny genetic mutations called single nucleotide polymorphisms (SNPs) in thirty-four strains of the bacterium *E. coli.* They found how random the mutation rate was in different areas of the bacterial genomes. At first glance, the results of their study [644] meant that key genes mutate at a much lower rate than the rest of the genetic material, which decreases the risk of such genes suffering a detrimental mutation. But contrary to what the researchers believe, I think that the mutation rate in all genomes is the same. However, when random mutations occur in key, evolutionarily old genes that conduct the basic functions of all cells, such as producing energy and important biochemical functions, randomization of the genome happens with much higher intensity than if mutation occurs in the rest of the genetic material. When the crucial parts of the genome are damaged, the legion of the new tens of thousands of molecules is produced by chaotic work of genome.

Evolutionarily more recent fractal functional structures of the genome occur on the basis of fractal structures of the early evolution of the genome. The destruction of the wave function of the mechanical pulsations of evolutionarily early structures of the genome leads to a chaotic work of genes in this area. But because these genes also manage many genes in the later acquisitions of the genome, they transfer chaos to these parts. Total chaos in such cases is much bigger. On the other hand, the destruction of the wave function of the evolutionarily late structures affects the genome less at the basic level and thus generates a lower level of chaos and, consequently, a smaller number and fewer types of molecules, which often cannot eliminate the mutational damage. It does mean that bacteria and other cells did not evolve a way to "safeguard" the crucial genetic material by some unknown mechanism, as suggested by authors of the study [644], but rather evolved the way to effectively repair the damages in it.

Another genomic study [645] at the Hebrew University of Jerusalem is confirming our conclusions about the study [644]. Researchers have identified two distinguishable sets of genes in every genome: those that produce very abundant biochemical products in the cell and function properly in most biological processes (evolutionarily early structures of the genome), and a flexible subset (later acquisitions) that might have abnormal function in a disease. The "hard core" of the genome network conducts the basic functions of cells. The biochemicals produced by the later acquisitions of the genome are less abundant in organisms, and their amount might vary significantly between different cancer cells derived from patients with the same type of cancer. This result coincides with the results citied above of the study [636]. In reality, I think it might be more than only two distinguishable sets of genes because it surely should be more than two fractal structures in genome structure organization.

The study [637] shows that not only mutations but even a single gene deletion in a non-coding area of the genome (junk DNA) may also affect the functioning of a cascade of many genes. This means that inherited information in the non-coding areas of the genome plays a strong role in the development of cancer. The researchers demonstrate that gene deletions called copy number variations (CVN) in either protein coding or non-coding areas of the genome play a crucial role in the aggressive prostate cancer. Other researchers have linked CNVs to Alzheimer's and Parkinson's disease, mental retardation, autism, schizophrenia, and neuroblastoma, a type of brain cancer [638]. But the presence of CNVs means a change in the mechanical properties of sections of DNA and change in their own mechanical frequencies of the pulsations. From this, *it is reasonable to assume that non-coding DNA regions under certain abnormal conditions* of CNVs by changing the mechanical state of chromatin in the nucleus contribute to the resonant excitation of the remote "bad" genes and the emergence of cascades of *on/off switches of many genes that cause disease.*

It should be noted that in principle it is the same mechanism described above that lies at the basis of work and evolution of the gel in the primordial soup, which we have reviewed previously in the analysis of life's origin. Thus, from the origin of life to the modern results of subsequent evolution, the work of a cell at all levels is controlled by the same mechanism of the viscoelastic gel mechanical pulsations and a relatively narrow set of BZ-type chemical reactions. The number of complex molecules in the process of evolution, of course, enormously increased with the complication and diversity of modern cells' functions, but the principles remain the same. This means that the influence of food and food metabolites on the chromatin structure and randomization processes in the nucleus is very important. If the influence is relatively small, it acts by reformatting the chromatin structure within the normal spectral range of possible states. That provides an epigenetic cell adaptation to new conditions of diet. If the influence is strong, fast adaptation at the genetic level with significant reformatting of the 3D chromatin structure to a new dynamic state happens. Moreover, this approach could help explain the acceleration of evolution in a period of intense changes in the environment compared with conventional low-enough rate of evolution in normal, more stable conditions.

Evolutionary biologists at the University of Toronto have found that individuals with low-quality genes may produce offspring with even more inferior chromosomes, possibly leading to the extinction of certain species over generations. Their study [640], published in *Proceedings of the National Academy of Sciences,* predicts that organisms with such genetic deficiencies could experience an increased number of mutations in their DNA, relative to individuals with high-quality genes. The research was done on fruit flies whose simple system replicates aspects of biology in more complex systems, so the findings could have implications for humans [641]. According to Nathaniel Sharp, author of the study, their research suggests that the problem is likely to compound over time, leading to a mutational meltdown that may devastate endangered populations and increase the risk of health problems in families in poor conditions [641]. Researchers examined the accumulation of mutations in the fruit fly *Drosophila melanogaster*, the genes of which are arranged on three major chromosomes. They introduced harmful mutations onto the fly's third chromosome and then observed how the presence of these mutations affected the fitness of the second chromosome over forty-six generations. According to the study, copies of chromosome two maintained in strains with poor-quality copies of chromosome three declined in fitness up to three times faster than those with good copies of chromosome three, suggesting that poor genetic quality elevates the mutation rate. The study's authors believe that the underlying mechanism of this process remains unknown. However, from my point of view, there should be a fairly simple explanation of the study's results. Introducing a harmful mutation to any chromosome leads to the randomization of that chromosome's work. This randomization leads to the production of a much wider-than-normal spectrum of RNAs and proteins. These RNAs and proteins, plus mechanical pulsations different from the normal wave function, bring chaos into the work of the other chromosomes and make them less capable of repairing DNA. That may generally lead the cell to a relatively ill state, or sometimes the elevated under chaotic condition of the genome mutation rate could also accelerate adaptation to environmental stresses and evolution of the cell.

We also have to remember that adaptation to environmental stresses is not the only driving force of evolution. In some cases, evolution may be driven by co-evolutionary interactions between species in any combination: cells in tissue, bacteria in colony, viruses, etc. In both cases, chemical signaling from outside of the cell changes the wave function of the genome as described above.

Of course, this scheme of the process includes a lot of a simplifications—as we know life is always much more diverse. But there are no assumptions that contradict the known data about gene mutation. The specific mechanisms of interaction of genes to enhance the joint survival are known in part, but I personally suspect that many more of these mechanisms are not known to modern biology. And the fact that in this example we have to consider the processes in the cells of the colony does not mean that they are not applicable to higher organisms. As biologists say, what is true for the cell is true for the elephant.

One aspect of gene interactions can be illustrated by results of recent studies [569, 570]. Researchers have uncovered a vast new gene regulatory network in mammalian cells that could explain genetic variability in cancer and other diseases. According to the half-century-old central dogma of molecular biology, scientists have thought that the only role of messenger RNA (mRNA) is to shuttle information from the DNA to the sites of protein synthesis of the cell. But these new studies suggest that the mRNA of a single gene can manage, and be managed by, the ensemble of mRNA of all other genes via their microRNA molecules, with up to a few hundred genes cooperatively interacting together. This means we have to consider here a few interacting ensembles of states of different networks in the nucleus of the cell. Namely, there are ensembles of genome states, chromatin matrix, mRNA, and microRNA, and perhaps some other proteins. And we have to remember that these interactions have a probabilistic character, and all conclusions made above have sta-tistical meanings.

Based on the above information, it seems to me reasonable to use the method of wave functions, widely used in quantum mechanics to describe the states of the sys-tems, for description of the above ensembles of probabilities of states. In this case, the interaction of these different wave functions for chromatin, genome, RNA, etc. in the cell nucleus is similar to the superposition of quantum states. In addition, the normalization condition of the wave function is crucial to biological systems. This means that there must be a stable life state of the biological system to which it as-pires in the process of evolution. In this case, the process of evolution is determined by a continuous process of renormalizing [222] the wave function described above for the jump across FV and earlier for the process of the origin of life due to the pres-ence of a viscoelastic feedback. Such an approach that uses concepts and mathemati-cal formalism of quantum mechanics can be quite fruitful for use in an analytical and in silico study of the probabilistic processes of interaction of ensembles of the states of various elements in the cell nucleus.

As shown in the study [569], in the case of some genes already expressed, dele-tions of their mRNA from total ensemble mRNA of all genes involved  appears to be acting as mutations of the gene itself. According to an essay [571] about this study, when two genes share a set of microRNA regulators, changes in the expres-sion of one gene affects the other. If, for instance, one of those genes is highly ex-pressed, the increase in its mRNA molecules will "sponge up" more of the available microRNAs. As a result, fewer microRNA molecules will be available to bind and repress the other gene's mRNAs, leading to a corresponding increase in expression. These studies reveal that mRNAs actually use microRNAs to influence the expres-sion of other genes. According to the author of the study it turns out that this type of microRNA-mediated regulation is commonplace in the cell, and thousands of genes are regulating one another through hundreds of thousands of microRNA-mediated interactions. The finding explains, at least in part, why all people with the same diseases often do not share the same genetic profile and why people with different genetic profiles have a similar reaction to the external factors from food, for example. The finding might help to explain how under the influence of external signals from food metabolites, the cell can alter the functioning of the chromatin matrix and interacting gene network.

In reality, not only two, but a larger number of genes may share a set of mi-croRNA regulators. By this "sponge up" mechanism, the wave function of the states of all mRNAs would re-normalize up with every single microRNA-mediated "mutation" until it comes to a steady state. Of course, this process will be influ-enced by the superposition of all renormalizing wave functions of the genome, the chromatin, various RNA, and proteins. This phenomenon of microRNA-mediated "mutations" can participate in the above-described process of "multiple simultaneous" mutations by participating in the interaction of the wave functions of chromatin matrix states and mutations in the genome. The superposition of ensembles probabilities of mutations of the genome and microRNA-mediated "mutations" theoretically may lead to a substantial acceleration of the evolution of these systems, since the frequency of mutation is complemented by an independent process of microRNA-mediated regulation. Additionally, in the case of low mutation rates in a system, it could as well improve system's ability to adapt because the possibility of adapting for some external influence is higher if two mechanisms are used rather than single one.

The existence of the RNA interaction network helps explain the influence of so-called "dark matter" of the genome on cell functioning. It is because, according to the study [570], the RNA regulatory network also appears to extend into the massive non-protein-coding region of the human genome. Thus, microRNA from non-coding RNA may play a role in microRNA-mediated interactions with expressed coding genes through their mRNA. Before this research, the function of that non-coding RNA had been a mystery. Despite the fact that in the studies described above, the processes were mainly studied as a mechanism of cancerous cell formations, they may be fully applied to the cell genetic and epigenetic adaptations for coming into the cell food derivatives.

One could ask, why should we pay so much attention to the details of the genome's work in a book on nutrition? I must say that our diet interacts with the genetic apparatus of cells in our body by many pathways. These pathways are in one way or another quite convoluted in the phase space of the reactions of our metabolism, but there are some very direct interactions of our food and the key processes of cells. It's no secret that our lunch messes with our biochemistry. Once food gets to your mouth, genes related to digestion have been activated and are causing the production of the many enzymes that help break down food substances. But a new study [572] suggests that the connection between our food's biochemistry and our own metabolism may be much more congested than we thought. RNA molecules usually found in plants have been discovered circulating in blood, and researches indicate that they are directly influencing the gene' network expression states. Therefore, all the processes described above are extremely important for understanding the interaction between food and our genes.

According to the study [572], when we eat natural food we are not only receiving proteins, carbohydrates, vitamins, etc., we are also receiving genetically modulated information from food. In the study, Chen-Yu Zhang's group at Nanjing University present a striking finding that plant microRNA could get into the host's blood and tissues via oral consumption. The microRNA molecules are not destroyed during digestion; they enter the body intact, find an mRNA target organ, and interfere with its functions. Moreover, once inside the host, they can elicit functions by affecting the target organ's genes and thus influence the host's physiology.

Researchers proved that this exogenous plant microRNA is present in the sera and tissues of various animals and that it is primarily acquired orally, through food intake. They demonstrated also that exogenous plant microRNA in food can influence the gene network's expression states and thus human physiology.

The significance of this finding lies in the fact that the transfer of genetic information is an important anthropic property of food that is almost completely lost in processed foods. And perhaps more importantly, that the products of modern agriculture conveyed to us through the food that we believe to be healthy bear genetic information that is evolutionarily erroneous for us. MicroRNA may be new and it has not yet been studied as to the type of functional molecules derived from food such as vitamins and minerals. I think that there are countless types of such molecules. Getting active RNA molecules of food means that the fine tuning of metabolic processes occurs at the level of direct exchange with the surrounding biocoenosis by genetic control signals. This also means that consumption solely of food in the state that retains the "live" properties of all food components ensures an anthropic diet.

But if the information can be transmitted in the form of RNA pieces, then why can it not be transmitted in the form of DNA pieces? This is especially important in the case of the exchange of information with the bacterial flora of our gut. There, the sharing of genetic information from the food's DNA and RNA might be much more intensive. Because in the bacterial world the rate of evolution is much higher than for mammals, it has surely already happened to humans. This is how modern disanthropic food leads to changes in our metabolism. Also, because the gut microbiome is the most active metabolic organ of our body, we have suffered, from the beginning of the Neolithic revolution, dramatic consequences for our health from genetically disanthropic food. And of course this study demonstrates the dangers that lurk in genetically modified (GM) food and is certainly the last nail in the coffin of the discredited "safety assessment" process for GM foods in the United States and worldwide.

Scientists still do not know all the details about the interaction with our cells of countless pieces of genetic information coming from the food, but the fact that they play an important role in the cells' biology is already proven. So findings in the study [592] clearly demonstrate that some miRNAs work as genuine cancer brakes that suppresses the malignant behavior of tumor cells. And those cancer brakes appear to fail in many types of cancer. When the endogenous production of these RNAs in senescent cells ceases completely or decreases, the RNAs from a truly anthropic food may help compensate for their shortage. This is why people who have a higher anthropy of diet have less chance of getting cancer.

It should be noted that the mechanism of foreign direct use of genetic information from plants in the human metabolism [572] and possibly genetic information from other products of nature, such as bacteria and/or animals, are endangered in the modern biocoenosis. This is related to biodiversity declining. This process leads not only to reduce the overall diversity of food in the food chains, but also to an overall reduction of the genetic material in nature. The lack of genetic information normally supplied to all species through the food chains (networks) leads to a change in the state of metabolism of all living beings in nature, including humans. As biodiversity declines, these metabolic changes become irreversible because their metabolism is increasingly affected. The biodiversity loss—i.e., the rapid declining in species number and the general degradation of ecosystems—is probably the greatest threat to the future health state of humankind. nature is losing species at a rate that is 100 to 1,000 times faster than the natural extinction rate. According to many scientists, we are now in the sixth mass extinction event on Earth, which is a result of a competition for resources between us and all other species. The previous five periods of mass extinction were driven by changes in climate as results of asteroid impacts, intensive volcanism, and change in atmosphere—the so-called Great Oxidation, the biologically induced appearance of free oxygen in Earth's atmosphere. The most dangerous thing in the biodiversity crisis for us is that we, like other things that lived during previous crises in the planet's history, do not have any guarantee or immunity from extinction.

## 3.8   Food, cancer, diabetes

Through the analysis of the behavior of mitochondria, the effect of viscoelastic cellular and extracellular protein matrixes on the translation process of signaling molecules and their aggregates, the link between metabolism and diseases becomes apparent. As the cells lose their normal function and show a decrease in ATP production from the entering food substances below the critical level, they are starting the process of apoptosis—i.e., suicide. In this way they exclude themselves from the body, allowing the remaining cells to exist in conditions of depleted nutritional supplies. In medical practice of disease development and aging, this looks like a shrinking of various organs and tissues of the human body. As already indicated, this is a natural process of aging. However, due to external chemical stress, from ingestion are not suitable for the body chemical compounds, when formation adverse metabolites and a lack of necessary substances occur, the above mechanisms accelerate the degradation of cells and organs, and lead, ultimately, to diseases or even premature death of the organism. Many examples of degradation are not related to natural aging: from the slow poisoning by disanthropic chemical compounds and substances to general household alcoholism. Malnutrition is between these two extremes.

However, the accelerated accumulation of classical or epigenetic mutations can lead to more dire consequences for the organism. After the accumulation of ten to twelve mutations in cell, originated from disanthropic living conditions and nutrients, a nasty outcome for the cell and the body may be a cancer. Cancer can, with a high degree of reliability, be considered a disease of the human metabolism that breaks down the feedback mechanism in internal biochemistry of the cells. Usually through a particular biochemical feedback signaling system, a cell receives signals from the other cells or from its own interior to self-destruct. However, in some cases, when the impact of disanthropic factors such as toxins is quite pronounced, those factors destroy these communication signaling pathways responsible for the feedback or, in other words, destroy some of the direct and backlash reaction in the phase space of metabolism. Simultaneous changes in metabolism in the cells and feedback mechanisms makes the cell cancerous - similar to invaded to the body bacteria, which does not obey well the signals from the body and behaves in a certain sense as a single independent microorganism.

The body cells become cancerous when the normal controls over cell growth and death go awry. This deregulation has traditionally been linked to DNA mutations of single genes or deletion of large sections of the chromosome. However, more recently it has become clear that gene silencing in cancer can also occur in the absence of changes to the DNA sequence but in the existence of epigenetic changes due to DNA methylation, which is one of the main epigenetic processes [374]. In cancer, the DNA methylation pattern of many genes changes. However, until now, it was believed that only individual single genes were silenced by methylation. But this is not necessarily the case. Researchers [374] have found that non-methylated genes that reside in a particular suburb near methylated genes are also silenced. Their physical proximity to the methylated genes affects their ability to function. But physical proximity between methylated and non-methylated genes can be due to modulation of nodes' positions in the spatial matrix structure of chromatin, as described in Section 3.7. But this modulation depends on environmental factors, of which the principal is the food consumed. The extent to which genes are methylated and acetylated is a means of controlling when, what type, and how much of a protein is produced from a gene. Monozygotic twins with the same DNA may develop different diseases because of expression and suppression of different genes. In cancer, the expression-suppression pattern of many genes changes. Genes that regulate normal cell growth can be silenced by the methylation process. But the same genes can be expressed by the acetylation process because epigenetic changes in gene expression is a reversible process and is a good target for anthropic-food-based cancer prevention therapies.

In a report published recently in *nature Genetics* [407], the investigators focused on a methylation, which silences genes. By comparing the epigenome of eight human tissue samples—three from noncancerous tissue, three from tumors, and two from early-stage colon cancer—the team found that in all the tumors the defining characteristic was a universally "chaotic" pattern of methylation. In noncancerous tissue, they found methylation occurring in well-defined places, either as small "islands" of methylation or huge methylated "blocks" [407, 408]. The researchers [407] noted that cells in their healthy tissue samples stayed methylated at around the 80-percent level for large blocks of the epigenome. By comparison, cells from tumors comprising those same huge blocks had no such stability and were much more variable in terms of methylation levels. In the cancer tissue, researchers "saw that the once-precise boundaries of the islands had shifted or disappeared altogether, and the start and end points of the sites appeared unregulated," says Andrew Feinberg, M.D., M.P.H., professor of molecular medicine at the Johns Hopkins University School of Medicine. "We also saw a loss of methylation, presumably increasing the randomness of gene function within them. What seems to define cancer at the epigenetic level may be simple and common, namely chaos that seems to be universal" [408]. The team [407] designed a custom test to compare about twenty noncancerous tissue samples to twenty samples from each of a variety of tumors as they investigated thousands of methylation sites for colon, breast, lung, kidney, and thyroid cancers. They found that, here again, methylation was well-regulated in the normal tissues, almost always occurring within a limited range of variability. However, in the very same specific places of the epigenome characterized by chaos in colon cancer cells, all the other cancerous tissues examined by the team showed distinctly variable and "chaotic" levels of epigenetic variation [408]. Thus it seems that the influence of external stress degradation of the highly organized 3D matrix structure of chromatin in the nucleus leads to chaos and degradation of the epigenome and in cell functions. One of the main sources of external stress is food, which includes in the present time a huge number of disanthropic chemical compounds and substances of different origin. Day after day, for many years, gradually changing the evolutionarily determined 3D structure of the chromatin matrix, which determines the normal type of methylation, these chemicals lead cells eventually to the state of chaos of epigenome observed in the study [407]. However, these changes are inherited during cell division and further aggravated by continuing disanthropy of external influence. Ultimately the accumulation of the changes may cause cancerous degeneration of cells. The fact that the inheritance of epigenetic changes is rather a normal process, not only in the cell, but also in higher organisms is shown in [409], where researchers have uncovered a mechanism by which the effects of stress in the fly species Drosophila are inherited epigenetically over many generations through changes to the structure of chromatin.

A healthy cell has a quite distinctive pattern of gene expression, in which some genes are active and others are suppressed, and an epigenetic derangement of this pattern contributes to cancer, in part by silencing genes whose activity would normally stop a cell from becoming cancerous. As confirmation of the fact that cancer may have epigenetic origins, we can consider the results of [581], which demonstrated that the survival rate of patients with lung cancer treated with epigenetic therapy was much longer than that normally expected. Patients showed signs of gene methylation reversal in at least two of four key genes, and some of them experienced dramatic tumor shrinkages. In one particular case, a man whose metastasizing lung tumors had spread to his liver, the epigenetic treatment cleared his metastases and markedly reduced his original lung tumor. Cancer of course also may arise through permanent genetic mutations, which cannot be undone with epigenetic treatment, but in many cases for prevention of cancer through epigenetic mutations, epigenetic prevention treatment by using a long-term anthropic diet may be very helpful. We also have to remember that very often the epigenetic mutations are a precursor to gene mutations leading to cancer.

From a very general point of view, transformation of healthy cells into a tumorous state looks like a genetic and epigenetic degradation of the cells' genome to the evolutionary level of the first cells that have begun to unite in multicellular organisms. Thus cancer could be looked as an atavism from the age of emergence of first multicellular organisms. But these first multicellular developments were not restricted by complex feedback loops, which limit the growth of the community of cells to a certain size—a body, an organ, or tissue. As a result of subsequent evolution, the genome of the cells with the development of higher forms of life became more complicated and the cells acquired a specialization, the ability to apoptosis, and information exchange aiming, among other things, toward limiting the size of an organ and a body. However, external influences such as radiation, toxins, lack of anthropic compounds in the diet, and so on sometimes disrupt the work of the genome, which in turn leads to disruption of its functions, manifested as degenerative diseases of various organs, including cancer.

Thus, cancer can be considered as the destruction of the functional part of the genome that has evolved in the course of genome evolution over the last 1–2 billion years. The cell in the metastatic state acquires the atavistic ability to divide indefinitely. As in every organ of the body, cells carry out certain functions—i.e., they express out the certain locis of the genome—then the effect of external stress appears to individually "erase" evolutionarily accumulated coding in some parts of the genome. This, in my opinion, explains why the many types of cancers are more or less specific for each organ, tissue, or body fluid.

The increasing complexity of organisms' genomes in the evolutionary process reduced their stability to external disturbances. This is similar to the stability of a column built from cubes not connected to each other. The higher the column (i.e., the more complicated the genome), the less resistance to external kicks the column has. Even quite small disturbances of the genome of a complex organism could change some functionality of the cells' or organs' tissue, leading to diseases. So from this point of view, cancer could be looked at as an atavism from the age of emergence of the first multicellular organisms. With stronger disturbances (high dose of radiation, poisons, etc.), an entire part of the genome, responsible for sophisticated cell functions acquired in the course of evolution, could be destroyed, and the cell could return to a more primitive state—i.e., become cancerous. The cancer cell is noticeably changed in its viscoelastic properties to a softer atavistic state. A softer cell (i.e., with decreased modulus of elasticity) acquires the ability to migrate within the tissue on the surface of the epithelium, similar to some type of ductile material able to flow like Dali's melting clocks. The already existing tumor starts indefinitely growing and spreading throughout the body. Thus, a cancer in some sense is the price we are paying for our organism's extreme complexity at the genetic level and is a sign of degradation of the cell genome. Relatively simple organisms in nature are almost never affected by cancer.

It is interesting to note, for example, that cognitive deterioration in Down syndrome due to a genetic defect is also associated with the frequency of cancer. People with Down syndrome have a markedly lower risk of most types of cancer. Some enzymes responsible for inhibition of cancer in cells are made by a gene on chromosome 21, which in people with Down syndrome is present in three copies instead of the normal two, causing the enzymes to be overproduced. It is difficult to say, though, whether the simplification of the structure of the brain is associated with the occurrence of cancer in the above sense. Degradation of the genome can occur under the influence of both epigenetic factors and genetic changes. As is mentioned above, degradation of normal DNA (genome) is a hallmark of cancer cells. Although it had previously been known that damage to normal cells is caused by stress to their DNA replication when cancerous cells invade, the molecular basis for this remained unclear. For the first time, researchers at the Hebrew University of Jerusalem in the study [399] have shown that in early cancer development, cells suffer from insufficient "building blocks" to support normal DNA replication. This means that an anthropic diet theoretically could halt this by externally supplying the building blocks, resulting in reduced DNA damage and significantly lower potential for the cells to develop cancerous features. Thus, hopefully, this could one day provide protection against cancer development. And it is how Mother nature used to protect us and most others animals from cancer via the abundance of anthropic elements in our everyday diet in natural life.

It is well known that during the anthropic lifestyle of our ancestors, cancer was an extremely rare disease, affecting only people with severe injury, burns, etc. The same applies to higher animals. The anthropy of their food did not stagger the work of the genetic apparatus of cells and caused no loss of functional efficiency. In the modern lifestyle, many disanthropic environmental factors, including diet, with a prolonged influence on the metabolic processes in the body lead to a much more frequent failure of cell functions. And this in turn increases the incidence of degenerative diseases in the population. Therefore, the task of preventing cancer in the current conditions of humanity pivots on the reduction of the disanthropic influence of environment. Diet plays a major role in that.

Besides normal aging, DNA is damaged and the body begins to deteriorate mainly because of oxidative stress [410]. As shown in the studies [411, 412] as you age, your whole body becomes more sensitive to this parasitic cancer mechanism, and the cancer cells selectively accelerate the aging process via inflammation in the connective tissue. This helps explain why cancers exist in people of all ages, but susceptibility increases as you age. If aggressive enough, cancer cells can induce accelerated aging in the tumor, regardless of age, to speed up the process. Lethal cancers show the same gene expression pattern associated with normal aging, as well as Alzheimer's disease. In fact, these aging and Alzheimer's disease signatures can identify which breast cancer patients will undergo metastasis. The researchers [411, 412] find that oxidative stress is a common "driver" for both dementia and cancer cells spreading. But if lethal cancer is a disease of "accelerated aging" in the tumor's connective tissue, then cancer patients would benefit from therapy with strong antioxidants. But in reality in most cases, tumors escape destruction by antioxidants from artificial sources such as medicine or dietary supplements DS. An "alternative" antioxidant therapy, such as Vitamin E or CoQ10 supplements in mega dose format, does not work. The task in this case is to keep the level of endogenous natural antioxidants in cells and tissues via the right anthropic diet over a long time, preferably the whole lifetime. Those endogenous antioxidants will eventually desert an aging body, as shown in one study [528]. But that happens at a much faster rate if the body has an intense disanthropic diet during a significant part of its life.

Once the cell degenerates to the cancer state, it is no longer satisfied with a relatively slow rate of absorption of vital compounds necessary for biochemical functioning. Cancerous cells begin to produce large amounts of so-called hormone growth factors to ensure the growth of blood vessels in tumors and thus improve the cell diet—an example of a cumulative stream of new cancerous cell metabolism (Section 2.7). Besides this, the cancer cells, especially in the initial stage of tumor growth, release into the surrounding tissues a wide variety of enzymes that degrade the tissue structures of the body and use the remains for their own nutrition and growth. Further, it starts the production and delivery of proteins to the membrane, which partially reduce the strength of the connection of cells in the outer layers of the tumor with each other. They eventually break away and spread through the body—the so-called metastatic stage.

The main difference between a cancer cell and a microorganism having an origin external to the patient's body is the fact that cancer cells have a set of proteins and other substances that are similar to the rest of the body's cells, particularly on the membrane. For this reason, the body does not consider cancerous cells alien and, therefore, the immune system does not attacking them. The effect of incoming to the body substances on the state feedback loops is critical. From this point, it is obvious why the cancer is so widespread in regions where the food and environment do not have certain biologically available trace elements (selenium, iron, magnesium, iodine, and others), which play a crucial role in feedback mechanisms through the formation of ion channels in cell membranes and other means. Thus, the metabolic disorders that promote cancer significantly alter the metabolism of cancer cells and the entire state of the phase space of the body's metabolism. The phase space of reaction is destroyed, and consequently there is a destruction of the body. In fact, cancer means that inside the body, another organism emerges (the colony of cancer cells) whose metabolism at the cellular and intercellular level is significantly degraded. At the intercellular level, as indicated above, this is manifested in the absence of normal responses to signals from the body. In the intracellular level in particular, according to [29], it is manifested as unusual for the cells' multipolar structures centrosomes—organelles of the cell—and thus as altered mechanism of separation of the chromosomes that does not exist in normal cells.

Cancer is a very dynamic disease that, during its development, undergoes many biochemical changes to evade natural defenses, adapting to new environmental circumstances and invading the organs to cause metastasis. As shown above, most of those biochemical changes manifest themselves through changes in the mechanical states of cell matrices. A tumor initially in a solid state, fixed somewhere inside of tissue, afterwards changing elasticity modulus and becoming a bit liquid or more flexible, go through the tissues to epithelial and cause metastasis. In this sense, drug treatments should aim at the mechanical properties of cells at every stage of cancer in order to try to reverse the process and move from an aggressive tumor form to a milder one, which could be better controlled in terms of the progression of disease. The importance of having healthy tissue structure for cancer prevention is demonstrated by a different angle of view in study [589]. Researchers have now shown that healthy spatial tissue structures may slows down the accumulation of genetic mutations, thereby delaying the onset of cancer. They have also shown that the destiny of cancer-causing mutations depends in part on where they occur and how much competition they are exposed to from other, similar mutations. This means that tissue with a strong spatial structure accumulates fewer mutations over a given period than tissue with unstructured cells. Cancer growth normally follows a lengthy period of development. Over the course of time, genetic mutations often accumulate in cells, leading first to pre-cancerous conditions and ultimately to tumor growth. For some cancer types, the accumulation process can take up to twenty years, and not everyone with pre-cancerous tissue will actually develop cancer; the formation of cancerous cells often has no medical consequences.

According to the study, in a tissue without any spatial structure or with mechanically weak spatial structure, genetic mutations can propagate and accumulate relatively quickly. In tissue with healthy and strong spatial structure, it takes much longer for cells to accumulate the critical number of mutations required for tumor formation, thereby delaying the onset of cancer. Thus, a well-structured tissue may prevent cancer cells' migration due to their strong mechanical elastic properties, and at the same time, such tissue prevents the occurrence of cancer cells by the scenario described in [589]. The importance of intact tissue metabolism is illustrated by the data from [24] on the increased frequency of deaths from the different types of cancer of survivors of previous successful cancer treatments. That is, even after cured cancer that already caused or was the result of such changes in the phase space of metabolism, the probability of cancerous failure in other parts of the metabolism remains high.

But this applies not only to cancer! For example, study [25] found that people with high levels of pesticides in the body tissue have high fasting blood sugar levels. This information establishes a link between diabetes tissue and metabolism altered by pesticides.

In the first half of the twentieth century, the point of view that cancer is a disease of the metabolism dominated among physicians and biologists. But later, this idea fell out of fashion when the study of genes and biochemical processes in the cells created the illusion that the molecular mechanisms of cancer and methods of combating it were about to be revealed.

The propaganda campaign of healthy lifestyles pursued by many governments is not able to moderate the unprecedented surge in cancer incidence worldwide. For example, in the United States, where the government actively promotes healthy lifestyles and nutrition, in particular, those efforts do not bring any noticeable effect. So the United States, despite a quite good system of health care, leads the world in the incidence of certain cancers, diabetes, and other degenerative diseases. The main reasons are the quality of food and general disanthropy of most people's modern Western diet. There should be continuous influence on the organism of anthropic food compounds in order to restore the people's weakened resistance to cancer. To prevent cancer in the adulthood of an individual, it is best to start with preventive measures of using anthropic nutrition in his or her mothers' diet and lifestyle for many years before conception. This can be achieved via prevention of congenital immunodeficiency, which plays an important role in the mechanism of cancer emergence.

A strange similarity between human embryonic stem and cancer cells has long been observed. This observation was the basis of the hypothesis that cancerous tumors occur from non-specific embryonic stem cells and their clusters that for decades remain in our bodies after birth but do not manifest themselves until a certain point. According to this theory, a predisposition to cancer is programmed in early embryogenesis. Later studies showed that these cells are in many people at any age, scattered throughout the organism from the time of fetal development. It is assumed that when these embryonic stem cells or their clusters are activated in human organs, a cancer begins. But its growth can take place in absolute silence over months and years of life. At the same time, the presence of these cells does not indicate a fatal inevitability of their activation. To activate them requires external provocation by pathogenic factors. Factors of provocation may be the following reasons: The first is the impact of chemical and radiation stress due to environmental pollution. For example, there are chemical pollutants in the environment, such as benzo-a-pyrene, which in experiments on animals cause malignancies to develop 100 percent of the time. The second reason is the abuse of drugs and the methods of modern medicine (antibiotics, hormones, chemotherapy, and diagnosis-related exposure to radiation). The third reason is a long-term metabolic stress under the influence of disanthropy of food. –The fourth reason is epigenetic factors.

The simultaneous combination of all factors leads to cancer virtually 100 percent of the time. But often, the third factor, which we identified as disanthropic diet, includes to a greater or lesser degree both the previous points and may interact with the forth. This means that the factor of disanthropy of food can have a critical influence on the occurrence of cancer in the body. If the body gets only a little essential nutrition from anthropic food for long enough periods, depending on the degree of disanthropy of food, there is a certain probability of violation of the homeostasis of stable metabolism, which may lead to a condition labeled in clinical practice as immunodeficiency. Depending on the degree of immunodeficiency, it can increase the risk of cancer. However, if a food contains the disanthropic elements that have biological toxicity, the likelihood of severe immunodeficiency states is much higher. Furthermore, disanthropy of food increases the likelihood of epigenetic mutations, and therefore, leads to a much higher risk of cancer.

According to the stem cell theory, all cancer cells descend from special embryogenesis self-renewing cancer stem cells. This view predicts that killing the cancer stem cells might suffice to wipe out a cancer. However, in reality the situation is much more complicated, with cancer cells able to interconvert between different types [514]. The common point of view, that the only way stem cells occur is by self-renewal, is possibly wrong [515]. The study [514] suggests an alternative possibility that cancer stem cells could be made, not only born. It means that cancer cells are probably not fixed at all, but that, at any given point in time, they exist (at least some part of them) in one of several phenotypic "states," and those states can interconvert [515]. These results have substantial implications for the treatment of tumors, in particular: while attacking cancer stem cells treatment is absolutely necessary, it alone may not be enough to fight cancer. The results also give substantial support to the point of view that metabolic stress from prolonged disanthropic diet could lead to more intensive interconverting between mentioned phenotypic states.

As I mentioned above, many types of cancer arise from widespread alterations not in the genetic code itself, but epigenetically—through the genomic chaos within cells relies on changes in the 3D structure of DNA and the proteins "community" alterations associated with wrongly expressed/depressed genes. What is the mechanism of action of disanthropic food that is at the heart of this process?

The chromatin matrix has regions in which change is quite easy in order for stem cells to become, say, kidney or brain cells, but at the same time, those regions are very vulnerable to changes that may ultimately lead to cancer. In my opinion, under the influence of disanthropic substances leading to chaotic biochemical alteration of the 3D structure of chromatin in those regions of normal cells, some parts of the chromosome could break off and fuse with parts of other chromosomes to create new fusion proteins. These fusion proteins subvert the normal function of the gene and theoretically could activate a set of cancer-causing genes. Usually, of course, it is not as simple as described above. In reality it takes many steps from the first fusion of the first wrong protein, which in turn makes a further alteration of the chromatin structure, and activates another set of genes, and produces a second set of fusion proteins (enzymes), and so on. Sooner or later after some n-fusion, we will get expression of cancer-causing genes. I have to point out again that those processes could be described the same way as above in Section 2 of this book as the interaction of the strange attractor of cell or nucleus metabolism with flicker-noise of disanthropic substances of food. Every disanthropic substance could generate its own cascade of n-fusion reactions probabilistically linked with few previous reactions of fusion and with reactions that originated from other disanthropic substances of flicker-noise spectrum. That could provide the basis for computational analysis of such processes in the cell. In addition, we have to remember that this cascade of reactions leading to cancer had started from disanthropic substances in our diet. Because the developments of such cascades of changes is a relatively slow process (years), it could be reversible in early steps of changing the 3D structures of chromatin. And correct anthropic food could stop development of n-fusion cascades and return the chromatin matrix from "chaotic" to unaltered healthy state. In a later state of "chaos" development in the nucleotide, after some critical point, return to a normal state is not always possible. The synergy of all the negative factors is also possible. So the lack of anthropic nutrition from the food leads to cracks in the phase space of metabolism. However, these cracks can be filled with cascades of reactions involving chemical and biological contaminants, which are a great many in today's food by cultivation methods in agriculture and processing technologies and storage.

The fate of embryonic stem cells remaining after the birth and of the newly formed type of stem cells [515] depends not only on provoking pathogenic and epigenetic factors, but also on the level of homeostatic protection of metabolism in a particular individual, which is largely determined by the degree of anthropy of his diet over a long time. Recently, researchers [216] discovered that the biochemical mechanism of cancer for the cells depends on activation of a particular type of microRNA. MicroRNAs are small bits of RNA within cells that wield enormous power. They influence virtually every biological process by controlling the expression of genes. The researchers found that the cancer could be developed by over-producing microRNA 380. That type of microRNA is very active while we are developing embryos, when cells need to divide very quickly, but after that it appears to get switched off. We still don't know why it gets switched on again in certain cancers.

But we can assume that provoking pathogenic dietary factors play an important role in this process—especially as [55, 572] showed the possibility of the presence of miRNAs directly in the human diet, and [216] also shows the effect of folic acid deficiency or an excess of arsenic in the activation of microRNA. Furthermore, it is known that a lack of natural sources of folic and methionine amino acids in food causes a disturbance in the processes of DNA methylation. It should also be noted that both these amino acids are not produced by the human body, but are vital for it.

But what is the mechanism of the effect of food on the characteristics of an organism such as resistance to disease? And is there any? To answer this question, we have to take into consideration the results of studies [422–424]. A stem cell that can morph into a number of different tissues is proving a natural protector, healer, and antibiotic maker, researchers at Case Western Reserve University have found [422, 423]. Mesenchyme stem cells (MSC) reaped from bone marrow have been hailed as the key to growing new organs to replace those damaged or destroyed by violence or disease, but have failed to live up to the billing. Instead, scientists who had been trying to manipulate the cells to build replacement parts have been finding the cells to be innately potent antidotes to a growing list of maladies [423]. The MSC "is a drugstore that functions at the local site of injury to provide all the medicine that site requires for its successful regeneration," according to A. Caplan, lead author of the paper [422]. Other scientists show that the cell's arsenal is even greater. They found that the cells produce a protein that kills bacteria including E. coli and Staphylococcus aureus, and enhance clearance of the microbes from the body [423].

Another recent work [424] addressed the hypothesis that MSCs represent a common mechano-responsive element upstream of osteoblast (cells that are responsible for bone formation) and adipocyte (fat cells) differentiation that could potentially be targeted for the control and treatment of both obesity and osteoporosis. They proposed that low magnitude mechanical signals (LMMS) could non-pharmacologically and non-invasively promote stem cell proliferation and thus an organism's healing and regenerative potential. In addition, researchers in this study assessed the impact of a high-fat diet on the resident stem cell population, as a possible contributing factor in the pathophysiology of diseases. The experiments and results [424] indicate that MSCs respond to LMMS by increasing proliferation. A developmentally mediated mechanism by which fat was suppressed and bone was enhanced was implicated and was linked to the mechanically based bias of the MSCs to preferentially differentiate towards osteoblasts over adipocytes. The mechanical promotion of the number of progenitor cells, as well as driving commitment choices, suggests a means to enhance an organism's regenerative capacity as achieved by exploiting stem cell sensitivity to mechanical signals.

Thus, from the results of [422–424], it follows that improper disanthropic diet with redundant or poor-quality fats affects many metabolic processes that control the body's production of natural endogenic protectors, healers, and antibiotic makers. If, as result of disanthropic diet, production of these is disrupted, obesity, osteoporosis, and other degenerative states of metabolism emerge. In addition, the results of [424] show that the mechanical viscoelastic state of cells, determined by anthropy of diet, and mechanical stimulation of viscoelastic stem cells influence their proliferation, and hence promote all properties associated with protection, healing, and antibiotic making.

In turn, good anthropic nutrition can do wonders in the treatment of various diseases. For example, the underlying cause of many cancers is that tumor suppressor genes have been switched off by epigenetic marks. To reverse this process, we must activate the silenced tumor suppressor genes to stop cancerous cell divisions. Doctors use some cancer drugs that work by de-methylation, meaning they remove methyl groups. But those drugs promote de-methylation of DNA non-specifically, causing severe side effects. The process of de-methylation in a cell is controlled by specific proteins [449] that come to the working field from digested food. For example, one study [450] found that extracts prepared from axolotl oocytes—the eggs of the axolotl salamander prior to ovulation—have a powerful capacity to change epigenetic marks on the DNA of human cells. It is very difficult at the current level of biochemistry to identify the proteins responsible for this tumor-reversing activity in axolotl oocytes. But in any case, these scientists [450] with extracts made from axolotl oocytes have brought cancer cells back under normal control by reactivating their cancer suppressor genes, and this could lead to a powerful new dietological platform for preventing at least some types of cancers. According the study [449] the cell uses proteins that normally repair DNA for a reversal of gene silencing by de-methylation. This means that therapeutic effect from axolotl oocytes could be extended to DNA reparation processes.

These studies form an excellent example of how food and the diet can be used to prevent cancer. Of course, for mass applications in the diet, it is necessary to develop the special new technologies of food production that could prevent degradation and oxidation of the wonderful molecules in axolotl oocytes that have such powerful epigenetic modifying activity. I am currently working to develop such technologies.

The fact that for cancer prevention it is better to start with the health of the mother of an individual suggests the need to consider how the mother's malnutrition undergoes epigenetic changes. As shown in the study [425] during intrauterine life and lactation, malnutrition brings about epigenetic modifications involving DNA and leading to metabolic pathologies at the adult age. Researchers have demonstrated for the first time in this study, such repercussions at the level of the leptin gene, the hormone that regulates satiety and metabolism. This work thus reveals the type of molecular process that takes place during the perinatal period and leaves an "imprint" in the genes of the fetus, lasting throughout the individual's lifetime [426]. But even organic food, if it does not contain the anthropic elements the body needs, is no guarantee of preventing cancer. In order to prevent cancer in an individual, you need to saturate the phase space of metabolism with anthropic compounds from food. The absence of epigenetic cracks of the phase space of metabolism prevents the appearance of a cascade of reactions not typical for a healthy metabolism, such as the activation of the "wrong" RNA, which is responsible for the development of the state of immunodeficiency and subsequent cancer.

As already mentioned in Section 3.6, with cancer significant changes in the viscoelastic mechanical properties of cancerous cells and surrounding intercellular space take place. The viscoelastic properties of the surrounding tissues is the only strong influence on the processes of tumor metastasis. But here I would like to point out the reverse: measuring the viscoelastic properties of cells and tissues in vivo may help to diagnose tumors in early stage and, therefore, to identify new ways to fight cancer by using drugs and/or a dietary methods of prevention based on anthropic food intake.

Study [381] described the regulatory molecules involved in formation of the protrusions that enable tumorous cells to turn metastatic—by degrading elasticity modulus of viscoelastic extracellular matrix and penetrating blood vessels and, ultimately, seeding themselves in other parts of the body. For many types of cancer, the original tumor itself is usually not deadly. Instead, it's the spread of a tiny subpopulation of cells from the primary tumor due the degradation of mechanical properties of the surrounding matrix to other parts of the body—the process known as metastasis—that mainly kills people. Preventing degradation of viscoelasticity of the extracellular matrix could help in fighting metastatic development of the tumor. In urgent cases of developed disease special medication should be used to induce the creation of cross-links between matrix polymers in extracellular space or possibly in the cell wall matrix (cortex) of the cells. In the long run, for avoiding the emergence of the initial tumor, it is best to have an anthropic diet which will help to support elasticity of all matrixes in the body tissues. But changes in the viscoelasticity within the cells themselves may also contribute to the development of metastases. Metastatic "fluid" cells migrate from the site of the primary tumor, through the stroma, into the blood and lymphatic vessels, finally colonizing various other tissues to form secondary tumors. As shown in [389], the tumor microenvironment is rich in highly contractile cells that are responsible for extensive remodeling of extracellular matrix, and enhanced invasion efficiency of metastatic cells requires the presence of the protein cofilin. It is a well-known fact that cofilin causes de-polymerization of actin filaments in cell, thereby preventing their reassembly. That leads to decreasing elasticity modulus of metastatic cell, making it more "fluid," with higher motility on extracellular space. Therefore, changes in viscoelasticity factor of the tumor cell [389] or extracellular matrix [381] alone make the tumor more aggressive.

The influence of the formation and dissociation of actin matrices of healthy cells on the processes of their migration has been studied in [419]. According to this study, each cell's migration along a gradient by crawling in the direction of the source of a chemo attractant is based on the ongoing reorganization of the viscoelastic cytoskeleton, a network of fibers made up of the protein actin within the living cell. In response to the presence of a chemo attractant, actin filaments are assembled at the front of the cell, causing the membrane to extend protrusions in the direction of locomotion. When cells reorient in the opposite direction in response to changes in the gradient during experiments [419], they go into reverse by disassembling the actin cytoskeleton matrix at the prior leading edge of the cell, and switching the site of viscoelastic matrix formation (protrusion) to the opposite end.

Analysis of the mechanical properties of cells and surrounding tissue could allow doctors to view how cancer cells behave under different mechanochemical stresses and to test different treatment strategies for each individual. Using mathematical models of the viscoelastic behavior of healthy and cancerous cells will significantly improve the prediction of tumor development and treatment. It should be noted that for the diagnosis of breast cancer, doctors already use methods of analyzing the mechanical viscoelastic properties of tissues [367]. A new ultrasound technique is proving valuable in distinguishing elastic properties of malignant from benign breast lesions in some patients. Ultrasound elastography as a valuable diagnostic tool provides specialists with elasticity scores, with lower scores meaning that the mass contains softer, less malignant tissue.

It can be argued that as we learn more details of metabolic processes related to the transport of molecules in the nanometer size range of biological systems, more attention will be paid to fine tuning of signaling processes due to changes in the viscoelastic properties of biopolymer matrices. I think it will reveal the essential role of resonance effects due to viscoelastic nonlinearity of the system in the control of relatively high-energy transfer processes of signaling molecules of proteins with molecular motors. In the presence of resonance, the polymer matrix of living systems performs the role of non-linear parametric filters or amplifies the translation of nano-size signaling molecules. The process of assessing the true impact of the fundamentality of viscoelastic properties of all living systems on their functioning and pathological conditions, including cancer, is just beginning. The number of studies related to the understanding of the processes of interaction of food metabolites and organisms within the framework of new ideas and paradigms of human nutrition will certainly grow in the foreseeable future.

I want to stress here that the mechanical viscoelastic properties of a cell's structures—and how they relate to the cell's molecular environment—is essential to understanding how healthy or unhealthy the cells are. For example, it is well known that tumors are mechanically different to normal tissues. And the difference is in the area of viscoelasticity. Probably the broad variation in the viscoelastic properties between cancer tissues in different patients and in different cancers may be the main reason for the variable success of chemotherapies that target the tumors. While most current medical and biochemical research focuses on the separate study of genes, signaling proteins, chemical suppressors and other molecular pathways, it might instead be the intrinsic change in cancerous cells' matrix and tissue viscoelasticity that leads to the change of molecular signaling, leading to the cancerous state of the cell.

# 4
# The Demon of Toxicity

## 4.1 Allergy and toxins

Interaction with a medicine has an entirely different character from the interaction with the spectrum of anthropic components in food. If an organism is interacting with a medicine, then instead of the two mutually interacting (like the multidimensional strange attractors of the organism and the food), we have the medical substance that sends a powerful controlling signal with a narrow spectrum to the body metabolism. Speaking simplistically, the majority of medicines are designed to change some single metabolic reaction or a part of the metabolic trajectory in the desired direction. However, because of an inability to make its effects discrete, the medicine simultaneously changes many other trajectories in a direction harmful to the organism. This is what is generally known as a medicine's side effects or toxicity. Often these side effects are greater than the positive effect of a medicine. For example, one study [42] has shown the risk of medicine for high blood pressure having a cancerous effect.

The impact of a medicine and the presence of any side effects depend on the state of the organism's metabolism. For young organisms with stable metabolisms, the side effects of a medicine are relatively little felt, while for organisms with metabolisms significantly altered by age and disease, drugs often lead to noticeable negative results. For example, a recent study [33] presents results of the impact of glucocorticoids medicine on patients in different age groups who suffered from a number of illnesses from arthritis to asthma. While young patients did not suffer significant side effects from this medicine, people who had disordered metabolism suffered a loss of bone mass and other undesirable results.

The same applies to the use of many DSs, a large part of "sports nutrition" that is a surrogate for food, and other supplements. Additional harm is caused as a result of these supplements often containing significantly altered substances the effects of which in no way resemble the effects expected by the user. In this case, the variety of disanthropic chemical substances contained in medicines or DSs acts like a bulldozer tearing through a forest, which, while making this or that route to a "specific tree" (to the cure of a specific dysfunction of the metabolism) also destroys other the "living trees" of the phase space of cascades of reactions.

If chemical compounds that become active in the organism are used as a medicine or a DS, this can provoke the appearance of cumulative streams, and as a result a strong and sufficiently quick change of the area of attraction of the metabolic phase space takes place—the transition to an unnatural stable state. Doping substances, hormones, and other supplements used in sports nutrition are often examples of this. Their prolonged abuse provokes an irreversible change in the organism's metabolism. Medicine has a positive result when its effect is accurately targeted at a specific part of the organism that requires correction. For example, for a medicine to have an effect on the water environment within a cell, it is preferable that its molecules entered the area around the cell diluted by the liquid substances of the organism and without chemical changes, and could penetrate the lipid bilayer into the water solution within a cell to carry out its effect. However, such an ideal working of a chemical compound used as a medicine happens very rarely. The medicine, diluted by the liquid environments within an organism, plays a part in various chemical reactions, and so changes itself and changes the trajectories of metabolic reactions in an undesired direction. As such, one of the biggest problems associated with the use of medicines is their delivery to the relevant part of an organism. Various methods are used to solve this problem, but none is universally applicable and in most cases have only a limited effect. This limitation on their effectiveness and precision of delivery leads to numerous side effects. So the use of medicine is in most cases a compromise between the benefit and the harm. I should note that the latest research on the use of nanotubes and nanoparticles to deliver medicine to various tissues and cells of an organism give cause for hope for certain future successes in this direction.

Overall, use of medicine can be justified under critical conditions or long-neglected illnesses, but DSs practically never are, due to their uselessness under critical conditions and their possible harm from prolonged use. The aim of the idea laid out in this work is the prevention of illnesses or the gentle correction of ill states through the correct diet, and increase in our life spans through factors associated with diet.

In general, speaking of disanthropic compounds, I should note that many of them are chemical substances, to this or that extent harmful for the human organism. Substances harmful to the human organism are called toxins. Inert chemical substances, however, are also not beneficial for the organism, as they replace beneficial anthropic ones in the mass of food consumed or require specific additional energy use to be removed from the organism.

Chemical substances that are absolutely disanthropic for the organism, that have no place in the natural human metabolism, are called xenobiotics. As a rule the increase of xenobiotics in the environment are directly or indirectly linked with man's husbandry activity. These include, pesticides, some cleaning fluids (detergents), radionuclides, artificial coloring, polyaromatic hydrocarbons, and so on. When these get into the surrounding natural environment, they can increase the frequency of allergic reactions, the death of organisms, alterations in hereditary traits, the weakening of immunity systems, the disruption of bodily processes, and the natural processes of the ecosystem.

Lipophilic or soluble xenobiotics are of particular current interest for ecologists and toxicologists, as these, accumulating in fatty tissues, can pass up the food chain into animal and human organisms, turning into more polarized and consequently more easily digestible or execrable substances (n-toxication). We sometimes hear about the deposition of toxins, in the form of the settlement of xenobiotic and other toxins and their products in an organism, that have their impact due to accumulation, and then their relatively constant presence in the organism, organs, cells, or tissues for a period from several days to many years. Deposition has four main causes:

1.   The active absorption of toxins by cells with their subsequent retention;
2.   The high chemical similarity of a toxin to certain proteins or to areas of their molecules;
3.   The dilution of xenobiotics in fats and ztheir retention there;
4.   The weak propensity of n-type xenobiotics to dilution and their slow metabolism, which lead to their accumulation in the space between cells or in fatty tissue.

The mechanisms through which toxins express their toxicity are fairly varied, as these take place on the molecular level with an enormous number of reactions interacting with many substances in the organism simultaneously, according to what is called the net principle. An organism might contain primary toxins that entered it from food, secondary toxins that formed during the processes of chemical interaction between toxins and the organism environment, and n-toxins, or n-th order toxins that formed as a result of subsequent misdirection of the trajectories of chemical reactions in the metabolism, provoked by toxins of the previous orders. That is, all n-toxins where $n > 1$ are metabolites of the organism, caused by the original toxin. Furthermore, toxins of the same order can appear on different branches of the tree of trajectories of metabolic reactions, but in essence these are different toxins that do nonetheless have the same origin. That is, at each stage there might be multiple parallel creations of toxins, and the sum intensity of their impact might strengthen what is called the principle of cascade intensification and parameter resonance.

The liver is among the most important of organs; it has a part to play in the transfer of bodily fluids and the decomposition of toxins. The blood that flows away from the intestine and contains substances that have entered the organism goes along the system of portal vein, first to the liver. A countless number of foreign compounds that are essential for the organism get into the liver and undergo metabolic alteration there. The liver is the organ of secretion. The mechanisms of substances entering the liver from blood and factors that affect this process have a number of notable features and are highly complex.

Toxins, as compounds alien to the organism's metabolism, broadly speaking, can impact metabolic reactions directly, but can also act indirectly during the process of accumulation in the cells and tissue of the liver, perverting the metabolism physico-chemically, and also through the mutation and changing expression of various areas of the genome of the cells' nucleus or mitochondria. A significant role of toxins and their metabolites lies in changing the viscoelastic properties of protein matrixes both in cells and in the space between them. This happens when molecules of an n-toxin provoke the bonding of parts of protein macromolecules or, on the contrary, the destruction (disassociation) of macro-molecules and matrixes. The result of this is the destruction of the vitally important natural mechanism of transporting signal molecules. This expresses itself either through the early onset of necrosis of tissue and cells, or through the premature apoptosis of the latter, or through other disorders.

N-toxins have a very important role in so widespread an expression of a perverted metabolism as allergic reactions. With a perverted metabolism, for example, of the intestine micro-flora disbacteriosis, entirely non-toxic compounds such as proteins can be a source of allergy. If the metabolism is disordered in the sphere of bacterial activity, for example in the intestine, protein might not be broken down to the level genetically appropriate to the organism and produces a reaction, for example the immune system. In this case, we see the appearance of toxins from a non-toxic environment. The various reactions of the organism and its parts—dermatitis, asthma, eczema—that often happen as a result of such allergies, are explained, in my view, by the situation of the consecutive appearance of n-toxins in the processes of the metabolism.

Applying the latest DNA-based technologies, researchers [587] demonstrated that gut microbial diversity is a crucial factor for allergy development in children. Their results show that diversity was significantly greater in the unaffected children compared to those who later developed allergies. And it is not *Bifidobacteria* that was used as a supplement in dairy products and advertised by manufacturers as an immune-enhancer, but the broad composition of intestinal microflora during the first weeks, months, and years of life that is critical to the immune system's development. In the absence of sufficient stimuli "training" from many different bacteria, the immune system may overreact against harmless antigens in the environment, such as certain foods. The frequency of developing asthma at school age for children afflicted by these allergies is five to six times higher [587].

Allergies often appear sometime long after the immediate impact of the al-lergen, which Fact is also, in my view, explained by the appearance of allergies to a toxin (allergen) formed on specific cascades of trajectories of metabolic reactions. Certain xenobiotics, after forming covalent links with protein molecules, change the structure of the molecules and their conformation (n-transformation), the me-chanical qualities of protein matrixes. As a result, some altered proteins take on the qualities of anti-genes for their own organism. This explains the often-observed al-lergisation of an organism by low-molecular compounds, which appear during their repeated interaction.

There are different methods of curing allergies: immune-therapy, the gradual "vaccination" through antigens and the injection of monoclonal antibodies; and medicinal treatment to reduce the sensitivity of allergy mediators. However, these methods are directed mostly at symptoms of the disease. A genuine cure can only be achieved through correction of the metabolism. The organism, under the in-fluence of toxins because of changes of its metabolism, tries to fight against the effects of toxins and can sometimes build a tolerance for certain ones. In this way, tolerance can be seen as a defensive reaction by the organism against the impact of xenobiotic and other toxins, because of which sensitivity to these substances is reduced, although it isn't rare for this to apply to only some effects induced by toxins. In this case, within the framework used in this book, this is equivalent to the displacement of the area of attraction of the phase space of the trajectories of reactions of the strange attractor of the metabolism. In some senses, a new fixed metabolism appears—a metabolism that is chronically toxified, that has certain stability with minimal entropy, but with nonetheless a higher level than the ge-netically determined metabolism. The kinetic reactions of toxins, as it were, be-come a part of the area of the trajectories of reactions in the altered metabolism and to some extent ensure its stability. The most colorful examples of such interaction are toxic mania, alcoholism, and drug addiction. The altered metabolisms of the organisms of those suffering from these illnesses begin to need the toxins, alcohol, or drugs to such an extent that abstention can sometimes end in death. So we can classify two aspects of the expression of an organism's chronic intoxication:

1.  The accumulation of toxins in the organism's structures, with a long time needed for breaking down or removal from the organism;
2.  A stably defective metabolism, expressed in the organism the form of the appearance of stable new defective trajectories of reactions, and defects of the structures of cells and tissue.

If deep changes take place due to intoxication, then the new stable state of the metabolism, along the second model given, begins to actually need toxins for its

existence, which is the equivalent of the dependence the organism develops on toxins during toxic mania, alcoholism, and drug addiction. In its weaker expressions,
this appears as an irresistible urge for this or that substance, like sugar or tobacco,
for example. Because they are widespread in the environment and in food, toxins can
impact the development of a population, its demographics, and so on, both through
the inner metabolisms of individual members of the population, and through their
metabolites in the environment around them. Speaking of food products, the main
source of their disanthropy is various types of toxins. These can be toxins of biological origin, organic and non-organic chemical compounds of natural origin, or
synthetic toxins that are used in the conservation, processing, and storage of food in
industry and agriculture, DSs, cosmetics, and medicine.

The heterogeneous and hierarchical nature of the human metabolism makes
it impossible for toxins to impact an individual part of the phase space of the metabolism—i.e., to impact this or that organ, tissue, cell, or a single part of them.
Due to the complexity of the metabolic processes in higher organisms and the enormous multitude of trajectories of reactions, the effects of the initial toxins and the
metabolites they produce result in a more toxic expression in the human organism, when compared to the effect of toxins on primitive metabolisms of bacteria
or plants given similar levels of concentration. The extent of a toxin's disanthropy
is determined by its atomic composition and molecular structure, including the
isometrics of molecules. The consumption of disanthropic food and the consequent
intoxication of an organism leads to the accumulation of toxins in the organism and
their disanthropic interaction with parts of the cells and the material between cells:
proteins, nucleic acids, lipids, etc. All this leads to changes, dependent on the level
of toxicity, in the phase space of the metabolism and, consequently, to reactions
flowing along distorted trajectories, which is the cause of illnesses in the organism.

All the disanthropic effects of toxic compounds listed above can be easily identified through a sharp impact on the organism. However, the modern diet of man is
generously filled with what are called weakly toxic compounds: free radicals, minerals, and other substances present in food or appearing from disanthropic methods
of preparing food: various coloring, preservatives, taste enhancers, pesticides, the
resultant products of the metabolisms of fertilizer, hormones, antibiotics, medicines,
and so on. The short-term impact of food with the disanthropic elements indicated
is usually not critical for the organism. But the use of the full menu for a prolonged
period of time has negative consequences for the metabolic processes, both through
their immediate impact, which is to increase the probability of classic and epigenetic mutations, and through the accumulation of toxic metabolites in the organism, the removal of which is in many cases a very complicated task. The inevitable
consequence of this is premature aging and the appearance of degenerative illnesses.
For example, as shown in recent research [427], several foreign chemicals (xenogens

such as bisphenol A, pesticides, insecticides, etc) that can accumulate to high levels of concentration in the human body have recently been linked to an increased risk of cancer and also to an impaired responsiveness to anticancer drugs. The presence of such substances was sufficient to enhance the cancerous characteristics of human colon tumor cell lines and primary human colon cancer tissue.

In addition, the prolonged intoxication of an organism through the chemical compounds in disanthropic food in the period leading up to pregnancy and, especially, during the pregnancy itself often leads to the appearance of allergies in the child. Changes in the mother's metabolism send incorrect signals to the child's metabolism and, consequently, provoke its metabolism to an inadequate response—changes in the trajectories of chemical reactions. These, in turn, interact with elements of food or their metabolites normal for most people and provoke a false reaction of the immune system, which in the majority of cases manifests as an allergic reaction. Over the last fifty to sixty years, the overall number of cases of food allergies present from birth has increased by around 20 percent every ten years and doesn't show even slight indications of slowing down its rate of growth. For certain allergies—for example, peanut allergy—there is evidence [70] of a three-fold increase over the last ten years. However, the consequences of changes in the metabolism of a mother and her child sometimes only become apparent after decades of the organism's life. One study [70] shows a link between children weak at birth and their propensity to type 2 diabetes in later adult life. This link is explained by the incorrect expression of key genes, the work of which is disrupted by changes in the organism's metabolism that began during the organism's presence in its mother's womb.

All of this is evidence of the gradual deterioration of the already difficult situation of mass disantropic nutrition in the modern world. Mothers who eat disantropically pervert their own metabolisms and the metabolisms of their future children. Later, during the period of childbirth, the process is made worse by the fact that the epigenetic changes now inherent in the metabolism of a girl after she has grown up interact with her disanthropic diet during her own pregnancy and pervert the metabolism of her own child even more. The girls of this generation, when they in turn become mothers, give birth to a generation statistically even more prone to various allergies.

Men too are affected by this. The disanthropic diet and incorrect metabolism of the mother leads to the birth of a weak progeny. This can be seen in the statistical increase of the number of men with defective sperm, which do not assist the creation of a healthy progeny. In this way, looking at various theories explaining the growth of allergies over the last decades, such as *the hygienic hypothesis* and *theory of incorrect food preparation,* we have to take a broader view of the problem than is taken by these hypotheses. The growth of allergies, in my view, is explained solely through growth in the disanthropy of the lifestyles of the post-war generations in developed societies. This includes disanthropic changes both in methods used in agriculture and in methods used in food preparation for mass consumption that are spread through supermarket chains. One confirmation of my hypothesis about the cause of allergies is that although the number of instances of allergy is growing for both sexes and for all age groups and races, a black child in the USA nonetheless has twice as a probability high as a white child to suffer from allergies [68]. Also, the probability that a child of Latin American origins will have some allergy or other is noticeably higher than for a white child. This is explained by the different quality of life experienced by these populations in the United States and, incidentally, these facts directly contradict the generally accepted *hygienic* hypothesis. The sole method of treating, not just alleviating the symptoms of, allergies in such cases is to correct the state of the metabolism through the anthropic diet.

## 4.2.   The evil of small doses

The situation is more or less clear with strong toxins that lead to the quick death of an organism. Super-strong toxins are what might be called a deadly poison for the organism. Over a relatively short period of time, from seconds to weeks, they destroy the metabolism of an organism. Some of these poisons, such as potassium cyanide, do their work over a matter of seconds. Such substances evidently create a self-contained set of high intensity, cumulative streams in the phase space of the metabolism, destroying practically instantaneously the vitally important trajectories of reactions. Disorders in the metabolism spread with an extremely high speed, either through the complete halting of part of the reaction or through the quick and powerful avalanche-like generation of high-strength n-toxins in many metabolic paths.

But the situation around the entrance of weak toxic in the form of the disan-thropic compounds in food, which do not kill the organism immediately but simply misdirect the trajectories of reactions somewhat, is much more interesting, especially as this, the slow intoxication of an organism by the disanthropic compounds in modern food, is the more widespread case. In this case, as indicated above, there can appear in the organism, during the process of metabolizing the initial toxins, toxins produced at the n-th stage along the trajectories of metabolic reactions. The production of these n-toxins takes place, clearly, along branches that are to this or that extent close to the system of bifurcate branching produced by the trajectories of reactions in a healthy metabolism. At the same time, these toxins also pervert in a certain way and metabolic feedback loops. Speaking generally, we can say that there appear in the organism a variety of n-toxins in different parts of the phase space of the metabolism. When we speak of different parts of the phase space, we mean that toxins can have their impact both in different physically distinct structures of the organism and at different stages of the metabolic reactions. Although the initial toxins, disanthropic compounds in their food and their n-order metabolites, might be relatively weakly toxic, the impact they have on metabolic processes, as classical flicker-noise, has been shown already. The flicker-noise characteristic of these toxins is because their action depends on the variety of previous reactions and the probability of their joining with each other along a single trajectory of reaction or on parallel branches of the trajectories of reactions. But as we know, for such systems as a metabolism, small flicker-noise can lead in the end to the disorder of the whole phase space of the metabolism. On the genetic level of the organism, both the initial toxins and their n-order metabolites can cause epigenetic changes. Small doses of a toxin can become a trigger for the process of the expression or repression of these or those genes. The cascades of reactions of an altered metabolism caused by this, as we have already noted, can mutually interact with cascades of reactions distanced in the phase space, and as a result significantly alter the structure of the metabolism as a whole. At the same time, we have to bear in mind that just as trajectories get distorted, so too do the structures of cascades of return links in the phase space of the metabolism.

In this way, the impact of many and prolonged weak disanthropic disturbances of the diet can in the end be the complete change in the state of the metabolism. What is the specific mechanism of this impact? Looking at mitochondria as an example, we know from the work of Nick Lane [27] that the culling (destruction) of defective mitochondria takes place after a certain number of changes in the functioning of these mitochondria. If the toxin is sufficiently strong, the cell dies or culls defective mitochondria and gradually returns to its initial state completely or partially. If on the other hand, the changes in the mitochondria are relatively small, below the floor at which effective culling takes place, a large number of these weakly defective mitochondria (relative to the norm) accumulate in cells, and their collective impact worsens the functioning of a large number of cells. This in turn leads to the degradation of the functioning of cells and tissues, which is expressed over the whole organism as an illness or prolonged weakness.

Hazard from damaged but still a life mitochondria appears as well as source of many neurodegenerative diseases [574]. Normally, when mitochondria go bad, they are destroyed by specific cellular agents. But when the damage is not so great, the containment system fails. And some forms of neurodegenerative diseases cause mitochondria to run amok inside the cell, spew excess amounts of toxic reactive oxygen species (ROS), and fuse to otherwise healthy mitochondria. Mitochondria are here just an example. Weak toxins can, and I'm sure do, have an analogous impact on other parts of the cell and metabolic processes in general. This certainty is based on many observations of live nature that show that *the principle of culling as a result of some limit of changes in functioning being reached* of some structure is sufficiently universal.

As confirmation of this idea, we can refer to the study which found that excessive DNA repair in cells may contribute to the development of breast and ovarian cancers [551].

Strong evidence for the existence of DNA "repair centers" and radiation-induced foci (RIF) in nuclei of human cells has also been found in study [586]. It showed the existence of non-linearity in RIF induction rate: the rate increases with increasing radiation dosage, whereas the rate at which RIFs disappear decreases. Researchers observed an absolute RIF yield that is surprisingly much smaller at higher doses measured in Grays (the Gray is the unit of the absorbed radiation dose): 15 RIF/Gy after a 2 Gy exposure compared to approximately 64 RIF/Gy after 0.1 Gy. This would seem to suggest that at low doses the influence of radiation exposure on the cell's DNA would be more effectively annihilated by large numbers of RIF. However, since the process of DNA repair is not 100-percent effective, there is a probability of the occurrence of defects. It is obvious that the number of defective regions of the cells' genome may statistically grow in a certain range of low-dose radiation. This means that as in the case of mitochondria, the number of surviving cells and the number of damaged cells among them may grow in some range of doses because of the lack of defective cells dying. But as we know, even a single damaged cell could lead to cancer. So the role of small doses of radiation should be studied further from this point of view, and we cannot exclude the possibility of some nasty surprises.

Consider also the consequences of the insecticide Kepon (chlordecone). This chemical remains in small doses in the soil and in human organisms for more than twenty years after the end of its use. Nonetheless, these small doses lead, after their prolonged impact on the metabolism, to the high frequency of cancer, and specifically, according to [50], prostate cancer. According to the research done in [60], the impact of certain chemicals on sea animals and on humans can continue for thirty to sixty years, even though their use has already ended thirty to forty years ago. As such, chemicals once used in industry or agriculture can continue to have a toxic impact in small doses sixty to eighty years in total after their use has ceased. One study [66] demonstrates that the exposure of the organism to phthalates at the lowest concentration possible reduces the level of testosterone in men, which in turn leads to weight problems and diabetes. Although phthalates have been used for fifty years in industry and in plastics, only recently have there been even weak efforts to limit their use in the manufacture of plastics.

I was personally involved in research into the toxicity of certain phthalates while on undergraduate placement in that distant year of 1974. The parts of this research connected to biochemistry were directed toward studying these toxic effects, which were entirely confirmed. The parts related to the physics-chemistry of high molecule compounds were directed to the discovery of non-toxic alternatives to phthalates for the plastification of Polyvinyl chloride, commonly abbreviated PVC. We also determined the potential replacements for phthalates then available. I lost interest in this subject afterwards. But just imagine my surprise when, while reading last year one of the technical journals dedicated to the chemistry of polymers, that it was only in 2003 that somewhat strict limits on the use of a small part of phthalates in plastics made from PVC that come into immediate contact with people were instituted. That is to say, it took around thirty years for the well-known facts about the toxicity of phthalates in polymers to result in some sort of action on the part of the regulators to prevent harm to people's health. I fear that there are many spheres I don't know about in which toxic substances are used and where the institution of even the smallest limits is proceeding with the same pace. The impact of phthalates and other chemicals in extremely small doses, but over prolonged periods, evidently is one of the main causes of diseases related to metabolism in later life. A growing body of research suggests that phthalates could be blamed in rising childhood obesity rates. As phthalates are used in a wide spectrum of products from cosmetics to pesticides, then the explanation for the fact that more than 75 percent of the population of the United States has a level of phthalates in their urine above the background level becomes obvious. And this is not because the United States is more infected with phthalates than other countries, but because there the government and private companies are conducting relatively more detailed studies of the health of the population than anywhere else in the world.

Speaking generally, the level of impact of whatever dose is determined by the evolved reactions of an organism's cells, depending on the intensity of the external influence. It is generally agreed that one of three things happen to cells in response to DNA damage: cells' growth is temporarily arrested so the damage can be repaired, cells go quiet and permanently stop dividing so they don't pass on the DNA damage, or cells simply die, a process known as apoptosis. With a small impact, the response is more often the repair of the genes damaged by the toxins. If the impact is large enough, then the cells either stop dividing or die. The response triggered by the signal of the external impact determines all subsequent developments on the level of the organism's cells. An interesting observation is made in a study [83] looking at an inhibitor of allergic interaction, the plant sugar arabinogalactan, a bio-polymer generously available in Meadow Foxtail. It was demonstrated that in small doses the pollen of Foxtail is the most powerful allergen. But in large doses its allergic properties do not manifest and, what is more, large doses provide immunity to allergy to it. So some principle of culling given a certain level of immune reaction is at work here too. This underlines yet again the importance of taking into account the influence of small impacts on the metabolism.

For example, it was recently shown in a study [79] that the process of DNA regeneration by converting genes in dormant cells leads to an unexpected outcome:

the amount of defective DNA increases. This takes place given precisely weakly toxic impacts on the cell structure of an organism, as strong impacts trigger the mechanism of apoptosis. A study [205] ties rodent data on the health effects of small doses of bisphenol A (commonly known as BPA) to predictions of human health effects from the chemical due to the use of everyday household products. Bisphenol A is one of the world's highest-production-volume chemicals, with more than 8 billion pounds produced per year. It can be found in a wide variety of consumer products, including hard plastic items such as baby bottles and food-storage containers, the plastic lining of food and beverage cans, thermal paper used for receipts, and dental sealants. The study provides convincing evidence that bisphenol A is dangerous to our health at current levels of human exposure. Mainly it is because bisphenol is related to the extremely hazardous endocrine-disrupting chemicals that produce adverse reproductive consequences on gene expression [204]. Another study [206] shows significant reproductive health effects in animals that have been exposed to bisphenol A at doses equivalent to or below the level that has been thought not to produce any adverse effects. Six environmental research studies reveal critical health risks with even very limited exposure to bisphenol A, phthalates, and flame retardants [207].

The authors of one study [236] argue that small doses of toxins not only directly impact some cell or organelle, but also impact the processes of interaction between cytosol's water and the structures of bio-polymer matrixes within the cells and membranes. Through this interaction, they change their characteristics, including their viscoelastic properties. Consequently, perhaps, changes in characteristics of the viscoelastic regulation of molecular transport have an impact on the functioning of cells as a whole under the influence of small doses of toxins. Bearing in mind that the interaction of water and the structure of protein matrixes have an entirely complex character, then the existence of anomalous effects of weakly concentrated doses is entirely possible. The idea, proposed in a study [235], of parametric resonance between processes of the reactions of signaling proteins with targets within a cell and the processes of the transport of proteins in viscoelastic matrixes, connected to the targets, explains perfectly the presence of extreme impacts in the diapason of limited toxin doses.

The impact of the majority of toxins in small doses can also have a delayed manifestation, worsened by the prolonged accumulation. For example, this happens when organic toxins accumulate in fatty tissue with individually small doses augmenting each other with time. Researchers [62] have discovered an important new mechanism through which cells can detect toxins and nutrients. This happens in the same way—and with the same effects—as when cells receive a message from a hormone. It is a well-known fact that, through receptor proteins on their outer surface, cells communicate with each other in extracellular space. By binding with different substances, such as hormones, molecules of food metabolites, and other chemicals, the receptors pick up signals from outside the cell and transmit the signals to the interior of the cells. This simplified picture is, in reality, a bit more complicated. When chemical signals from foreign substances, such as medicine, toxins, or metabolites from disanthropic food, come into contact with receptors, the cell first tries not to open them a way to the interior of the cell. If this succeeds, the substances accumulate in the extracellular matrix or go to somewhere else in a body, usually to fat cells. If they get into the cell interior, they could severely damage cell— immediately, in most cases. Part of these foreign substances stay in extracellular space and have a long-term poisonous effect on the body. But the biggest part, which goes to fat cells, stays more or less isolated from the body's metabolism with relatively little harm to the metabolism.

Yet they may not remain harmful indefinitely. The accumulation of these substances in fatty tissue over a long time leads to a high concentration, and if the body's store of that fatty tissue drops, those chemical pollutants, organic and non-organic, may be released. Recently, Duk-Hee Lee of Kyungpook National University in South Korea has shown that weight loss causes chemicals to be freed, leading to their build up in the blood [188] and dispersal to other tissues of the body. This relatively sudden dose of these substances can now have a more serious, and potentially dangerous, impact on the metabolism. The chemicals that enter the food chain from sources including pesticides, manufacturing, and so on have been linked to an increased risk of diabetes, cancer, and dementia.

As shown in a study [189], the same mechanics apply to cannabis users: stress or dieting might trigger "re-intoxication," resulting in a positive drug test long after you last used the drug. The main psychoactive ingredient of cannabis is tetrahydrocannabinol, and once in the body it is readily absorbed into fat cells. Over the next few days, it slowly diffuses back into the blood. Since tetrahydrocannabinol is taken up by fat more readily than it diffuses out, continual intake means some residual quantity can remain in the fat cells. It has been suggested that stored tetrahydrocannabinol can be released at a later date in situations where the body's fat is rapidly broken down. This is based on anecdotal reports of spikes in blood cannabinoid levels in people who have not taken the drug recently but have experienced rapid weight loss. This does not mean that we should take no action against excess weight, but it does meant that we should work to prevent the buildup of dangerous substances over time.

The disanthropy of our diet is also expressed in the smaller physical volume of food consumed by modern man. Modern man consumes fewer calories compared to ancient hunter-gatherers due to an inactive lifestyle and the greater concentration of disanthropic substances in food that has been processed. Consider that the main function of food is replenishing energy that has been used. In distant times, when man plowed dry land, sowed, hunted, stayed "on the job" from dawn 'till dusk, his energy consumption amounted to 3.5-5 thousand kilocalories per day. Social and scientific-technological progress has reduced this number one and a half or twice over. To replenish our strength, we no longer need a side of lamb and a multitude of roots and plants. But with a smaller volume of food, we don't get that amount of vitamins and the whole complex of other bio-organic compounds that are essential for the organism. To reach the goals of an anthropic diet, a smaller volume of food can be compensated for by food with a greater density of bioactive components. The classical example of this is superfoods from live nature.

The facts laid out above about the dangers to the organism of excess weight demonstrate how non-linear is the interaction of various factors of the environment determined by diet and the inner processes within a human metabolism—and also how unexpectedly the incorrect diet and pollution during food processing can impact our health. Those fact also show that it's practically impossible to solve the problems connected to correcting our metabolisms on the basis of a linear approach. Rather we have to use specialized methods to reduce the concentration of disanthropic substances in fatty tissue and to force them out through the prolonged detoxification of the organism through an anthropic diet. In addition, small levels of pollution are not all that small. Currently there are 700 different disanthropic substances and chemicals in the waters and soil of most regions of intensive agriculture and developed industry that got there as a result of the modern methods used in agriculture and of industrial waste. The synergistic impact of these toxins has not been studied due to the size of the task and the complexity of the problem. However, their sum impact can be seen in the growth in the overall level of illness in all age groups that has been noted by many researchers.

## 4.3.   Radioactive life supplements

I cannot neglect to mention the case when small impacts of radiation from food products infected by nucleoids, have the sum effect, especially given prolonged periods of time, more harmful than stronger sum impacts, that take place when you are subject to external or internal sources of radiation. During the multi-year study of the last meltdown at Chernobyl, the fact (paradoxical at first glance) was established that the accumulation of small doses of radiation received from products grown and processed in the affected zone had a significantly stronger effect on individuals' overall health and on the appearance of illness connected to radiation, than doses received from a single exposure of radiation that was several times larger. Up to a certain point, of course! The solution to this apparent paradox is that in this case the same *principle of culling as a result of some limit of change of functioning being reached* applies. Strong doses of radiation activate the evolved mechanisms of an organism's defense, which lead to the culling of the impacted structures of the metabolism (the same mitochondria, for example), while weak doses, by causing relatively small changes, ensure the prolonged continuation of their distorted lives, causing significant harm to the organism. As a result, the number of defective structures of small impacts can ultimately be much larger than with a relatively stronger impact that activates the mechanisms of culling. A large number of functioning defective structures gradually lead to much larger blows to the organism, both through the length of their impact, and through their number.

In a study [253] published online in *PLoS Genetics, 2010*, researchers at the Stanford University School of Medicine have artificially removed a protein, called Perp, important in the most common variety of skin cancer. They found that Perp loss promotes cancer fiercely via the increased survival of cells that in normal conditions would have committed apoptosis in response to radiation. This experiment provides evidence for the stated hypothesis that disruption of the culling of stricken cells—i.e., their survival—in reality does nothing good for the organism. Also, researchers from Indiana University-Purdue University Indianapolis (United States) and Umea University (Sweden) report [341] that a method by which cells repair breaks in their DNA, known as Break-induced Replication (BIR), is up to 2,800 times more likely to cause genetic mutation than normal DNA synthesis. When one or many cells repair themselves using the "efficient" BIR method, accuracy is lost.

The activation of the *principle of critical culling* as a result of radiation is connected, it appears, that the evolutionary paths of many living organisms included mutations, caused by the natural radioactive environment of the planet including cosmic radiation. During the early stages of human evolution, this environment was significantly more radioactive than it is now. Evidently, we have inherited a relatively high limit at which defensive reactions against radiation are triggered, as a sort of atavism. Indeed, there was no reason why a lower limit would evolve alongside the fall of natural radiation on the planet. A lower limit gave no advantage in competitive struggle for billions of years while the environment became less radioactive, and our ancestors did not encounter the consequences of nuclear station catastrophes and radiation. Moreover, there is the suggestion that, in order to maintain the rate of mutations in an organism during reduced radiation doses, as often happens in living systems, there may have evolved a compensatory, non-linear mechanism that would increase the frequency of mutations and give low doses of radiation to maintain the rate of evolution. The precise mechanism that causes this acceleration is not yet known. However, while looking at the frequency of mutations in the cells of mammals given relatively high doses of radiation, a study [30] found that the dose-effect (in terms of mutations induced) graph has a very clear non-linear character. Incidentally, such results have been obtained not only with radiation, but also when looking at the effect on the organism of small doses (concentrations) of many disanthropic substances—chemical pollutants, of which there are now so many in our environment. It is entirely possible that the *principle of critical culling* is a universal principle that determines the behavior of cells and organisms under the impact of the majority of external disanthropic factors related to small doses and concentrations in our environment.

If we suppose that there is a non-linear, power-law relationship between small doses of radiation or toxins of any sort and the accumulation of classical and epigenetic mutations, then the conclusion is that the cause for amplification of effect in the small-doses range is the non-linearity indicated above providing the work of the non-linear amplifier. Looking at the intermediate zone between small and large doses, here we see the distinction of evolved responses in action, as the number of persistent mutations falls due to a higher mortality of mitochondria and other organelles or cell structures. From this point of view, even small concentrations of disanthropic substances in the human diet can lead to uncontrollable consequences for health. Practically every week, the scientific world discovers some chemical compound or other that appears to be toxic in the lowest possible doses. At present, thousands of such chemical compounds that have gotten into the soil, water, and air due to man's economic activity are known. More than 80,000 industrial chemicals are in use and about 600–700 new chemicals are introduced to the market each year in the United States alone, according to the U.S. government. Therefore, increased study of low-dose response for a general risk assessment of those innumerable chemicals is crucial for the population's health.

What is most worrisome is that the effect of small impacts of both disanthropic toxins and radioactivity on the human organism is virtually unstudied when looking at long periods of exposure. The most horrifying example of this is the universal introduction to airports, in connection with the war on terror, of devices for the radioactive scanning of passengers. These devices have been intensively introduced to the surveillance policies of the majority of airports worldwide in the wake of 9/11. In my opinion, all of their developers' claims about these devices' safety do not stand up to scrutiny, as they were never before used on a mass scale. The small dose of radiation is not sufficient criteria, as, first, the dose is not that small, or otherwise, due to natural background radiation, it would not be possible to see through clothes; and second, because small doses that are nonetheless in excess of the natural background ones harbor their own particular threat, which is virtually unstudied. Artificial, brief, but repeated doses in excess of the natural, which have been the source of the majority of the mutations in our cells over the course of our evolution, must be uniquely dangerous. These could be the trigger for a multitude of additional classical and epigenetic mutations in the genome of the nucleus and mitochondria of a cell. Even now, there are no attempts to gather statistical information on the relationship between illness and the frequency with which a person is scanned.

I think governments worldwide are afraid to do this as they don't want to deprive themselves of a potent weapon in the fight against potential terrorists. The overall health of the population is of no concern for politicians of whatever hue. However, the impact of small doses of radiation on systems of the organism, especially given its frequent repetition for groups of the population who fly frequently, and bearing in mind the principle of limit culling outlined above, holds a potential and completely unstudied risk to our health. I myself have seen small children and women being scanned in many airports across the world, especially in Russia. But the unstable, developing metabolism of a child up to and including the age of puberty is uniquely vulnerable to the external disanthropic impact of these devices' radiation. This is even more relevant for women of child-bearing age, who, unsuspectingly, could have been pregnant for up to –one to six weeks at the time they were scanned. But it is well known that the embryo is maximally vulnerable to epigenetic changes at precisely that age. My view is that, in these cases, the probability that the children and embryos' metabolisms will develop defectively is unacceptably high. Bearing in mind that millions of passengers undergo such a procedure every year, there must be tens of thousands of such women among them. In addition, considering that the majority of passengers go somewhere and back again within around a week on average, then the impact is doubled over a relatively short period of time.

In addition to millimeter scanners, more than 500 so-called backscatter scanners have appeared in seventy-eight U.S. airports so far, and the federal government ultimately has plans to triple the number of them by the end of 2012. Backscatter scanners are different from millimeter wave scanners, also used in airport security, which image the body using a different type of energy than ionizing radiation. But it is not safer based on any standard. Because backscatter scanners use X-rays to penetrate clothing, image the human body, and reveal hidden articles underneath, they have drawn public scrutiny about concerns of potential health risks. How certain can Americans be that there are not as yet unknown safety risks? Even the defenders of these devices caution [372] that their analysis of their safety is based on the assumption that the backscatter and millimeter wave scanner devices work perfectly and are used as designed. But there is always potential for software glitches, human errors, or mechanical malfunctions of equipment that could cause the scanners to change their design specifications. It could be a change in the width of the emitting spectrum, or the appearance of potentially harmful higher harmonics with different frequency X-rays or millimeter waves and the exposure of people to higher levels of radiation. Given how many of these everyday pieces of equipment will be used in busy airports and how many people will be exposed to these machines, nobody could guaranty that error or malfunctions could not happen, or that they would even be noticed quickly. Even if routine checks happen every day, it is still possible that tens of thousands people would be exposed to malfunctioning devices before the problem is spotted. Even more worrisome is that the US government is buying custom-made vans packed with something called backscatter X-ray capacity equipment. This X-ray technology works by bouncing narrow X-ray streams off an object like a car and then decoding the scatter rate of the reflected rays in order to locate objects in a car that could be bodies or bombs.

Moreover, millimeter wave radiation has already been used for weapon technology. According to study [372], specialists argue that millimeter wave scanners are still dangerous because of the beam concentration and can increase the rate of tumor growth. Studies show an increased rate of tumor growth in animals after exposure to millimeter waves. Studies on mice have also shown that millimeter waves influence the immune system . It is also stated that because of the shallow penetration depth of millimeter waves, thermal injury to the eye and the skin is most likely in the case of an accidental stronger exposure. Experimental studies on mice show that millimeter wave radiation induces irreversible brain damage and may cause fatal damage. Some experimental studies have also shown irreversible eye damage (blindness) after exposure to millimeter waves in the same threshold area as most waves used in airport millimeter wave scanners.

Male fetuses of mothers who are exposed to small doses of radiation during early pregnancy may have an increased chance of developing testicular cancer [619]. When the researchers gave small doses of radiation to female mice in the first half of their pregnancies, all the male offspring developed testicular cancer. During the past fifty years, the incidence has tripled in young Caucasian men throughout the world. But Caucasian males are living mainly in the developed world with many nuclear stations, airport scanners, and other sources of radiation.

The scale of the potential problem is very high: around 250–300 million passengers take flights in the United States, Europe, and Asia each year. Implementation of scanners could significantly increase the lifetime risk of cancer for travelers, and especially for children and youngsters, who will take more flights over their lifetime because they are expected to live longer than adults, so their risk of cancer would be proportionally higher. Even an extremely small percentage of increase in cancers over ten years of airport scanning, say 0.01 percent, would kill 30,000 people— many more than terrorist attacks have killed worldwide in the last ten years in the air transport industry, including the events of 9/11.

The consequences of utilizing such scanners, whether mobile or stationary, are unknown and potentially dangerous. What is more, the danger lies in what the developers of such devices regard as their virtue: the low doses of radiation. No one has yet proven the reverse, while the increased harmful consequences of small radiation doses are known. The use of such devices is a certain road to a personal hell of loss of health, conscientiously assured for you by the good intentions of bureaucracy and the ostentatious concern of politicians. The war on terror is of course a good intention, but not at the expense of our health. So one of the recommendations of this book, which goes beyond its subject of healthy eating, is the warning of this threat to my readers' health: do not believe bureaucrats and the management of companies that want to literally illuminate you right through. Instead of this radiation of you and your nearest, of children and women especially, demand a conventional inspection, which they do not yet have the right to refuse you.

Besides their security applications, the use of procedures with low-dose ionizing radiation in medical practice, such as computed tomography (CT), angiography, and nuclear scans, is increasing, which has led to mounting concern in the medical community that patients may be at increased risk of cancer [331,332]. "We found a relation between the exposure to low-dose ionizing radiation from cardiac imaging and therapeutic procedures after acute myocardial infarction, and the risk of incident cancer," writes Dr. Louise Pilote, the author of the study [331]. We have to take into account that in the last twenty years, the number of CT scans obtained has quadrupled. In addition, the use of myocardial perfusion scans—which accounted for the single most frequent test and the highest radiation dose—increased by more than 6 percent per year over the course of the last ten years. Medical exposure constituted nearly half of the total radiation exposure of the U.S. population from all sources [335]. Recent estimates suggest, for example, that as many as 2 percent of cancers could be attributed to radiation during CT scans alone [333,334]. But there are many other procedures with low-dose ionizing radiation in modern medical practice. In total, the attribution rate could be as much as 10–15 percent or more of all cancer cases. Even small doses of radiation exposure during early childhood significantly increased the risk of adult solid cancers, according to a study of survivors of the Hiroshima and Nagasaki bombings [336].

Apart from patients, these small doses are dangerous for doctors as well. Studies [521, 522] published online in the *European Heart Journal* found that among interventional cardiologists who were regularly exposed to x-rays for their work, levels of some chemicals in cells rose, especially glutathione, an antioxidant that protects against cell damage from reactive oxygen species (ROS), and hydrogen peroxide, which indicates the amount of oxidative stress caused by the ROS. The authors of these studies believe that these changes indicate that the radiation was inducing potentially harmful changes at the cellular level, hence the three-fold increase in hydrogen peroxide, but that this in turn activated a protective response, reflected by the two-fold increase in the antioxidant glutathione. The response of cells to increase the antioxidant level is the work of the mechanism that prevailed in the environment of smaller doses of radiation—that is, trying to repair cell damages—but might potentially lead to the cancerous transformation of cells. In addition, the studies registered an increased susceptibility of white blood cells to apoptosis, which could be the body's response of killing off severely damaged and potentially cancerous cells. This second type of response actually may protect from the development of cancer.

So exposure to a level of radiation that is considered "safe" by regulatory standards for cardiologists can induce profoundly dangerous biochemical and cellular changes. And it is very difficult to say what dose actually is safe due to the nonlinear relation of radiation damage and dose given low doses of radiation. But if the principle of limiting culling is working here, the first danger will lie in the valleys of low doses, and the second one in the mountains of high doses.

Another source of the risk of receiving small doses of radioactivity is that many countries are starting to grow agricultural produce on lands infected by radioactive isotopes during the Chernobyl catastrophe. The meltdown on the Chernobyl nuclear station, which took place on April 26th, 1986, led to the radioactive pollution of land on the territory of Belorussia, the Ukraine, Russia, and a host of European countries. Due to the wind direction prevailing at the time, the main cloud of radioactive ash was carried to Belorussia. And it is precisely Belorussia, one of the republics that appeared on the territory of the former Soviet Union, that is beginning this process. From the economic point of view, the re-cultivation of lands stricken by the meltdown makes no sense. Radio-nucleoids have not gone anywhere from the soil affected by Chernobyl pollution. There's plenty enough of both cesium-137 and strontium-90 there. And these are especially dangerous and, moreover, long-lasting radio-nucleoids that can accumulate in an organism. In the opinion of experts, the polluted territories will only return to their natural state in, at the earliest, 20–25 thousand years' time. Yet Belorussia delivers agricultural produce totaling $2.5 billion onto external markets every year, and hopes to increase this to $5–7 billion by 2015. Of course, Belorussian products do undergo inspection for radioactive elements on the border; however, companies, according to experts, have long known how to neutralize it. For example, radioactive milk from one region can be mixed with entirely clean milk from another. The result has a level of radioactivity just below the maximum level allowed. The same applies to meat: radioactive carcasses are made into mince, and then mixed with other, clean mince, and made into dumplings or sausages. What is more, Belorussia has one of the lowest levels of population density in Europe, and so there's an abundance of agricultural land. The whole process of re-cultivating and utilizing these lands is aimed at increasing profits, by delivering to other countries products with a radioactivity, on average, just below the maximum safe level. But the process of mixing biological products is rarely uniform. In both meat and milk, there can be clusters with a greater level of radioactivity, which can cause irreversible harm to the consumer, especially to children. Moreover, the prolonged consumption of products at the maximum allowed radioactivity can itself be dangerous. As small doses of radiation are still considered safe in most countries, bureaucracy is in no hurry to invest resources into studying the anomalously strong impact of small doses on health. The equally corrupt dictatorial regime that rules Belorussia has not displayed any great concern for the consumer of its produce.

Twenty-five years after the Chernobyl disaster, experts from Greenpeace have found a high level of radiation in milk, mushrooms, and berries in many Ukrainian regions. Overall, 114 varieties of milk, berries, mushrooms, and vegetables were tested. Virtually all varieties (93 percent) of milk had an unacceptably high level of radioactive cesium-137. Its concentration exceeded the norms specified for child products by 1,2 - 16,3 times over. The acceptable level of radioactive pollution for children in milk is 40 Becquerels. In the samples tested by Greenpeace, this level exceeded 665 Becquerels. I must note that the Ukraine has stopped taking lands polluted by the Chernobyl catastrophe out of agricultural activity over the last few years. Radioactive cesium enters plants, milk, meat, and human organisms from the soil.

Eating such products is harmful both for children and for adults. Contrary to common beliefs, the risk of cancer associated with radiation exposure in middle age may not be lower than the risk associated with exposure at younger ages, according to studies [260, 261]. The results [259] suggest that for adult exposures, radiation risks do not decrease with increasing age at exposure. A biology-based model draws attention to the importance of low-dose radiation exposures [260]. The first large-scale study of the relationship between low-dose radiation and birth rates [417] shows that, among populations living in the proximity of nuclear facilities (within twenty-two miles), the sex odds (change in male births relative to female births) increased significantly in both Germany and Switzerland during the running periods of nuclear power stations. In addition, there was a significant increase of sex odds in Europe in the year 1987 following the Chernobyl catastrophe, whereas no such effect was detected in the United States, which was practically unexposed to the consequences of the catastrophe. Taken together, these findings show a long-term impact of low-dose radiation exposure on human sex odds, proving cause and effect [418]. What is less clear from the study is whether this increase in male births relative to female births is the result of a reduced frequency of female births or an increased number of male births. The authors estimate that the deficit of births and the number of stillborn or impaired children after the global releases of ionizing radiation amount to several millions globally [418]. These results contribute to disproving the established and prevailing belief that low-radiation-induced hereditary effects have yet to be detected in human populations. The authors [417] found strong evidence of an enhanced impairment of humankind's genetic pool by small doses of artificial ionizing radiation. It is very likely that in a few years, similar research will show the similar results on human populations due to the airport screening programs we have now.

This prediction is supported by the fact that a slight impact of ultraviolet rays on the organism leads to undesirable changes in the skin, as laid out in a study [36]. It demonstrated that the most extensive mutation took place not in the outer layers of skin, but those at a certain depth—i.e., those that were impacted by a lower level of radiation. Under normal circumstances, the accumulation of defective mutant cells is balanced by their apoptosis. However, given a prolonged exponential growth of their number, the life duration and number of the mutants' descendants grow, and so a change that will lead to the appearance of cancerous cells becomes more likely. This explains the medical observation that malign skin tumors appear on inner, less exposed skin layers. The inner and outer skin layers are protected by more or less the same level of concentration of a specific protein. However, its protective effect against the mutagenic factors of UV radiation is lower for inner layers of cells than for outer. This research also demonstrates that the use of various sun protection creams might be useless, and on occasion might even do additional damage to the cells. The reduction of UV radiation caused by sun protection creams could increase the overall depth of a weakened UV impact, which is the most mutagenic for skin cells. As a result, the number of defective cells might grow in comparison to "unprotected" skin and increase the risk of the appearance of cancerous skin cells. For most people, tanning and protection from UV damage seems a simple proposition. A person lies in the sun for hours, using protective sun creams and hopes to end up with a bronze tan and without UV damage. But the reaction of skin to ultraviolet light is more complicated. In a recent study [573], scientists found that melanocyte skin cells detect ultraviolet light using a photosensitive receptor previously thought to exist only in the eye. This ability of skin to photo sense light triggers the production of melanin more quickly than previously thought, in an apparent rush to protect against damage to DNA. Until now, scientists thought that melanin production occurred days after UVB radiation had already begun damaging DNA. But this finding about how the skin responds to and protects itself against UV radiation could lead to a revision of how sunscreen creams are used. That is, sunscreen cream use may slow the immediate response of cells to UV (the production of melanin) and thus cause increased damage to DNA. This study demonstrates the ambiguity of the use of creams, which may have in fact just the opposite effect to what they were for.

Doctors from many clinics in northern countries are reporting a dramatic rise in skin cancer in young adults. Mayo Clinic researchers recently conducted a population-based study using records from the Rochester Epidemiology Project, a decades-long database of all patient care in Olmsted County, Minnesota. They looked for first-time diagnoses of melanoma in patients eighteen to thirty-nine from 1970 to 2009. The study found the incidence of melanoma alarmingly increased eight-fold among young women and fourfold among young men. The lifetime risk of melanoma is higher in males than females, but the opposite is true in young adults and adolescents [635]. We know that young women are more likely to use indoor tanning booths than young men. But the use of sunbeds only gives a 74-percent increase in risk for developing melanoma. Where is another 600-percent increase in women coming from? We do know that young women are more frequently sunbathing with the use of sun protection creams. I think the carcinogenic effect of tanning under the sun while using sun creams is the reason for that 600-percent increase in girls and 300 percent in boys.

As such, I would not recommend to my readers to rely entirely on sun creams to protect their skin from UV radiation. The best approach is to reduce the time your skin is directly exposed to the sun. This is even truer for tanning booths, which use UVA-type UV, which, although it has less energy per photon, nonetheless leads to a statistically more harmful impact on the skin. This is clearly explained by the mechanism described above. Support for such approach can be found in the results of a study [59] on the changes of cells in the deepest level of the epidermis, mela-nocytes, which had never previously been looked at with regard to UV radiation specifically. Studies into the impact of UVA-type radiation show that melanocytes are less stable in the face of the carcinogenic effects of UVA than the external cells of the epidermis. Solar radiation in the part of the spectrum studied provokes changes in the DNA of melanocytes. As the cells cannot reconstruct themselves, mutations take place. At the same time, due to the relatively weak level of radiation, apoptosis does not take place. The existence and accumulation of mutant cells leads to the development of melanoma—one of the forms of cancer hardest to cure. The impact of the same radiation on external cells leads to their death, either from the immedi-ate radiation or apoptosis. Cancerous growths do not occur, even though the cells are exposed to a much harsher radiation and receive much larger sum doses over the course of the exposure.

Another mechanism of radiation's effect might consist of the fact that a small radiation dose can relatively easily impact the protein structures of cells rather than the core, which is evolved to be better defended. In this case, by causing inter-linking chemical interaction with macro-molecules, the radiation can impact the viscoelastic qualities of the cell structure and consequently change the balance be-tween viscoelastic phase separation and the rate of the processes of nano-diffusion (translation) of proteins and aggregates of a relatively small size. At the same time, apoptosis of the cells does not take place, but the protein exchange of a cell's core is disrupted, which subsequently causes those changes that lead to illness.

Of course, we need a more detailed study of the effects of UV radiation in gen-eral, and with the use of sun creams specifically, through experiments on model animals. These effects are complicated by many processes, for example the chemical reactions and dissolution of the fats in creams under the impact of UV radiation, the interaction of the substances created by the dissolution with the skin, and the substances secreted and their infiltration into the skin. The impact of these processes can intensify the harm done by UV radiation non-linearly.

Most sun creams contain dibenzoylmethanes (DBZM), one of the UVA-absorbing substances. Some of the DBZM degradation products turned out to be highly allergenic. This kind of allergy is called photo contact allergy. Women use cosmetics on an enormous scale. But the requirements for the safety of cosmetics products are much lower than those applied to medicines. Virtually no one controls them. Many cosmetics companies, in order to attract customers, proudly announce that some cream or other cosmetic product was not tested on animals. It's completely clear that in this case our women are the guinea pigs that the cosmetologists companies use to test their products on and expect them to pay for the privilege. Can you imagine the same situation in any other sphere producing products that would be directly applied to human skin?

Government regulators virtually do not regulate at all the various additives and scents used in products. The sweet smell of fragrances in cosmetic and household products may contain a sour note. Widely used fragranced products—including those that claim to be "green"—give off many chemicals that are not listed on the label, including some that are classified as toxic. A recent study [261] led by the University of Washington discovered that twenty-five commonly used scented products emit an average of seventeen chemicals each. Of the 133 different chemicals detected, nearly a quarter are classified as toxic or hazardous under at least one federal law. Only one emitted compound was listed on a product label, and only two were publicly disclosed anywhere. "We analysed best-selling products, and about half of them made some claim about being green, organic, or natural," said lead author A. Steinemann, a UW professor. "Surprisingly, the green products' emissions of hazardous chemicals were not significantly different from the other products." More than a third of the products emitted at least one chemical classified as a probable carcinogen by the US Environmental Protection Agency (EPA), and for which the Agency sets no safe exposure level. Manufacturers are not required to disclose any ingredients in cleaning supplies, air fresheners, or laundry products, all of which are regulated by the Consumer Product Safety Commission. Neither these nor personal care products, which are regulated by the Food and Drug Administration, are required to list ingredients used in fragrances, even though a single "fragrance" in a product can be a mixture of up to several hundred ingredients. Researchers found that all products emitted at least one chemical classified as toxic or hazardous. Eleven products emitted at least one probable carcinogen according to the EPA. These included acetaldehyde, 1,4-dioxane, formaldehyde, and methylene chloride. But because product formulations are confidential, it was impossible to determine whether a chemical came from the product base or the fragrance added to the product.

It is a well-known fact that every year the average woman receives 10 grams of poisonous lead and 100 grams of aluminum with her cosmetics, and pours up to 6 liters of oil by-products and an uncountable number of other mineral and organic compounds alien to humans in toxic quantities into her hair with her shampoo. In addition, because of the processes of the acidification of cosmetic fats on the skin, kilograms of free radicals are sucked up through the skin over the course of a year. We must also bear in mind that the majority of creams acidify intensively even before use due to prolonged or incorrect storage in cosmetics stores. The action of free radicals can also lead to bonding (the creation of cross-links) of proteins and change their viscoelastic properties in cells. A glaring example of dangerous reductionism is the use of the chemicals called PFCs in mundane plastics and chemicals. A study [369] suggests exposure to them may be associated with earlier menopause. But PFCs are contained in the products of several cosmetic companies that produce so-called "oxygen cosmetics." Based on primitive assumptions about skin metabolism, these companies suggest that oxygen, diluted in creams, has a rejuvenating impact on the skin. However, instead of rejuvenation, women who buy this pitch get premature aging.

We can also say that antioxidizing additives to creams and other cosmetic products do not protect from acidified substances in the skin. Overall, the effect of free radicals cannot be so easily compensated through antioxidants. The reason for this is that radicals and antioxidants are usually entirely different molecules with different characteristics and rates of diffusion. As a result, they are often in different phases of the cell environment and might not interact in the way that the simplistic views of those who advocate the use of antioxidants as panaceas against all illnesses suggest. So just because an excess of free radicals (ROS) is bad, it does not follow that an excess of antioxidants is very good. What is important is that the antioxidants impact the right place—in that phase of the cell structure where an excess concentration of ROS appeared. What is also important is that in the right place, in the right amount, the right molecular spectrum of antioxidants appeared which would have the ability to recombine with precisely the spectrum of ROS in excess there.

For example, researchers [432] have demonstrated that some antioxidants can even damage DNA and theoretically may turn cells to a cancerous state instead of protecting them. This means that attempts to boost the body levels of antioxidants, especially by taking DSs or vitamins such as A, C, E or beta-carotene, may be harmful. nature is much more delicate than the defenders of the antioxidant theory of universal curing allow, and so, when it comes to preventing illness, it is better to rely on nature itself, using its components with their anthropic qualities. True, antioxidants in some cases may have a certain positive, but limited, impact on the state of many organs. However, the mechanisms of this impact are not entirely clear. We can merely note that only natural antioxidants in the anthropic form of natural food have a genuinely healthy effect.

# 5
## The Nonlinear Code of the Gut

## 5.1 Bacteria files

Another interesting aspect of the intoxication of an organism is intoxication under the impact of microorganisms. The impact of toxins can show itself or not show itself depending on the presence and state of these or those microorganisms, which themselves do not have a toxic effect. There can be situations where a virus can, by interacting with bacteria, have a toxic impact on a human. Incidentally, there is a significant difference between the viruses suffered by bacteria and by mammals. Generally speaking, bacterial viruses cannot strike an animal or a human, and vice versa. However, the pathways of how bacteria, viruses, and higher organisms interact are often highly convoluted in nature. For example, a study [50] looked at the interaction in a mammal organism of the bacteria E.coli. This bacteria can be infected by a virus that codes for the production of Shiga toxin and of the simplest eukaryote Tetrahymena, which usually consumes E.coli. But if the E.coli is infected with a virus that generates the Shiga toxin, Tetrahymena dies while consuming these bacteria. Or alternatively, the infected cell start to release the toxin into the surrounding environment and so kills the Tetrahymena. In this way the population survives, as the conditions of life for even the uninfected bacteria are made easier because of the reduced number of predatory bacteria in the environment.

But at the same time, the battle between the two bacteria leads to the intoxication of the host organism. Curing it is made extremely difficult by the fact that using antibiotics kills the large bacteria and, as shown in the study [47], leads to the production of a large number of toxins. It is interesting that, given a low concentration of Tetrahymena, the harmful impact and presence in the organism of the virus that is producing the toxin might not show itself in any way. That is, therapy with the aim of detoxification has to first be directed against the presence of Tetrahymena.

I offer this example at the start of this chapter to underline the unique complexity of interactions that take place in the human intestinal tract.

There is evidence from scientific studies that changes in the intestinal microbiome typical of modern man lead to the spread of such a seemingly unconnected condition as depression. One piece of evidence is that rates of depression in younger people have steadily grown to outnumber rates of depression in the older populations, and researchers think it may be because of a loss of healthy bacteria. In an article [301] published in *Archives of General Psychiatry*, Emory neuroscientist Charles Raison and colleagues say there is mounting evidence that disruptions in ancient relationships with microorganisms in the soil, food, and the gut may contribute to the increasing rates of depression. According to the authors, "the modern world has become so clean; we are deprived of the bacteria our immune systems came to rely on over long ages to keep inflammation at bay. Since ancient times benign microorganisms, sometimes referred to as 'old friends,' have taught the immune system how to tolerate other harmless microorganisms, and in the process, reduce inflammatory responses that have been linked to the development of most modern illnesses, from cancer to depression."

Trillions of bugs known as gut microbiome live symbiotically in the human gut. They play a key role in many of the processes that take place inside the body. Different people have different types of gut bacteria living inside them, and abnormalities in some types have recently been linked to diseases such as diabetes and obesity. Because of the huge influence that bugs in the gut have on people's health, targeting gut microbiome with specific diets of anthropic products, rather than only concentrating on the chemistry of the human body that are the current focus of most drugs, could provide an array of new possibilities for preventing disease. Much research is still needed to untangle the roles played by each different type of bug and food in bug–food interaction. The important thing about this is that it should be easier to use the anthropy of food to change the bugs than to change the metabolism and signaling pathways inside the cells of the human body. And it is a well-known fact that it is possible to alter the makeup of bugs in a gut, affecting the metabolism, by using different food. These kinds of makeup of microbiome will mean a more holistic approach to metabolism correction through change in nutrition. But bugs in the gut not only benefit their host by helping with digestion; their presence also prevents the more pathogenic bacteria that may be in the gut from proliferating. When antibiotics kill all non-resistant bacteria, including those residing in the intestines, the usual balance of beneficial versus harmful microbes could be destroyed, leading to problems ranging from diarrhea to infections with dangerous microorganisms.

Researchers [393] found that the combination of microbes in the human intestine isn't random. "Our gut flora can settle into different types of community," says Peer Bork, one of lead authors of the study. After analyzing gut bacteria of 278 individuals from America, Asia, and Africa, researchers found that all these cases could be divided into three groups, based on which species of bacteria occurred in high numbers in their gut: each person could be said to have one of three gut types, or enterotypes. Like blood groups, these gut types are independent of traits like age, gender, nationality, and body-mass index. In my opinion, because gut bacteria, according to [394], interact with the human immune cells for mutual benefit, these three types of microbiome could be related to differences in how people's immune systems distinguish between bacteria. And as we age, we become less efficient at processing different foods' nutrition, because of the changes in our immune systems. So in order to survive in the human gut, microbiome have to take up the task and modify bacterial composition, but varying depending on which group they belong to. Within each group of gut microbiome, bacterial diversity still could be very wide, depending on diet, lifestyle, and the general health state of the individual.

## 5.2. The food digestion

We will briefly look at the first border post of the strange attractor of the human metabolism—the digestive system, which consists of the masticatory system, esophagus, stomach, and intestine. Each of these parts has its own specific function in the process of food digestion and the absorption of the bio-organic and chemical substances contained in food into the organism's metabolism.

Fermentation, digestion, and absorption of elements of food takes place in all parts of the digestive tract. For example, during the process of mastication (chewing), food is mixed, made homogenous, and enriched by saliva, which contains various ferments, enzymes, and other compounds needed for the biochemical transformation of food. At the same time, the process of absorbing elements of food into the organism is already beginning in the mouth. This process continues in the esophagus. Then in the stomach, the mixing, grinding out, and chemical decomposing of the substances in food is carried out through specific musculature. The stomach is a highly specialized organ, which is designed to digest a certain type of food that is natural precisely for man as an omnivorous organism. An intensive process of absorbing the substances originally in food and the metabolites produced also takes place in the stomach.

Then the prepared food enters the intestine—the main place where the chemical transformation and absorption of proteins, fats, and carbohydrates takes place. In the intestine, the bacterial environment and liquid secretes of the liver and pancreas chemically transform the food that has already passed part of the digestive tract and ensure the absorption of all substances essential to the organism in the correct quantities and proportions. In the last sections of the intestine, the reabsorption of water, non-organic ions into the organism takes place. The temporary (until a periodic removal) storage of undigested parts of food also takes place in this section.

Despite how it might appear from this simple schematic description, in reality the metabolism of food digestion is uniquely complex. It includes thousands of varieties of bacteria and reactions with substances different in their atomic and molecular contents—proteins, peptides, amino-acids, lipids, micro-elements, and so on. Subsequently these substances enter the blood, lymph, and cells of the organism.

The metabolism of the majority of plants and animals has certain traits that are paradoxical at first glance. One of these traits, in my view, is the limited utilization of the metabolic scope (MS) of food. The question arises: why has evolution not led to the development of mechanisms for the 100 percent utilization, or a degree close to this, of food? For example, from the results of a study [44] looking at fish metabolisms, used in many cases as analogous to the human metabolism, it follows that even with an active metabolism, the degree of utilization of the metabolic contents of food is about 77 percent.

To me, this seems connected to the fact that the metabolism of any organism, of a fish or a human, needs the maximum amount of all substances from food, but in specific proportions. However, these proportions are different in different varieties of food. As such, the complete utilization of food did not have a positive evolutionary effect for animal organisms and did not give a competitive advantage. In addition, the complete utilization of the metabolic contents of food evidently demands large energy expenditure that is not justified by the remains usually unutilized. It is true that the correct functioning of the natural state of any organism requires precisely specific amounts of the elements involved in the metabolism. The excess of any substance, evidently, creates problems as the organism's metabolism "does not know how" to utilize it. It begins to play the role of ballast or a toxin, disrupting the course of biochemical processes along the trajectories of metabolic reactions. As they say, there are no non-toxic substances—there are only non-toxic doses. In addition, an excess of unutilized nutritional substances begins to accumulate as fat. This underlines the need to balance our diet so as to ensure that it contains anthropic elements in anthropic proportions. In this way, to some extent the anthropy of a diet is determined not only by its inclusion of anthropic substances, but by the anthropic proportions in which they are absorbed by the organism. Food utilization can be considered anthropic if the organism can derive sufficient anthropic material from the food and remove the non-utilized parts without the excessive energy expenditure.

Let's look also at the process of how the metabolic contents of food are used with reference to its non-linear nature [43]. If we label the end level (normed to the initial amount) of utilization of the metabolic contents, and the initial normed level of MS of the food that has entered the organism is in this case equal to 1, then we can say that:

Lf = 1/

where is the Feigenbaum constant equal to 4.669.....

Given these assumptions, = 0.785. This is unexpectedly close to the level of utilization of 77 percent or, as a number, 0.77, which has been derived from experiments on zebrafish in study [47].

It is hard to say if this is more sheer coincidence or evidence of the universality of the nature of non-linear dissipative processes, first discovered by M. Feigenbaum [43], for strange attractors. If the theory is universality is true, then the metabolism of the organism of a zebrafish can be almost precisely seen as a strange attractor with a doubling of its period and a rate of compression towards the area of attraction equal to the Feigenbaum constant. Incidentally, it is known that the evolution of a population's strange attractor takes place precisely according to this law. We can speculate that the behavior of the strange attractors of a population and a metabolism follow universal laws, due to their hierarchical interaction.

The possible maximum degree of the utilization of MS limits the existence range of the strange attractor of the metabolism and its stability. In practice, this means that the organism, if this limit is exceeded, would not be able to exist, due, for example, to excess energy expenditure on food absorption. However, the Feigenbaum scenario of dynamical interaction, in my opinion, has a deep heuristic value in the science of the interaction of food with the organism. The basis for the iterative process that creates the Feigenbaum period-doubling cascades is entering the solution to a non-linear equation into the equation again and so on in each step ad infinitum. However, this precisely that universal and fundamental process of analogue iteration that has a place (in cells, organs and systems) in the strange attractors of the metabolisms of real biological entities. This gives the opportunity to base our system of counting the MS of a food and its utilization in the organism on something other than the calories typically used now. Its foundation can be erected on the principle of determining the nutrition of a food on how much MS is utilized. The actual real-world level of MS utilization can be found through experiments on model organisms and, in their safe parts, directly on people. However, this work is made difficult by its scale. If it is completed, we will acquire data that will allow us to precisely ascertain, for example, what part of the MS of food goes to maintaining the vital functioning of the organism, and which to generating fat through over-eating.

For any organism whose metabolism is working correctly, the absorption of food follows the Feigenbaum scenario, with the organism itself determining the amount of food needed by comparing this to the energy expended to maintain vital functioning. With large energy expenditures, food consumption increases and with small ones it falls. However, the normalized level of MS utilization evidently remains constant. In a disrupted metabolism, where the signals meant to indicate satiation do not reach their target, the organism, nonetheless, continues to utilize food with that same constant normed level of utilization. The excess energy is converted into fat deposits.

The further biochemistry enters into the metabolic processes on the molecular and quantum levels, the more convoluted and chaotic the picture of biochemical processes becomes. At the same time, the knowledge we can acquire about individual metabolic reactions, of which there are probably virtually limitless, might have a limit connected with the cognitive limitations in an information processing. At the same time, we must underline that this is not a defect of biochemistry as a science, but a consequence of the grand complexity of nature itself. It was due to these limitations of the analytical science about the human metabolism, perhaps the most complex object in the universe, that the idea of a nonlinear anthropic phenomenological approach, presented in this work to describe the impact of food on the metabolic human processes, was born.

Returning to the question of food utilization, we have to note that food enters animal organisms at an impulse tempo, only when they eat. As such, animals have normalized mechanisms to prevent the disruption of nutrition, sometimes prolonged ones, by storing excess fat and carbohydrates. However, the organism cannot store for long, or store very many, minerals and other chemical compounds that in excess quantities are toxic for it and disrupt the metabolic processes. As such, the proportions in which the elements of food are absorbed are crucial. That is, it is important to balance the mineral and organic contents of food.

## 5.3   The Anthropy of the Gut

The metabolism of the intestine depends on many  factors. For example, it was recently discovered [639] that how a person was born—whether naturally (through the vaginal tract) or by Caesarian section—is among the factors that determine his metabolism. It was discovered that, given different methods of birth, the bacterial environment in an infant differs by which type of bacteria predominates. As such, there are certain risks for the subsequent health of individuals born through Caesarian procedures, which, in the main, consist of these children being observed to have a high susceptibility to certain pathogens and a higher rate of allergies, asthma, and similar ailments. The difference between these two methods is that in the first case, the infant has a direct contact with the vaginal micro flora of the mother during birth, and in the second, contact with the micro flora of the hospital ward and the doctor performing the operation. During a natural birth, the mother's micro flora works on the child as the first injection, which determines the state of the child's micro flora for the entirety of his subsequent life.

This example demonstrates that the metabolism of food during its consumption and the human metabolism as a whole is determined by a multitude of factors connected to his life, beginning with his development in the womb and emergence into the world, and finishing with the whole history of his food consumption. The importance of the state of intestinal micro flora is demonstrated by the results of a study [290] that looked at the difference in the processes of milk absorption by infants and adults. This study shows that infants are more efficient at digesting and utilizing the nutritional components of milk than adults are, due to a difference in the strains of bacteria that dominate their digestive tracts.

Many illnesses, from cancer to allergies to obesity, are connected with the appearance of incorrect bacteria in the intestinal flora. Earlier we noted the vicious circle formed by interaction between the state of the intestinal micro flora and allergies. A similar vicious circle exists between the state of the micro flora and obesity. More than 1,000 different strains of bacteria and organisms called archaea co-exist peacefully in the typical healthy bowel. In normal conditions after breaking down food compounds, the gut microbes excrete remnants that are used in the body's metabolism. But when the balance is altered, by antibiotics or other causes, some bacteria strains can become dominant. These changes in the balance lead to diarrhea, inflammation, etc. Poor intestinal flora is believed to trigger obesity and diabetes. In the same way, healthy gut flora could reduce the risk. Diabetes and even obesity, as well as Parkinson's disease, might be cured, according a study [466], just by replacing the bacteria in your intestines.

The state of the micro flora is mostly determined by our diet—whether or not it contains substances and micro-organisms that are essential for the intestinal flora to function fully. Nowhere in the organism do the side effects of medicine show themselves as clearly as they do in the state of the intestinal micro flora. It is widely known that the use of antibiotics burdens the micro flora and leads to serious illnesses. As antibiotics are used by a significant proportion of the population for entirely trivial reasons, the sum harm they do to the metabolisms of organisms through their destruction of micro flora might significantly outweigh their positive effects. The number of bacteria in the microbiome (flora) of our intestines greatly exceeds the number of cells in our organism. Taken together, the number of bacterial cells in the human microbiome is tens of times greater than cells of the organism itself. Any change in their number and variety, which are caused by the use of antibiotics, has the strongest impact on the state of our metabolism. As demonstrated in a study [395], the administration of antibiotics to mice killed off the big part of the intestinal microbiome and dramatically changed the metabolic landscape. The most profoundly altered pathways involved steroid hormones, eicosanoid hormones, sugar, fatty acid, and bile acid. These important substances control the immune system, reproductive functions, mineral balance, sugar metabolism, and almost all other aspects of either mouse or human metabolism. Intestinal microbes help us digest our food, provide us with vitamins that we cannot make on our own, and protect us from pathogens.

The restoration of micro flora (bacteria and archaea) requires an element of therapy after the use of antibiotics, the need for which is unfortunately often underestimated by doctors and patients, especially in developing countries. Unfortunately, a state of incorrect bacteria is even now still viewed by many specialists as a minor change in the organism. However, we have to admit that disruption in the intestinal microflora impacts in the strongest way the health of a person in general over a prolonged period of time. It exceeds in its impact many chronic illnesses and significantly reduces the life span of the individual. Bacteria naturally present in the human intestine could produce substances that help to protect against colon cancer and provide therapy for inflammatory bowel disease [539].

However, apart from diet, the intestinal microbiome is also determined by external stresses, according to the findings of one study [364]. The study showed that stress can change the balance of bacteria that naturally live in the intestine and alter the immune system of the body. Exposure to stress led to changes in the composition, diversity, and number of intestinal microorganisms. The study reveals the dynamic interactions between multiple physiological systems, including the intestinal microbiome and the immune system. The bacterial communities in the intestine not only became less diverse, but also had greater numbers of potentially harmful bacteria, such as *Clostridium*. This research suggests that not only does stress change the bacteria levels in the human gut, but also that these alterations can, in turn, impact our immune system.

Another source of the destruction of the intestinal ecosystem is the charlatan procedures, widely used in certain clinics. It is dangerous to disrupt any part of the gut bacterial biotope (microbiome). Scientists have known for decades that various microbes in the colon collaborate to ferment undigested food and degrade and dispose of the byproducts of fermentation [485]. The researchers of one study [486] found that the colon biotope harbored three important types of hydrogen-consuming microbes: methanogens, which convert hydrogen to methane; acetogens, which make acetate from carbon dioxide and hydrogen; and sulfate-reducing bacteria, which expel hydrogen sulfide gas. Disruptions of the colonic biotope can have profound implications for human health. Studies have already demonstrated that hydrogen sulfide, a major byproduct of bacterial fermentation in the gut, is genotoxic (damaging to DNA) and may lead to colon cancer [485]. That is why the widely used procedure of colon cleansing has absolutely no benefits but could lead to many complications in a healthy state. This procedure, sometimes called colonic irrigation or colonic hydrotherapy, often involves the use of chemicals followed by flushing the colon with water through a tube inserted in the rectum. It has ancient roots, but was discredited by the American Medical Association in the early 1900s. Yet colon cleansing has staged a comeback [487]. Georgetown University doctors say there's no evidence to back the claims of any benefits for such a procedure. In fact, their review of research in this field [488] demonstrates that colonic irrigation can cause side effects ranging from cramping to renal failure and death.

The microbiome of the intestine is extremely different for each individual and includes thousands of varieties of bacteria, divided, as we have already noted, between three main types. The homeostasis of the intestinal microbiome of any type is uniquely vulnerable to the use of antibiotics, which get into it not only during their direct consumption as medicine during therapy, but also from a wide range of products; these include meat and fish, which contain antibiotics in significant amounts due to the barbaric practices common in animal feeding. Disruption of the intestinal microbiome, the creation of the incorrect bacteria, has, as we have already noted, a fundamental impact on human health and life span. However, the restoration of a disrupted microbiome, despite the advertised promises made by producers of prebiotics and probiotics, is virtually impossible with the help of such substances. Not one of the well-known prebiotics or probiotics achieves the promised long-lasting and stable effect in restoring the biological diversity to be found in a normal, healthy intestine. The transplant of a healthy microbiome to an ill person is a relatively promising method. However, studies into this are at their earliest stage and have certain difficulties connected to the possible issues of compatibility, similar to the issues in tissue transplants.

The most appropriate method is to restore microbiome through an anthropic diet, which contains elements of nutrition and bacterial flora evolutionarily close in their biological diversity and types of bacteria to those in the human intestine [18]. An example of such food are raw cacao beans (ones grown in the wild), the shell of which contains a variety of bacteria remarkably close to the microbiome of the human intestine. Thanks to the dense structure of the shell, the bacteria easily pass the acidic environment of the stomach and multiply in the intestine, recreating the disrupted biological diversity.

Many studies have shown that compounds known as polyphenols, present in high content in cacao beans and green tea and in lesser amounts in grapes, possess neuroprotective properties, binding with toxic compounds and protecting the brain cells. When ingested, the polyphenols are broken down to produce a mix of compounds. Digestion is a vital process that provides our bodies with the nutrients we need to survive. But, says Dr Okello in a study [321], it also means that just because the food we put into our mouths is generally accepted to contain health-boosting properties, we can't assume these compounds will ever be absorbed by the body. What was really exciting about this study was that researchers found that when green tea is digested by the enzymes in the gut, the resulting chemicals are actually more effective against the key triggers of Alzheimer's development than the undigested form of the tea. In addition to this, the study also found the digested compounds had anti-cancer properties, significantly slowing down the growth of the tumor cells that scientists were using in the experiments. There are certain chemicals that we know to be beneficial, and we can identify foods that are rich in them, but what happens during the digestion process is crucial to whether these foods are actually doing us any good [321]. However, the effective digestion of whatever food components requires the presence of the right enzymes. Only in the presence of the right enzymes do the beneficial qualities of mineral substances and organic compounds in anthropic food show themselves in full. This is true not only for the specific example of green tea, but for all food we consume. However, the creation of the majority of specific enzymes is the work of our intestinal micro flora, which is on the whole quite individual. As has been noted, it is significantly dependent on the person's diet. Some people produce the enzyme needed to digest green tea in their intestine, and others don't. Perhaps this is due not only to the absence of whatever bacterial strains are necessary, but also to genetic and epigenetic changes in the bacteria of our intestines.

Jeremy Nicholson, a biological chemist at Imperial College London, points out, according to Michael Eisenstein [322], the tremendous diversity of the intestinal microbial flora, citing a report in 2010 which showed that Europeans each carry a complement of at least 160 bacterial species, with more than 536,000 bacterial genes between them—well over twenty times the human gene count. "It actually should be thought of as a multicellular organism with a very large genome," he adds. Nicholson has already found some compelling evidence that genes expressed by the gut flora have effects that reach far beyond the digestive tract. "We've found deep compartmental connections between microbial status and bile acid metabolism," he says, "[and] there are some staggering connectivities between blood pressure and gut microbial metabolites. Certainly less than 10 percent—and it might even be less than 2 percent—of the bugs that are in you are also in me," says Nicholson [322]. In other words, the variation in the bacteriological contents and in the genes expression of in the trillions bacteria's of the intestinal microbiome can reach truly cosmic proportions. This in turn puts in doubt the possible or declared successes of a discipline such as Nutrogenomics, which attempts to study the interaction between the diet and genomes of a person. To take into account so many variations, never mind the practical application of these, exceeds the calculating capacities not just of existing computers but of any we might imagine.

Until now, scientists did not know how individual variations in gut bacteria might influence specific health issues. But a recent study [330] shows a strong relationship between the complex microbial community in the human gut and the medical condition known as fatty liver. As was mentioned above, a multitude of different kinds of benign bacteria inhabit our gut, and these populations can vary widely between individuals. It is a well-known fact that fatty liver can be caused by alcohol abuse, obesity, hormonal changes, and/or diabetes. Researchers [330] have suggested that diet is also important, with strong indications that deficiencies in the essential nutrient choline might be partially involved in some incidences of the condition. Choline deficiency also implicates genetics, since many people lack the genes to efficiently produce choline in their metabolism. Now, a new bioinformatics finding [330] shows that the abundance or scarcity of certain types of bacteria in the gut may also help predict susceptibility to non-alcoholic fatty liver. The researchers noticed that among the numerous classes of bacteria present in each patient gut, variations in the populations of *Gammaproteobacteria* and *Erysipeoltrichi* seemed to correspond with variations between patients in the degree to which they developed a fatty liver during the period of dietary choline depletion. The implication of the finding is that these groups of bacteria may be influencing the body's ability to properly use the choline available in food. What does this finding mean for people in terms of their diet? Even when doctors put the patients on exactly the same diets, because everyone's gut microbiome is different, the nutritional outcome of digestion is different.

Variations in intestinal microbiome may be a cause of many other diseases. Researchers [602, 603] have analyzed the multitude of microorganisms residing in the human gut as a complex, integrated biological system, rather than a set of separate species. Their approach has revealed patterns that correspond with different diseases. The results [602] indicate that the communities of microorganisms that reside in the gut of autistic children with gastrointestinal problems are different than the communities of non-autistic children. Whether or not these differences are a cause or effect of autism remains to be seen. Researchers [603] observed that obese and lean people have differences in their gut microbiome and that difference leads to the difference in the genetic makeup of gut microbe networks. Consequently, it affects the metabolism of the gut and the health state of the human host.

Another recent study [604] suggests that the types and levels of bacteria in the intestines may be used to predict a person's likelihood of having a heart attack and that manipulating these organisms may help reduce heart attack risk. The researchers hope to set up a laboratory pipeline in which we can deliberately manipulate collections of human intestinal microbes from people of different ages and dieting cultures. They think it will give them the opportunity to identify groups of microbes for transplant that may be beneficial in various therapeutic settings. Some researchers have grown bacterial microbes in the laboratory before for transplantation, but there's still no reliable way to know whether communities captured in a

Petri dish mirror the diversity of the bacterial collections that exist in the particular habitats of the intestine. The problem is that there are so many types of bacteria that live in the human gut, as well as big differences in these collections from person to person, that it is impossible to get the diversity and richness of microbial communities when the researchers try to grow them in the laboratory. In reality, this approach makes it impossible to obtain personalized microbial communities for people. A person's overall nutritional and gut microbiome status can be improved only by general anthropic diet interventions.

However, in healthy people, substantially different starting states (in the system of the interaction among food, microbiome, and the body's metabolism) lead in the end to more or less the same final state. Such behavior by this system is the absolute opposite of moving towards thermodynamic equilibrium (as the end state is minimally entropic), and can be assured only if this end state is determined by the macro-state of the organism. That is, the demands of the organism's metabolism to the intestine are determined by the ensemble of the micro-states of the intestinal microbiomes. To reference the studies on this point [222, 223], the macro-state of an organism's metabolism renormalizes the probabilities of the various micro-states of the intestinal microbiome. Here we have to remember, however, that the ensemble of the micro-states of the intestinal microbiomes includes in itself the ensemble of the micro-states of food and the ensemble of the micro-states of chemical signals from the organism. The precise biochemical process that leads to such a result from the interactions in the food-microbiome-body metabolism system is unfortunately unknown for now.

It is my firmly held conviction that despite the large differences between the contents of their intestinal microbiomes between people, overall the intestinal microbiomes of all healthy people has the same end result. This means that the food-microbiome-body metabolism system contains mutual compensatory mechanisms that allow the nullification of the negative consequences of specific distortions in the processes taking place in the gut. This has evolved under conditions of an anthropic diet. So the consumption of anthropic food is the only way to correct the metabolic state of an individual through his diet. Any creation of isolated signals by using medicine or food supplements has a quite patchy impact on all people, and is completely useless for a part of the public. What is more, practical experience based on the medical study of the human intestine shows that the disanthropy of any part of the food-microbiome-body metabolism system changes the end result of the functioning of the intestine in an undesirable direction.

Most likely at least part of the change in the intestinal bacteria appeared, according to studies [171, 322], because of the horizontal transfer of genetic material. For example, Japanese individuals can digest seaweed carbohydrates more efficiently. This was made possible by an ancestral gene transfer event from kelp-borne bacteria that endowed their gut flora with the capacity to produce porphyranase and agarase enzymes [322]. According to Michael Eisenstein, "microbe-watchers like Nicholson suggest that this study could be a strong indicator of the future, as the research community begins to come to terms with the extent to which human genetic effects on diet might be overwhelmed by the bacteria we carry" [322].

We must note that, returning to the example of the absorption of green tea, if your organism does not produce the necessary enzyme, because you have never or rarely drunk green tea, then you cannot digest it completely and derive the full benefit. However, the prolonged consumption of natural green tea brewed with fresh (or correctly preserved) plant leaves can make possible the expression of genes that code for the production of the correct enzyme and ensure that you derive the maximum health benefit from green tea. This is one of the principles of an anthropic diet, that consuming some anthropic variety of food can epigenetically modify the basis of the human metabolism.

Moreover, even if the components of green tea and other plants are not directly digested by the human body, they still may well be involved in the metabolism of the intestine and by other means to metabolic regulation in animals and humans. Generally, plant food can affect the metabolism in three ways. First, for example, ginseng is rich in many steroids glycosides, dubbed ginsenosids. But that ginsenosides have no any significant direct effect on humans because the gut walls cannot absorb them. After partly breaking down ginseng to ginsenoids and digesting it, microbes in the gut leaving the remnants that used by the body with well-known highly beneficial results for it.

The second way is that some plants can influence the balance of microbial species living in the gut. A good example is Ginkgo leafs that increase the abundance of beneficial bacteria and probably archaea. The third way is when ingredients of some plants and genetic information [572] move directly through the gut walls to the body's blood stream. So green tea and other plants may affect human health in any of these three ways or in any combination of them.

Looking at the question of the transfer of genes in the intestine, I would like to note that many researchers, even now, think that the horizontal transfer of genes between organisms, especially from eukaryotes, is the rarest of eventualities. But this process is completely typical if we look at bacteria.

Since the Second World War, the widespread use of antibiotics has led to a situation in which virtually every single bacterium on Earth, from the South Pole to the North, is resistant to the majority of antibiotics. I am certain that this is the result of the genetic transfer of genes that cause stability in the face of antibiotics. No one knows how much time was needed for these horizontal transfers. This has been discovered only recently, after sixty years of the mass use of antibiotics. Possibly, this process of gene transfer has a much quicker rate and happened over the whole planet just a few years after the development of the first bacterial strain became stable in the face of some antibiotic. Incidentally, it would be interesting to attempt to experimentally determine this rate of transfer by using a new type of antibiotic that bacteria are not yet stable in the face of. If we constantly analyze certain bacteria to be found in live nature on different continents for the appearance of this stability, then we can determine the rate at which this stability spreads. I suspect that doctors will be unpleasantly surprised by the speed of this global process. But even now, it is already obvious that the process is extremely quick compared to many other processes that have an impact on the evolution of the organism.

In a human, the intestinal microbiome is one of the key parts of the metabolism. At the same time, it is where the bacterial world and the eukaryotes world meet and exist in relatively close contact. Needless to say, the bacterial environment of the intestine consists of trillions of bacteria of hundreds of varieties. The processes of horizontal transfer can be carried out in their case not only through highly mobile genes, but also accidentally through genes that are not so highly mobile. For this reason, the dangers of using genetically modified products is several times greater than is assessed by many scientists who state that the transfer of genes between bacteria and eukaryotes is virtually impossible. During the trillions upon trillions of interactions of bacteria among themselves in the human and animal microbiomes, the opportunity for the horizontal transfer of genetic material on a wide scale that is thought impossible is in fact realized. During this, certain changes in the metabolic trajectories in the human as a whole take place, which as we know leads to the general weakening of individual organisms and the population as a whole. For example, one study [337] has for the first time confirmed that genes can transfer from a human organism to a pathogenic micro-organism.

But if the process can happen in one direction, then there is no reason to think it cannot happen in the opposite. So for example, scientists have found that the intestines of the inhabitants of Japan contain special enzymes essential to the digestion of seaweed, which is used in many Japanese dishes [171]. In the human digestive tract, carbohydrates are dissolved by a group of enzymes under the general name of glycosides, which includes only 261 substances. These enzymes are not produced by our organism's cells, but are created by the intestinal micro flora, including the bacteria of the type bacteroides. Each of these ferments dissolves a specific type of carbohydrate that enters the organism alongside plant food. It is well known that the consumption of raw products has an effect on the variety of enzymes created by the intestinal micro flora. Glycosides, which are involved in the digestion of red seaweed, were isolated in the bacteria Zobellia galactanivorans, which inhabit the surface of these plants. During a comparative analysis of the genomes of these bacteria and samples from the intestinal micro flora, genes for the enzymes that digest seaweed were found in the bacteria Bacteroides plebeius, which inhabit the intestines of the Japanese, while the equivalent bacteria, inhabiting the intestines of North

Americans, either lacked these genes or these genes did not express themselves. In the opinion of the researchers, samples from the intestinal microflora of Japanese people received these genes as a result of exchanging their hereditary information with bacteria that inhabited the seaweed that are used in many Japanese dishes, including in many sorts of sushi. The researchers did not pinpoint when precisely the gene exchange between the bacteria took place.

Given a close interaction between bacteria themselves and with the carrier organism, we cannot exclude the possibility of relatively rare cases of transfer between bacteria and the organism's cells. Any slight chance becomes a possibility given more than a trillion attempts every day. So genetically modified products, the carriers of modified genes, can, through the billions of various metabolic trajectories of live nature, have an unknown and unpredictable impact on the human genetic apparatus through the products consumed. What is important is that you can personally not consume genetically modified products at all, but modified genes will still enter you through countless paths from the surrounding environment and, of course, through food. Scientists currently have an ambitious and dangerous research project to develop an *in vivo* biological cell-equivalent of a computer operating system. The "success" of the project to create a "re-programmable cell" could revolutionize modern biology and would pave the way for scientists to create completely new and "useful" forms of life. Long-term consequences of such experiments are unpredictable.

What's the solution? I can see only one solution—categorically ban the use and creation of genetically modified *anything* that has any part in the cycle of substances in nature. Perhaps mankind has already let the genie out of the bottle. Perhaps not yet, and we still have a chance.

The development of biology over the last decades shows, with increasing clarity, that it contains no hard and fast rules nor the iron correlations we are used to. There are different methods of evolution, which, speaking generally, counteract each other. For example, the lengthy debate between the followers of Darwin and Lamarck ended with the unexpected result that both methods of the functioning of the evolutionary mechanism exist and work simultaneously, at least in bacteria. But we have to note also that scientists can grow no more than $1/1,000$ of all known varieties of bacteria in cultures and observe the evolution of their strains. If we take into account that modern biology has classified approximately $1/1,000,000$ of the bacterial sphere on our planet, then we can understand the limited extent of knowledge in this area. Perhaps, apart from the mentioned Darwinian and Lamarckian ones, there are tens, if not hundreds or thousands, of other evolutionary mechanisms awaiting their own discoverers.

The interaction between various bacteria and with the organism's metabolism sometimes leads to surprising results. For example, one study [46] looked at the peculiarities of the interaction between colonies of Streptococcus pneumonia and

Haemophilus influenza in the organism. Both types of bacteria usually show no symptoms of their presence in an organism and perhaps have some sort of metabolic symbiosis with the "host" metabolism. However, given certain changes in the metabolism of the host organism, as a result of which the populations of these organisms start to compete for living space, H. influenza begins to have an unusual impact on the metabolism of the host organism, producing an immune response to S. pneumonia. In reply, the latter changes its metabolism to improve its defensive qualities against the immune system of the host organism. Its improved defenses against the attacks of the host organism's immune system allow S. pneumonia to enter the bloodstream and cause various illnesses, from pneumonia to meningitis. Several thousand of such interactions between the bacteria living in an organism and the organism as a whole are known. Moreover, who can be certain that there do not exist, apart from the three-way interaction described here, n-order interactions, as a result of complex changes in the bacteria's and organism's metabolisms, with interwoven interactions at different levels, having a significant impact on the state of health. The more we learn about the human metabolism, dieting, and nutrition, the closer we get to being able to prevent diseases—and the more complicated our understanding of disease itself becomes.

It is known that bacteria communicate with each other, and this exchange of signals helps bacteria attain greater viability and resistance to antibiotics. Studies [76, 78] have shown that bacteria can form, in the intestine included, a shield (biofilm) against the organism's immune system and antibiotic molecules, through communication with each other. The extracellular matrix of structured microbial communities, usually called biofilms, constitutes the framework that holds the component cells together.

Although the presence of cell-to-cell interconnecting matrices appears to be a common feature of structured microbial communities, there is a remarkable diversity in the composition of these matrices [493]. The microbes in biofilms live in a self-produced matrix of extracellular polymeric substances that form their environment. Polymeric substances of biofilms are mainly polysaccharides, proteins, nucleic acids, and lipids; they provide the mechanical stability of biofilms with the ability to reversibly deform and form a cohesive, viscoelastic 3D polymer network that interconnects and transiently immobilizes biofilm cells. The biofilm matrix also acts as an external regulatory system by regulating the rate of extracellular enzyme release to the cells, enabling them to participate in processes of cell metabolism. Viscoelastic mechanical properties and constituents of the matrix are what make biofilms among the most successful form of life on earth.

The impact of biofilms forming in the intestine is enormous. For example, according to recent research [376], microbes in the gut could fight flu by triggering an immune response that suppresses infections. The study suggests that our diet affects our ability to fight viruses by altering the composition of our gut bacteria population. Researchers don't yet know which bacteria trigger those pathways of our metabolism, but suspect that this effect is caused by *Lactobacillus* species residing in the gut. In experiments, after introducing antibiotics, the populations of these bacteria were significantly diminished. However, if bacteria form biofilms, then the impact of antibiotics is completely different—they cannot effectively impact the bacteria. This underlines yet again the importance of controlling the use of antibiotics, which, through their effect on the intestinal flora, can impact in countless ways the organism's ability to resist viral diseases.

The unique importance of the bacterial composition of our intestinal tract and the food consumed by us on the state of so seemingly unrelated an organ as the brain has been shown in one study [82]. It demonstrated that certain bacteria that live in the intestine play a significant role through the immune system in the development of multiple sclerosis. Indeed, doctors have long known that the appearance and development of multiple sclerosis are connected to environmental factors—single-egg twins, who have identical genomes, have only a 25-percent chance of suffering from this illness if the other one does if they live in different environments. However, before this study [82], no one knew which environmental factor may causes this illness. Over the last decades, the population of developed countries consumes, as a significant part of its daily ration, products that have undergone strong thermal or chemical processing, and hermetical sealing to prevent any bacteria from getting in. As a result, the balance of pro- and anti-inflammatory bacteria in our intestines changes, which, as its final result, leads to the development and complication of such auto-immune illnesses as multiple sclerosis, among others.

Besides this, researchers at McMaster University [369] discovered that the "cross-talk" between bacteria in our gut and our brain plays an important role in the development of intestinal diseases and probably other health problems as well, including obesity and diabetes. They show in research that gut bacteria influences anxiety-like behavior through alterations in the way the brain is wired and that the bi-directional communication of the body and the brain may produce metabolic disorders, such as obesity and diabetes. The study results allowed the researchers to put forward the interesting hypothesis that the state of our immune system and our gut bacteria—which are in constant communication—influences our personalities.

Recently, an alternative mechanism of evolution was observed [69], whereby the organism borrows, as it were, the genome of another creature. Up until now, it had been thought that any form of evolution is based on the principle that the metabolism of an animal or plant adapts genetically (through its own genes!) to assure its viability in changing circumstances. However, researchers have discovered a surprising example of bacteria that infects an animal and gives the animal reproductive advantages, which are then inherited. This symbiotic relationship allows the animal to receive a ready-made system of immunity against bacteria, which spreads throughout the population through natural selection, in the same way that genetic mutations useful for survival are spread in general. The results of studies of the interaction between drosophila and parasitic worms that destroy the reproductive capacities of the flies show that flies infected by the bacteria spiroplasma maintain their reproductive functioning. This bacteria, which is useful for the survival of the population, quickly spread across the fly population over the whole of North America, as the bacteria is inherited. These useful bacteria went from being present in 10 percent of the population in 1980 to 80 percent by 2008. This newly discovered method of evolution, where one organism uses the genome of another organism (in this case, bacteria), almost certainly took place many times in the evolution of the human digestive system. This also illustrates the possible complexity and ambiguity of symbiotic relationships in one of the main organs of the human metabolism—in the bacterial environment of the intestine with its trillions of bacteria. Symbiotic relationships during the process of evolution evidently included a three-way symbiosis of bacteria, the organism (individual organs), and incoming nutrition. Changes in any component of this "holy union" produce the strongest changes in the phase space of the strange attractor of the human metabolism.

So according to research [363], bacteria in the human gut may not just be helping metabolize food but could also be exerting a certain level of control over the metabolic functions of the liver. These findings show the importance of the symbiotic relationship between the whole human metabolism and the metabolism of their parts (gut) and how changes in the gut microbiome can impact the functions of other organs. The diversity of the microbial community in the gut could enhance the metabolic capacity of the body for processing food nutrients and modulate the activities of multiple pathways in a variety of organ systems, according to the authors of this study [363]. Research on mice, as modeling organisms, shows that there exists a strong association between liver metabolic functions and microbial populations, determined by a family of bacteria and hepatic lipid metabolisms. These results provide insights into the fundamental mechanisms that regulate body-gut microbiome symbiotic interactions. Moreover, in that work researchers found that gut colonization strongly stimulated the gene expression and activity of some essential enzymes in drug-detoxification pathways. Both these findings confirm the importance of an anthropic style of diet for the regulation of these most complex of interactions in the gut-body system, and in general its extreme importance for nutrition and for metabolic and systems biology.

Microbial communities thrive not only in our digestive tract but elsewhere in the body. And it is clear that their net genomic expression is inextricably linked with the body's metabolic function. The composition of our mouth and gut microbiome is a direct by-product mostly of our diet. As I mentioned above, the gut microbiome should be viewed as probably the biggest metabolic organ of the human body because it process components of our diet and their metabolites and profoundly influence our inner body metabolism. In addition, the smaller, but maybe equally diverse, microbiome of our mouths and nasal passages certainly play an important role in food digestion.

I would also like to speak here about one topic related to the bacterial state of modern man's intestine: specifically, the presence of incorrect bacteria. The various producers of medicines, DSs, probiotics (which contain live bacterial cultures), pre-biotics (substances that cannot be digested by humans, but which are nutrition for a number of bacteria in our intestines), and, especially, lactic acid food products are at the moment piggy-backing on concerns about incorrect intestinal bacteria. Let's look at probiotics first. In reality, the bacteria in probiotics are not natural for the human bacterial micro flora. That is, the claims about the capacity of the specially grown healthy bacterial cultures (probiotics) to colonize our intestines are lies and rubbish. In reality, even if these bacteria do overcome the barrier of stomach and pancreatic juices and reach the intestine, they will still only be able to become what's called transitional colonists [378], with a maximum duration for their presence in the intestinal micro flora of around one to two weeks. And we cannot regard such a transitory colonization of the intestine as normal or desirable for the organism, as the ultimate aim of all dietary methods of restoring the intestinal micro flora is the restoration of a healthy intestinal microbiome natural to precisely this specific individual.

Nonetheless, we do have to note that some probiotics, while not especially effective, have shown themselves as quite useful in treating certain sharp diarrheas found among infants. Probiotics can also be of some not-especially-significant assistance in states when it is essential to use large doses of antibiotics. A weak conflict between the probiotics and the human organism can, strange as it may be, have something of a positive effect, due to the activation of an immune response both in the intestinal metabolism and in the metabolism of the human organism as a whole. However, it is clear that the impact on the immune system of a person of his own healthy micro flora far exceeds the modulating impact of any probiotics. Moreover, the alleged anti-carcinogenic effect of probiotics has no evidence in its favor.

Probiotics in yogurts also secrete a microbe lactose ferment, which breaks down milk sugar, into the milk product. It is precisely milk sugar that leads to digestive complaints among people who have, with age, lost their own ferment. The source for this ferment is for a large population of the planet, especially in Asia, products such as natural yogurts. However, in this case the best solution is for people with a deficit of endogenous ferment to limit their consumption of milk products. Many of the studies published in the professional medical literature over the last twenty years have not shown a certain, noticeable impact of milk-based probiotics on the health of children or adults [378]. As such, the enormous exaggeration of the effectiveness of milk-based probiotics is mainly connected to the advertising campaigns of milk and pharmaceutical companies. If you have good reason to think that your intestinal micro flora is deformed in its number and composition of bacteria, then instead of using whatever miraculous medicines and probiotics you hear about, you should restore the micro flora the natural way—through an anthropic diet. You should include in this super foods, wild-grown products of live nature in their anthropic state (live or moderately fermented). For example, a small quantity of organic cacao, that has not been heat processed and with a moderate fermentation (three to five days long), consumed in its natural state of beans still in their natural covering will restore an intestinal microbiome ruined by antibiotics within three to seven days.

As we already noted, the individual contents of the intestinal microbiome differs significantly from person to person. It differs even more between populations living in different countries, and under different geographical and climate conditions. That is why many people (children especially) suffer from "traveler's diarrhea" while traveling abroad. This type of diarrhea is a method the organism uses to adapt to the new bacterial environment. If the diarrhea is not strong, there is no reason to combat it with any sort of medicine. You have to remember that, through the constant emission of fecal matter, which is 70-percent bacteria, and due to the bacteria's capacity for intensive multiplication, the intestinal microbiome quickly restores itself and modernizes and adapts its contents, through the new types of bacteria, to the new external conditions. Even when the external destabilizing factor is eliminated—i.e., you return to your habitual environment—the new composition of your intestinal microbiome is maintained, at least for some time. After your organism undergoes this adaptation, it will enrich the contents, variety, and future stability in the face of negative factors of your microbiome. Among other things, your immune system will probably improve. Another important factor is taking care of your intestinal flora. In the case of illness, this should be entrusted only to highly qualified, specialized medical professionals.

The usual practice of analyzing the bacterial composition of the intestine is to base it on an analysis of the feces. However, specialists have long known that the correlation between the flora composition in the feces and its actual proportions in the intestinal microbiome is weak. The main reasons for the absence of the correlation noted above are:

1.  The analysis of intestinal flora is based on the method of growing bacterial cultures in nutritional environments. The problem is that the various types of bacteria found in an intestinal microbiome grow at varying rates in artificial nutritional environments. Some don't grow at all. As a result,

the quantities and composition of the bacteria we get in a Petri dish differ significantly from what we see in reality.

2.   The composition and numerical proportions of bacteria in different parts of the intestine differ significantly. The feces consist of an ensemble of bacteria completely different to the ones in the "work" zones of the intestine.

3.   How a sample is collected and worked with matters. If the bacterial sample spends a long time under the impact of the oxygen in the air, dries, and so on, then you will get a highly distorted picture from an analysis of it.

Modern specialists are aware of these problems and are more or less capable of combating them. However, not enough attention is paid to these factors in the majority of medical centers.

Speaking more generally, the way to take care of the intestinal microbiome is to restore its anthropic mode of existence. This can be achieved only through an anthropic diet, and not through futile attempts to supplement the microbiome with milk bacteria alien to it. Your lifestyle, and one of its most important aspects, your diet, determines the qualitative and quantitative composition of your micro flora. At the very least, this is the most important external factor that is susceptible to our influence. The usual recommendations of modern dietology do not allow us to restore our intestinal metabolism to a highly anthropic level. As already stated above, even the intestinal microbiome of a healthy modern man who has the perfect diet from the point of view of modern science (specifically, a man in the Mediterranean region), has a bacterial variety several times lower than a healthy man living in a relatively "wild" state in Africa or South America. As such, our ration has to be supplemented by highly anthropic elements of live nature. One of the main sources of such a diet are superfoods, but, with a few exceptions, not in the commercial form in which they are to found on the market currently.

This is explained by the fact that the bacterial flora in our intestines is formed, starting from childhood, under the influence of diet, among other things. Food modulates the way the intestinal micro flora's genes express themselves. During their subsequent functioning, the bacteria have a modulating epigenetic impact on each other and an epigenetic impact on the whole organism through the epithelial genes, creating a stable metabolism for the intestine and the whole organism. The microbiome that has appeared over the area of the intestine makes impossible (or at least more difficult) the penetration into it of alien forms of bacteria and its colonization by them. Such a microbiome has a high capacity to adapt to new foods and to new bacteria. This is of course the ideal. The microbiome of a modern Western city dweller does not have even a tenth of the variation in bacteria, that it should have had based on the biological nature of man. Only an anthropic diet may produce a positive shift in gut microbial metabolism lasting for a long time.

A demonstration of the significance of interaction between genetic factors and diet is celiac disease, an autoimmune condition disorder of the small intestine that occurs in genetically predisposed people of all ages and afflicts roughly 1 percent of Americans. It is caused by a combination of genetic and diet factors. Researchers from Spain [534] show that the level of risk of celiac disease influences the composition of an infant's gut microbiome, and confirm earlier studies showing that the type of milk feeding—breast versus formula—also influences the species distribution. Obviously, breast feeding is the best anthropic diet for infants. This is for two reasons. First, scientists cannot imitate in "formula" composition most parameters of a mother's milk influencing the child. The secondary metabolism of infants is especially sensitive to disanthropy of feeding. The research shows that infants at high genetic risk of celiac disease have a high prevalence of certain bacteroides that are different from the population in those at low genetic risk. Researchers found that the type of infant food influences the gut species composition, in particular with breast feeding favoring the prevalence of a species associated with the low-risk genotype, and reducing differences in bacteroides species composition between the two genetic risk groups. This is a good illustration of how dietary intervention helps to modulate the intestinal microbiome in subjects at risk of developing genetically predisposed celiac disease. In addition, according to the study [591], intestinal flora are involved in the emergence of multiple sclerosis and many other diseases. This means, figuratively speaking, that the gut is place for conversation between your body and your microbes. When that conversation gets out of whack because of disanthropy of your diet, and if you have the genes that are susceptible to diseases, then disease can occur.

Take into account also that our digestive system is home to an almost endless number of viruses. In a recent study [529], researchers have investigated the dynamics of virus populations in the human gut, shedding new light on the gut "virome" and how it responds to changes in diet. Obviously, these interactions between viruses, bacteria, and the human host have significant influence on the gut and human health, through changes in the gut microbiome. By analyzing DNA sequences from viruses and bacteria present in the gut of the patients, researchers found that although the largest variation in virus diversity observed occurred between individuals, over time dietary intervention significantly changed the proportions of virus populations in individuals on the same diet, so that the viral populations became more similar.

Even diabetes may start in the intestines. Scientists at Washington University School of Medicine in St. Louis have made a surprising discovery about the origin of diabetes. Their findings, reported in the journal *Cell Host & Microbe* [620], suggest that difficulties in blood sugar control—the hallmark of diabetes—may begin in the intestines. This study may upend long-held theories about the causes of the disease. Because insulin is produced in the pancreas and sugars are stored in the liver, many scientists before have looked to those organs for the underlying causes of diabetes. Researchers studying humans with ulcerative colitis had previously observed that colon biopsies from these patients have low amounts of fatty acid synthase. According to the authors of the study, the fatty acid synthase is required to keep that mucosal layer intact. Without it, pathogens invade wall-cells in the colon and the small intestine, creating inflammation, and that, in turn, contributes to insulin resistance and diabetes.

To finish this section, I would like to note that the processes of intestine formation are determined by the viscoelastic characteristics of its tissue. The editor's summary in the article [508] is worth quoting at length: "The human intestine, much longer overall than the human body, is tightly looped within the body cavity in a pattern that is very similar between individuals, and characteristic of species. A study of gut morphogenesis in the chick, combining cellular and developmental biology, biophysics and mathematical modeling, shows that the looping complex shape of the vertebrate gut is a simple consequence of mechanics. As a body grows, the gut inside grows faster. It is anchored at each end and suspended by a muscular sheet called the mesentery, so is forced into loops. The looping pattern is determined solely by the elasticity (of viscoelastic gut), geometry and relative rates of growth of mesentery and gut, but the various twists and turns and loops are very reproducible, occurring with the same number in the same location from individual to individual."

# 6

# Decoding the Truth

## 6.1 The apocalypse of modern food

Modern plant and protein food, consumed in developed countries in general, does not have anthropic qualities for human nutrition, due to the methods and conditions of how it is grown, the processes used during its processing and culinary preparation, and the genetic qualities of plants and animals. The diet of animals, filled with artificial feeds made out of genetically modified products with enormous concentrations of "vitamins," growth hormones, antibiotics, and other medicines and supplements, in combination with the unnatural methods of their keeping and the barbaric chemical processing of carcasses, leads in the end to meat and meat products not simply of low quality, but with characteristics dangerous to human health. An enormous number of studies reveal the unimaginable monstrosity, in terms of known and potential consequences for the lives of billions of people, of the mass production of meat and superfoods.

Contrary to popular belief, birth control pills account for less than 1 percent of the estrogen found in the US drinking water supplies, scientists have concluded in an analysis of published studies [346,347]. A report [347] suggests that most of the sex hormone—a source of concern as an endocrine disruptor with adverse effects on people—enters the drinking water supplies from other sources, mainly (90 percent) from animal manure.

Take pig husbandry as an example. Under modern husbandry methods, pigs spend their entire lives in enclosed spaces. They achieve the desired condition, a weight of 120 kg, within six to eight months. People have been wondering how to put them on a stream feeding as long ago as 1920, when it was discovered that, if animals are supplied with vitamin A and D, then they can do without sunlight or movement. However, another problem had to be solved. Piglets raised in cramped and dirty conditions often fell ill, which prevented weight gain. The invention of antibiotics changed the situation, and now substantial amounts of antibiotics are used in modern animal husbandry. Feed antibiotics are used with two aims: to combat illnesses and as a stimulator. In the latter case, they provoke the accelerated division of cells and increase the animal's adaptation to unfavorable living conditions. The danger to healthy people of the substances used to make an animal grow quickly is that these stay in the tissue of the animal, and then enter the human organism. If the piglets have been filled with a particular sort of antibiotic for a long time, then resistant strains of bacteria form in them, and these can be stable even in the face of antibiotics from another group. Resistant bacteria can spread from animals to humans, mainly through the food chain. Three of four recently emerging infections in humans originate from animals: avian influenza H5N1, severe acute respiratory syndrome (SARS), and most recently the Salmonella outbreak in Germany. As a result, certain medicines might then simply have no effect on the person. They can also stop the action against the pathogenic bacteria. For example, a strain of the potentially deadly antibiotic-resistant bacterium known as MRSA has jumped from livestock to humans, according to a study of Northern Arizona University researchers. The most powerful tool in evolution is selection. Modern farming has supplied a strong force of selection through the excessive use of antibiotic in animal production. That inappropriate use of antibiotics is now coming back to haunt us [624].

In pig complexes, everything is done so that piglets eat as much and move as little as possible. Their physical activity is entirely replaced by feeding and sleep. For this, depressants are added to their feed. Domesticated pigs, due to inbreeding, are prone to stresses and the development of infarct. Because of this they are given a heart medicine, beta-blockers, during transportation from the farm to the slaughterhouse to prevent their death from infarct. These medicines, which have numerous negative side effects, enter the human organism through the pigs. The use of such substances is dangerous for the consumer, as their trace remains might have an undesirable impact on the human central nervous system. Growth hormones have been banned in the EU as their negative impact on human health was demonstrated: for example, they can lead to oncologic illnesses. But they are not banned in the United States, and this puts the lives and health of American consumers at risk of illnesses connected to these hormones appearing.

The productivity of animal husbandry has also increased the use of nitrogen-containing protein-vitamin concentrates, wheat, bacterial and seaweed proteins, urine, and synthetic amino-acids, all to fatten animals up. As a result, non-utilizable chemicals, specifically Benzedrine, and also lipids not natural for traditional sources of nutrition, can enter our organisms through meat. At the same time, meat contains proteins that are involved in all the vitally important processes of the organism. And pork, as the cheapest variety of meat, is very popular among the public.

To a slightly lesser extent, the same applies to the production of fish. The majority of fresh fish available in stores—salmon, trout, sea bream, sea bass, carp, and other—are reared under artificial conditions with the use of artificial feeds and supplements to accelerate growth, with an enormous concentration of fish in the water. Even in those cases where the fish is reared in enclosed spaces on the edge of a sea and is, seemingly, washed over by clean, running sea water, its meat has a heavy metal content that exceeds by three to eight times the maximum concentration recommended by specialists—no, not based on dietology, but on medical experience in treating poisoning. This happens because the places where fish are artificially reared do not change their location for at least five to ten years, and in many cases, for example Norway and Scotland, for twenty-five to thirty years. In addition, many fish farms work simultaneously in places of mass commercial fish rearing. But fish excrement, which accumulates on the sea floor, has a relatively high concentration of heavy metal salts, which enter the fish organism from sea water and food under conditions of high intensity feeding. The biochemical processes of fish's and other sea creatures' internal, multi-cellular metabolism works in such a way that they expel these excess salts out of their organism in the form of the byproducts of vital functioning. Over the course of high tide days, an enormous amount of excrement accumulates beneath sea farms and, consequently, an enormous concentration of heavy metal salts builds up in the sea water. These salts enter the fish organism again from the water, but their metabolism is not up to the task of dealing with these salts in such concentration, and as a result they accumulate in the fish's tissue.

For example, specialists in food metabolism allow such fish to be consumed by school-age children only if the farm remained in the same place for no more than three years, and in a quantity of 100-150 grams once every three months. If a large quantity is consumed, some degree of poisoning is possible. The same specialists underline that consuming such food is generally undesirable.

The same applies to the production of other sea foods, in particular in Asia, where norms about the concentration of foreign substances in a quantity of water, the number of sea farms on a certain distance of coastline, and so on, are often disregarded in all thinkable and unthinkable ways. Incidentally, the same happens in such places as France, Ireland, and other countries during the rearing of oysters.

Another compelling piece of evidence is that the chemical compound triclocarban—an antibacterial ingredient in some soaps and the source of environmental health concerns because of its potential endocrine-disrupting effects—has a tendency to bio accumulate in fish, as presented in a study [375]. Accumulation occurs when any organisms take in chemicals faster than their bodies can metabolise it down and exert it (or if they cannot metabolize it at all). If a substance can be bio accumulated, especially in the case when it cannot be metabolised by the body, even minute and seemingly harmless amounts in the water can build up to toxic amounts inside the body. "Due to its widespread usage, triclocarban is present in some amounts in 60 percent of all rivers and streams in the United States," said study [375] leader Ida Flores, of the University of California-Davis. "Fish are commonly exposed to triclocarban (TCC), even though much of it is eliminated by waste water treatment plants. And the fish quickly accumulated TCC. The levels of the TCC in the fish soon after exposure were about 1,000 times higher than the concentration in the water."

Bear in mind that we are speaking of fish and other sea foods that are recommended by modern official dietology as something to be consumed often, and which are hailed as uniquely healthy. Over the last decade, the import into the United States of fish from Norway and Scotland and sea foods from countries of the Pacific Rim was frequently banned. Yet the restrictions were most often lifted as a result of political agreements and the threat of counter measures against US agricultural produce, which also are not without sin.

We have to bear in mind also that producers use supplements to fish feed which contain the same substances as are used in the commercial production of meat (growth hormones, antibiotics, etc.), plus, in the case of salmon for example, coloring to improve the fish's appearance on the supermarket shelf, which is entirely harmful to humans and has mutagenic qualities. However, growth hormones in fish and animal husbandry are themselves used by regulatory organs as proof of the safety of the fish similarly reared. The FDA's statement is priceless: "The growth hormone itself presents no specific risk, as we consume growth hormones in all meats we eat" [211]. That meat filled with these hormones presents no threat is, in my view, far too optimistic a viewpoint. The most we can say is that the specific risk of growth hormones in meat has not yet been demonstrated. But the existence of a yet-unspecified risk is entirely probable. This means that people suffer different illnesses based on the differences between their individual metabolisms. The FDA's assertion looks especially arrogant against the background of a sharp rise in oncologic and other degenerative illnesses in the last decades. Scientists have not yet been able to find a certain cause for these, but the use of growth hormones in animal husbandry stands under the most suspicion.

As with the raising of livestock for mass consumption, the normal practice in the fish-rearing industry is the use of enormous amounts of antibiotics in the feeding of fish to prevent possible infections. The amount of antibiotics compared to the weight of the fish is so monstrous that it cannot be digested by the fish's metabolism and is contained in large quantities in the tissue of the commercially available fish on supermarket stalls. At the same time, mutant forms of pathogenic bacteria and organisms stable against antibiotics appear in the actual fish bodies, as well as pathogenic bacteria that actively break down antibiotics. In a study [169], the interaction between the dove population and human population in a seaside area was used to show the frightening similarity in the antibiotic resistance of bacteria found in both populations. The transfer of pathogens from fish to the human organism through the food chain leads to irreversible consequences and is one of the most underestimated threats to modern mankind.

The situation is made more difficult by the fact that the rearing of seafood given contact between its sperm and eggs, as happens during the rearing of the majority of them on the shores of the seas in permeable enclosures, impacts the genetic fund of wild varieties of the same fish through the transfer of genetic material from artificially reared fish to wild ones. So wild salmon near the shores of Norway in the North and Barents Seas becomes less viable, with widespread mutant varieties appearing in the wild population. This has already been cause for an official diplomatic complaint to Norway on the part of Russia. Clearly, the fish farms of Scotland and other countries in the region also contribute to this. Precisely the same is happening in Asia, where such phenomena are either entirely ignored or do not have sufficient attention paid to them.

However, there is more. To prevent the transfer of genetic material from farm-raised salmon and increases in the rate of growth, scientists have started to genetically modify fish, which now have, instead of the two sets of chromosomes that are natural, three sets [92]. As a result, such fish do not breed and grow quickly. It has reached a point where you can easily distinguish a wild salmon from one reared on a farm through simple chemical methods [93]. This shows that the deviation of the metabolism of fish reared on farms from the metabolism of wild varieties has reached improbable proportions. The interaction between fish farms and the wild population is leading to a dramatic decrease of the latter [94]. In this way, all these tricks used in the cultivation of salmon, apart from threatening the genetic health of people, also reduce the availability of natural wild fish for consumption.

But the reality around the genetic manipulation of salmon and trout, a million tons of which all together are produced globally every year, is more frightening still. A study [95] proudly announces that experiments in the trans-genetic transformation of trout have led to the six-fold increase in the fish's musculature. The external view of such a fish brings to mind a bodybuilder and makes you shrink away from its ugly appearance and from the thought alone that someone might cook and offer you this monstrosity on a plate. These monsters can already be found on supermarket stalls and the completely unaware customer purchases it in dissected form, thinking that he's buying a healthy fish. No one on earth can say what the long-term health consequences are of the mass consumption of genetic freaks. In essence, this is a genetic experiment carried out on the human population.

Unfortunately, even fish and sea animals that live wild in nature often cannot be seen as entirely safe for human or animal consumption. For example, killer whales that inhabit the open seas near the shores of Norway have a concentration of chemical toxins in their tissue which is a record for sea-dwelling creatures. The concentration of various chemical compounds in them exceeds the concentration of these same poisons in the bodies of seals that inhabit the extremely polluted Arctic ports. This is evidence that certain types of metabolism are more vulnerable to the impact of chemical pollution that enters their organisms through the food web and directly from sea water. The cruel joke was played on the killer whales by what is a large evolutionary advantage of these sea mammals, their ability to survive on different foods. Incidentally, in this they are similar to man. What is interesting is that the normal killer whale life span is also close to the human: about ninety years for females, but dropping constantly over the last decades with a worrisome prognosis for the future. The carcasses of dead whales also enter the food chain and in turn increase the level of pollution in the organisms of animals and fish, which in turn increases this concentration even further in new generations of sea dwellers, including the killer whales themselves.

The extremely high concentration of polychlorinated biphenyl (PBC) and chlorine pesticides in the tissues of whales is caused by the fact that their main source of nutrition is herring. In herring we can observe a relatively high level of concentration of those chemicals that accumulate in the bodies of whales. Despite the many years' long bans on using these chemicals in industry and agriculture, there are still plenty of them in the water of the seas, rivers, and groundwater runoff. But whales of this type consume a significantly broader selection of sea organisms, apart from herring, by comparison to their sibling breeds. As a result, the concentration of these chemical substances in them is on average is more than twenty times higher than in other sea animals.

The same picture can be observed in the wild salmon population in the same region, as it also feeds on herring fry. According to one study [60], this situation, the poisoning of killer whales and salmon, will continue for the next thirty to sixty years, even if the use of pollutants dangerous to their lives is stopped.

The chemical pollution of sea water leads to serious changes in the metabolic processes in female salmon, which over the last decades have shown themselves in the form of syndrome M-74, or the increased mortality of fry from acid stress. The impact of acid stress shows itself in the changes various chemicals cause in the trajectories of fish's metabolic reactions, through their impact on the expression of genes connected to this illness. This example is relevant for our analysis in two ways. First, it demonstrates the extreme importance of pollution in changing the food chain that culminates in man and other animals, and second, and no less importantly, it shows the impact of the diet of living organisms on how the genetic apparatus of cells expresses itself. Clearly, the latter process has a universal character in nature as a whole, and is equally true for fish and for other organisms, including man.

The results laid out in a study [72] show how a changing environment leads to a cumulative growth in the concentration of chemical compounds in fish organisms, even if their concentration in the water is relatively low. Using the example of ecological systems with various amounts of algae, it was shown that the accumulation of methyl mercury (the product of the mercury metabolism in the organisms of sea animals and plants) in fish goes up to the highest concentrations, despite the fact that the overall mercury content in the water is relatively low and does not exceed typical safety norms. This accumulation is caused by the increasing concentration of mercury in seaweed alongside a decrease in the seaweed's volume in the aqua-space. This increased concentration transmits itself into plankton, which feeds on seaweed, and subsequently into the fish organism, which feed on plankton.

Mercury is but one example of this non-linear mechanism of the accumulation of concentrated chemical elements and compounds in animal organisms under ecologically disturbed environments, but it doubtless can be applied to the majority of other substances. Clearly, this hyper-concentration of, looked at earlier in killer whales, appears as a result of the indicated process of non-linear accumulation in the whale's organism due to a disturbed balance of components in their habitat. The precise details of this imbalance are not yet known, but it is known almost for certain that man is the causal factor. I would like to repeat yet again the extraordinary fact that due to this mechanism, chemicals accumulate in the organisms of wild animals despite their entirely safe levels in the water. This is what is scariest of all! I am sure that we will discover a great multitude of such undesirable long-term consequences of the growth, after World War II, of the chemical industry and methods of sea animal nutrition and cultivation over the coming years. To understand the scale of the problem and how much aquatic cultivation has entered our lives, we can look at the findings of the United Nations, which show that at the present time around half of all fish and sea foods (73 million tons per year) are reared artificially. Sixty-three percent of all aquatic cultivation in the world takes place in Asia, where it is almost impossible to monitor the quality of produce or the conditions under which it was raised.

Looking at plant products, the situation is no better and in many cases is even worse than in other spheres of agricultural production and animal husbandry. For example, grain sown on the best land and with the use of only natural fertilizer (manure, etc.) cannot be denser than 15-17 centner per hectare. The reality is different: in some European countries, the average is 85 centner per hectare. The same applies to absolutely all other agricultural cultures in the whole world, which are grown, mostly, on impoverished lands that have long lost their natural fertility. The situation around cultures grown in irrigated agriculture and with the use of irrigation is especially monstrous. Let's consider rice, more than 558 million tonnes of which are produced yearly across more than 100 countries and which is critical for more than three billion people. In regions where rice has been traditionally cultivated, the concentration of arsenic in the soil exceeds the usual level by seven to twenty-five times over. The same applies to many other heavy metals harmful to people. As a result, in countries such as Bangladesh, the level of mortality as a result of arsenic grows by 20 percent across the country as a whole, and by 70 percent in some regions [52].

This is connected to the fact that water is the most important economic component in irrigating land. As a result, it's usually re-used. However, arsenic and other harmful compounds accumulate in it from the large amounts used in chemical fertilization during its re-circulation. On the enormous territories of India, Vietnam, Bangladesh, and other countries, virtually all the underground water in the areas of rice cultivation are poisoned by super high concentrations of arsenic and other harmful compounds. A large quantity of arsenic enters rice from such water, and spreads virtually across the whole planet.

Precisely the same situation applies to rice in developed countries, such as the United States, where certain researchers assure the public that the arsenic contained in US rice is in its less deadly form {53}. In truth I can scarcely conceive of what a less deadly form of arsenic actually is. In addition, we have to note that water from third-world rice fields is often used to rear fish, which I of course recommend that no one eats. In this way, 90 percent of commercially available rice on the planet has a heightened concentration of arsenic, lead, mercury, and other minerals, in combination with hundreds of different organic pollutants. Japan is something of an exception, where the practice of processing and selecting rice and attention to ecological issues is head and shoulders above the rest of the world. The situation is the same with all other cultivated plants grown to feed man.

In addition, the recent comprehensive global review of the data regarding stature and health during the agriculture transition, published by the journal *Economics and Human Biology* [456], shows that when people started turning to agriculture around 10,000–13,000 years ago, in all locations around the globe and all type of crops, the health of the populations significantly declined. The studies included populations from areas of China, Southeast Asia, North and South America, and Europe. Farmers experienced nutritional deficiencies and had a harder time adapting to stress, because they became dependent on particular food crops, rather than having the significantly more diverse diet of hunters-gatherers. In the last century, following the industrialization of food systems and better medical service, the trend toward shorter life span reversed, and average health for most populations began improving. But it is important to note that the number of degenerative diseases in the population grew at the same time. That is, the path to better health followed during the twentieth century turned out not to be applicable to a significant part of the world. This data, taken in conjunction with other studies, confirms the idea of declining health and increasing nutritional diseases as societies shifted from foraging to agriculture.

Some economists and other scientists are using the rapid physiological increases in human stature during the twentieth century as a key indicator of better health. "I think it's important to consider what exactly 'good health' means," says A. Mummert, the leading author of one study [456]. "The modernization and commercialization of food may be helping us by providing more calories, but those calories may not be good for us. You need calories to grow bones long, but you need rich nutrients to grow bones strong."

Over hundreds of thousands of years, humans have selected and bred plants for traits that benefit us economically—bigger, juicier, and easier-to-harvest fruits, grains, and so on. They never paid attention to their health benefits. For example, the evolution of perennial plants under human influence has resulted in large changes to their reproductive biology, and in many cases perennial crops have reduced fertility under cultivation [562]. Reduction in fertility almost certainly means that those plants became less healthy. They certainly lost many benefits for humans that co-evolved during million of years of normal evolution, because you cannot expect the same benefits from diseased or degraded species. According to one study [460], the planet's soil is under greater threat than ever before. In some parts of the world, losses due to erosion are greatly outstripping the natural rate of soil formation, and the intensity of human activity is impacting the ability of the soil to produce food, store carbon from the atmosphere, filter contamination from water supplies, and maintain necessary biodiversity. Because of growing demand for food, intensification of agriculture alone will put a huge strain on soil over the next few decades, and climate change adds to the challenge [461]. If we add the wide spread of genetically modified cultures and the horizontal transfer of genetic information that takes place as a result of cross-pollination between different varieties of plants, then we can see a picture of a genuine food Armageddon in the modern world forming quite clearly. It is very naive to expect health benefits from such products.

The consequences of the latest direction in genetic modification, the cisgenic modification of agricultural cultures and animals, are also unknown. The essence of this method of genetic manipulation is to swap areas of the genome between sibling species that can, in principle, interbreed. The evangelists for this method state that they are only accelerating the natural evolution of these species and so their experiments are much safer than ordinary genetic modification. But who knows? In my view, given the current level of knowledge and technological inability to predict the consequences, any genetic manipulation has a chance to produce monsters that will always remain with us. Ordinary genetic modification has already left us with only genetically modified forms of soya and other cultures across the whole planet.

What is more, in its attempts to improve the commercial qualities of its products, the food industry extensively processes raw food, exposing it to temperatures of between 140–300 degrees Celsius, to high pressure, and to sheer stress—dissolving it, refining it, and exposing it to the action of alkalis and other chemical impacts. As a result of these actions, the healthy qualities of raw food disappear almost completely and new disanthropic chemical compounds, unquestionably harmful for man, appear. The problem is worsened many times over by barbaric practices that exist in the majority of world countries, such as the addition of physiological solvents, artificial coloring, and preservatives to raw meat and fish, and the even more monstrous additives used in preparing salami, pates, and other fish and meat-based products.

Sometimes even preparing a food at home can cause harm. For example, according to data [77], frequent consumption of the traditional Ukrainian soup, the borsch, is dangerous to your health and can become one of the causes of the development of oncological illnesses. This is explained in the study [77] by the fact that all the vegetables used to prepare this Ukrainian dish ordinarily contain an enormous number of nitrates. Moreover, while consuming some of these vegetables in a raw state doesn't hold a serious threat to the organism, during this soup's lengthy preparation, the nitrates in the vegetables undergo complex chemical reactions and become highly dangerous, including becoming nitrites. It goes on to state that we shouldn't trust even vegetables we've grown in our own garden without using fertilizer, as in our time the whole soil and the underground waterways in most developed countries are filled with nitrates.

But some researchers irresponsibly claim that the nitrates in vegetables and plant cultures grown on fertilizer are even beneficial. A recent study [339] claims that there is a benefit to nitrates and the nitric oxides that stem from them. The authors of this study insist that increased dietary nitrate can have a beneficial effect for muscles. But they don't make clear what might happen to people who consume higher levels of such disanthropic substances as inorganic nitrate over long periods of time. It is clear enough that the long-term impact on their health will be negative, mainly because nitrates and nitrites are very toxic substances in themselves and because they tend to form chemicals called nitrosamines, some of which are carcinogenic. Such a large intake of nitrites and nitrates poses big health risks in the long term.

Because of everything indicated above, irrespective of whether he wants to or not, modern man consumes food produce (meat, fish, plants, and their produce) that don't have their initial healthy qualities or have only a small proportion of them. Indeed, they are often so harmful that they are virtually a slightly poisonous substance. In other cases they are harmful because their organism's genomes differ from the genomes of the natural population, which can lead, through the expression of the wrong part of cells' genomes, to the synthesis of undesirable toxic compounds in animal organisms, and to their subsequent bio-active interaction with the human organism with entirely unpredictable long-term consequences.

The various changes in protein nutrition of plant or animal origins impact the human organism through the cumulative effect of numerous simultaneous disruptions in the trajectories in the phase space. This applies both to the initial toxins that come from disanthropic food and to their interactions, described in the preceding chapter, with the organism that lead to cascades producing n-toxins. The impacts on the organism's metabolism of the appearance of second-order interactions within the web of second-order toxins, which undoubtedly exist, have hardly been studied at all. The intoxication of the organism which takes place as a result of prolonged (over the course of the individual's life) consumption of such products leads to the full spectrum of the results of slow poisoning, the main ones of which are weak, sickly descendants, a relatively low life expectancy, genetic changes at the level of cells, heart diseases, diseases of the inner organs, and cancer. For example, as was shown in one study [42], a childhood diet based on the frequent consumption of burgers leads to asthma, allergies, and so on. However, the disanthropic contents in many children's diets in the modern world do not end with burgers and fries. This is why, it seems, various childhood diseases are spreading across the planet at the rate of pandemics.

For example, according to research by Finnish scientists, sperm quality has worsened significantly over the past few decades, and simultaneously the incidence of testicular cancer has risen. According to a study [371], impaired semen quality and testicular cancer may be linked through a testicular digenesis syndrome of fetal origin. The incidence of testis cancer has been shown to increase among Finnish men born between 1979 and 1987. Scientists have focused on Finnish men as it was found in the past that their sperm has the highest concentration of spermatozoids in the world. The researchers came to the conclusion that the blame lies with the impact on men of negative factors in their environment, and specifically with industrial and day-to-day chemicals that get into a mother's organism during pregnancy and then into the child's organism  through food, among other things. A greater incidence of testicular cancer was found in men born in the 1980s compared to those born in the 1950s. As a result of the nooevolutionary activity of man, huge changes are taking place in the planet's bacterial world. Global resistance to antibiotics is just the visible part of this iceberg. The overall impact of the planet's bacterial flora on the food chain that ends with man is absolutely unstudied. But it is obvious that this impact has to be very great indeed.

Another powerful and potentially very dangerous source of the disanthropy of food is the industry around nanotechnology. Scientists are having significant success in creating nano-materials: nano-paints, metals, and other chemical elements and compounds in nano-form. The essence of this process is the creation of stable and size-differentiated substances, with an area of between 1 and 1,000 nano-meters, that can easily enter the organism. However, the size-differentiate we have in nature is such that any one size is also differentiated by a spectrum of other factors. The sum size of substances in each nano-diapason is relatively small. However if the entire volume of a substance is concentrated in the same area, as happens if all the particles have the same nano-size, then their sum surface area grows in proportion to the number of particles involved. This, firstly, changes the area of interaction with their environment and increases the rate of reaction at least in proportion to growth in the area of interaction. In addition, nano-particles enter entirely different areas of the organism due to their strong tendency to disperse. Finally, some size diapasons have a greater impact on health than others.

All this can lead to the strongest disanthropic impact of even relatively inert or overall anthropic substances because accelerated reactions, say by 1,000 times, change the situation significantly. This is even truer for toxic materials. Moreover, the toxicity of materials in nano form can exceed their toxicity in their normal state several times over. Also, the appearance of new dangerous toxic substances, related to the toxicity inherent in small particles, is likely. So in a study [360] published recently in the journal *Environmental Health Perspectives*, scientists showed that, in people with diabetes, breathing in nano-particles can activate platelets, cells in the blood that normally reduce bleeding from a wound, but can contribute to cardio-vascular disease.

At the present time, nano-particles are used in various areas of technology and in day-to-day life. Automobile and other paints based on them are already in fairly wide circulation, and nano-particles are used in food technologies, agriculture, and so on. Overall, nano-particles are already involved in more than 500 day-to-day products [116], including food [117], with 3–4 new products entering the market weekly [118].

A recent study [621] examined how large doses of polystyrene 50 nm nanopar-ticles—a common, FDA-approved material found in substances from food additives to vitamins—affected how well the gut absorbed iron, an essential nutrient, into its cells. The researchers tested both acute and chronic nano-particle exposure us-ing human gut cells in petri dishes as well as live chickens and reported matching results. They chose chickens because these animals absorb iron into their bodies similarly to humans, and they are also similarly sensitive to micronutrient defi-ciencies. High-intensity, short-term exposure to the nano-particles initially blocked iron absorption, whereas longer-term exposure caused intestinal cell structures to change, allowing for a compensating uptick in iron absorption.

Companies producing food often do not inform the general public about the use of nano-particles in their food products [119]. They nonetheless state, without even blushing, that in making food on the basis of nano-technology, they will be able to effectively deliver the beneficial ingredients in food to the right organs and tissues in the consumer. It is entirely possible that they will manage to deliver some ingre-dients to some organs. But along the way, they will also deliver them to where they are entirely unnecessary. Moreover, at the current time, we lack precise knowledge about the benefit of these or those specific ingredients, and are becoming increas-ingly aware that the healthiness of food comes from the unbroken interaction of all its ingredients with the numerous metabolic processes in the organism, which are still not entirely understood, and the full understanding of which is still a long way off. The impact of nano-particles, which can penetrate deep and unpredictably into our organisms, is absolutely unstudied. It is clearly mostly negative, due to the dis-anthropy of their size and their physical and chemical interaction with the organism. But the experiment in vivo on the whole of humanity has already begun. This greatly resembles the development of the chemical industry in the twentieth century.

Alongside economic and technological development, the chemical industry has also given us an enormous negative impact on people's quality of life across the whole world, due to ecological problems and our use of chemical products. Entering sea water alongside the byproducts of human life, nano-particles, after going through the oceanic metabolism, can take on truly monstrous qualities, of the sort we have not yet witnessed. The lesson is this: we have to learn to design technology for the application of nano-particles more intelligently and to take adequate actions in order to prevent potential hazards. For example, researchers at Brown University found [518] that nano-particles of nickel activate a cellular pathway that contributes to cancer in human lung cells. Getting down to the nanoscale made the nickel parti-cles more harmful than in the usual state of this metal and gave them cancer-causing properties. The reason obviously is that for the same mass, smaller particles in total expose a much bigger surface, and that makes them more chemically reactive. The results of the study [518] show a big difference in how nickel nano-particles and nickel oxide nano-particles react with cells. The nickel oxide nano-particles are so lethal that the cells exposed to them died quickly, leaving no opportunity for cancer to develop. Metallic nickel particles, on the other hand, were less likely to kill the cells, but still did some damage. Metallic nickel nanoparticles caused sustained activation, but they are less toxic, so damaged cells survive. As such they are more likely to be transferred to the cancer state. That is the reason for, as established in Section 4.3 of this book, the principle of critical culling—less toxic particles' ac-tion in the cell might, paradoxically, transform cells to a much more deadly state. The dangers of using nano-particles do not end here, moreover. We are now bearing witness to the birth of a new sphere, known as nano-agriculture, which uses nano-technology to boost the productivity of food plants or plants for animal feeding. But as scientists admit now, there exist huge gaps in knowledge about the effects of nano-particles on food crops and consequently on humans. Researchers from the University of California Center for Environmental Implications of Nanotechnology, and colleagues at The University of Texas at El Paso, note [491] that nano-particles are used in products ranging from medicines to cosmetics. The particles also end up in the environment, settling in the soil, especially as fertilizers, growth enhancers, and other nano-agricultural products hit the market. Some plants can take up and accumulate nano-particles [492]

Food that contains nano-particles and their metabolites could turn out to be uniquely dangerous for man and animals. According to a review [519], scientists at the Centre of Cancer Biomedicine in Norway showed that uptake and accumulation of nano-particles in cells can disrupt important intracellular transport pathways. The researchers found that the particles interrupt the transport of molecular substances in and out of a cell, causing changes in the cell's physiology and disrupting normal cell functioning. The likely explanation is that nano-particles of a certain size either cannot enter via the very thin tubes in the cell organelles, or if they enter, they then lodge inside and plug them up. Researchers at the National Institute of Standards and Technology (NIST) have provided the evidence that engineered nano-particles are able to accumulate within living cells and damage their DNA [643].

The impact of such disanthropic factors as nano-particles, among others noted above, shows itself in different ways on the level of the population as a whole. So, for example, the widespread hormonal imbalances of a disrupted metabolism that lead to insomnia also lead to heightened mortality for sufferers from this ailment [38], and consequently have a significant impact on the state of the population in general through the increased frequency of car crashes and other catastrophes caused by human error. There is no number to such impacts, the majority of which are unknown to us. Beside this, there exists a strong tendency to use nano-particles in medicine. But the challenge is that even clinical studies carried out on patients with chronic diseases will not provide the whole truth. The negative effects of a nano-medicine may not show up in a short-term study, but patients who use that medicine over many years to fight a chronic disorder may end up exhibiting an over-occurrence of certain cancer types due to the nano-particles being incompletely excreted, disrupting transport in the body's cells [519].

I should also note that the characteristics of human metabolism are inherited from much more remote ancestors than is usually assumed. A study [577] of evolutionary alterations in gene expression in chickens, frogs, puffer-fish, mice, and people has revealed surprising similarities in several tissues. Researchers have shown that expression in some not-highly-specialized tissues (i.e., with a limited number of specialized cell types) is strongly conserved, even between the mammalian and non-mammalian vertebrates. The authors studied 3,074 genes that were present as a single unambiguous copy in each of the five genomes; they noted among them clearly strong evolutionary constraints on tissue-specific gene expression. In many genes, these constraints go back so far as to show conserved human/fish expression. This suggests the existence of a basic ancestral pattern of expression in each tissue, the so-called inner fish in all of us.

But what does all this mean in terms of human metabolism? I think it certainly means that the details of metabolism have very deep evolutionary roots. Therefore, the deformation trajectories of metabolic reactions, developed over hundreds of millions of years, solely have a very strong effect on functional status through modern disanthropic, and hence the health of cells and tissues of the body. It is hard to imagine that gene expression patterns formed over a hundred million years can be adapted to all innovations of nooevolution, such as chemical and industrial pollution, genetic and biochemical changes from natural status in the products of modern agriculture, and so on.

I should also mention the potentially harmful effects of a huge fraud on the public health. The Italian newspaper *La Republica (December 26, 2011)* claims Italy's €5 billion olive oil business is a fraud. Citing an ongoing investigation by different police agencies in collaboration with the Italian agriculture trade group Coldiretti, it says 80 percent of the country's oil carrying the label "Extra virgin" and "Made in Italy" is mixed with lowest quality Spanish, Moroccan, Tunisian, and Greek oil (with a 25-Euro-cents-per-litre price tag). Some producers, when exporting to not-so-sophisticated markets, such as ex-Soviet countries, may also mix in "Lampante oil," which is olive oil not suitable for use as food. Italy exports 250,000 metric tons of oil per year, but imports a total 470,000 tons. Last year oil imports rose by 100,000 tons, prompting authorities to ask where it ended up. Characteristically, the recent scandal in Italy surrounded a large scale lie about food that was claimed to be ecologically clean. According to the information from law enforcement, criminal groups were exporting these products from Italy to many other countries, and profited 220 million Euros from this scam over several years.

## 6.2 Superfoods' glitter isn't gold?

As has already been noted in the preceding chapters, nutrition—i.e. the metabolism of food—is the complex process of transforming elements of food in the organism with the aim of acquiring the energy needed to maintain vitality and the substances necessary to the construction and functioning of the organism. The ideal food contains anthropic substances essential to the metabolism in the correct phase states that ensure that the genetically determined minimum level of entropy is attained. As a result of the complexity and variety of the metabolic processes in the organisms that are a source of food for humans, it is virtually impossible to determine anthropy through a reductionist approach by simplifying all the various interactions of thousands of substances and their phase states to the assortment of chemical compounds to be found in any one food. The data presented in preceding chapters about vitamins and DSs allows us to conclude that such an approach does not lead to any noticeable steps in the direction of healthy eating. The anthropy of a food shows itself in the broad spectrum of material signals that enter the phase space of metabolic reactions and augment the trajectories of reactions that are, for the most part, determined by the organism's genotype and, to a slight extent, the phenotype of its lifestyle, including aspects of its environment. The food entering the organism is also a strange attractor, both as a whole and as its component elements. Apart from this, the anthropy of a food is also determined by the interaction of all its elements and metabolites in the inter-linking cascades of biochemical reactions in the phase space of the strange attractor of the organism's metabolism.

This theory is well illustrated by the situation around the production of certain medicines and substances that are vitally important for man. For example, diabetics use insulin produced by bacteria. All modern biochemical knowledge is still incapable of synthesizing a consumable product under laboratory conditions using purely chemical methods. Turning this function over to bacteria, and using them as a bio-reactor, the detailed working of which is still unknown to scientists, nonetheless allows the production of insulin of a fairly high quality, close to the human. Moreover, a study [89] showed that more complex molecules, for example human proteins, have to be produced in bio-reactors that are more complex in their set-up—i.e., organisms more complex than bacteria. Protein, which exists in the human organism in the blood's immune system, can be effectively produced, using methods developed in this study [89], in bio-reactors based on specifically created mosses. In this case the strange attractor of the mosses metabolism can be used to produce compounds suitable for the strange attractor of the human metabolism. I think this is a majestic way to create new effective medical products—even not knowing the details of how the strange attractor of the mosses' metabolism works, scientists can still make good use of it. And in this case, "details" means thousands of the most complex molecular transformations that modern science will not be able to discover, generalize, and most importantly re-create within the foreseeable future.

As already noted above, it is most accurate to look at the combined strange attractor of the human metabolism and the incoming food. Because of the unimaginable complexity of the processes of biochemical interactions, it is impossible to single out individual strands to effectively control the metabolic processes of such an attractor. However, this does not mean that affecting the state of the organism with some sort of chemical substances is entirely futile. For example, in the case of medicine, it is entirely justified to use them to correct certain extreme states of the organism. However, such an approach is entirely fruitless when it comes to maintaining the health of the organism at its natural level and to correcting illnesses caused by a disanthropic diet and lifestyle.

Given the actual conditions of modern life, we consume foods that, taken together, do not have many anthropic compounds and, at the same time, have a high content of disanthropic ones. This means that the phase state of the attractor of the incoming food or, put differently, of the nutrition is significantly altered. This altered attractor alters the state of the strange attractor of the human metabolism and leads to the development of ill states, which lead to degenerative illnesses and an early death. Prolonged—i.e., over the course of many years—incorrect nutrition leads to an especially wasted state of the human metabolism, the pollution of the organism with toxins. This state can be corrected if we consume highly anthropic food in combination with leading an anthropic lifestyle.

So what can be a source of highly anthropic food for modern man? Sadly there are very few such sources left in most areas of the world. In earlier sections of the book, we dealt with the systematic problems in the production of meat and fish, and the creation of agricultural cultures. But this is only one side of the coin. Modern agriculture over most of the planet will not be able to produce anthropic food, even if we immediately cease using medicines, hormones, and the entirety of agricultural and industrial chemicals that are poisoning animals, plants, and the environment. The reasons are as follows:

1. The prolonged (up to fifty years) impact of trace chemicals in soil and water leading to toxins in agricultural produce.
2. The overall impoverishment of the majority of the soil in traditional farming areas.
3. The insufficient speed of metabolic processes in the soil and surrounding environment, and the relatively brief yearly vegetative period in the mid-latitude band, where the majority of the agricultural production in modern developed countries takes place.
4. The insufficient productivity of sea resources due to their excess exploitation over the past few decades.

It is my deep conviction that only the products currently called superfoods can assure the provision in the diet of the maximum possible complex of anthropic food elements. What then are superfoods? In this work, this term is applied to all consumable products of live nature that have the bio-organic and mineral contents needed for an anthropic diet. During the time of hunter-gatherers, and indeed for modern peoples who get their food in a similar way from untouched nature, all foods consumed were and are superfoods under this definition. For example, bushmen living in wild nature under extremely unfavorable conditions nonetheless consume a diet that allows them a long life span and a capacity for physical exertion far in excess of what modern representatives of the developed world are capable of.

Or consider a study [182] that demonstrates another example of the miracle of wellbeing on a very restricted diet. This study recently found new proof of that miracle in the nomadic Maasai people of Kenya in Eastern Africa. The nutritionist, from the Friedrich Schiller University Jena (Germany), analyzed the diet of a nomadic tribe in the Kajiado District. The surprising results of the field study show that the Maasai are in good health in spite of a very limited diet.

The development of humanity and the need to feed a growing population, resulting in the farming of relatively infertile lands, plants grown very densely, or animals reared in large numbers, has led to the four main reasons outlined above why humanity cannot obtain anthropic food. However, scientists noticed long ago that the populations in certain regions of the planet have an average life span much longer than in the developed countries. Analyzing this information has allowed the scientists to pinpoint what the common factor is among the various long-living populations. It turns out that their diet that is key; this category of people consume much more of the products labeled superfoods. The richest and most stable sources of superfoods in wild nature at present are the tropical and sub-tropical regions of the continents and oceans. This is connected to the fact that the combination of a high intensity of solar radiation and the great biological diversity in these regions accelerates the natural metabolic processes of the environment several times over compared to the mid-latitude band. Strictly speaking, though, the biological diversity is itself the product of the intensive metabolism of the environment in the broadest sense, including the virus, bacteria, plant, and animal worlds of these places.

Solar energy, given sufficient moisture, transforms living material in numerous ways, and this leads to its diversity in the tropical forests. The Amazon basin has the planet's broadest diversity and variety of all five forms of life: animals, plants, fungi, protista (eukaryotes), and monera (bacteria), plus the viruses inhabiting them. The tropics are the motor of biological diversity on Earth. The biological diversity in the tropics is maintained thanks to relatively high temperatures. Due to the kinetics of chemical reactions being determined by temperature, the speed of metabolism in organisms and in ecosystems overall is much higher than in the organisms and ecosystems of the mid-latitude band regions of the planet. This leads to quickened reproduction and, consequently, to quickened evolution of new species and of new traits within species at all levels of living matter. It also leads to the large sum activity of the vital processes and an increase in the amount of nutritional material, which is sustained by the large number of food chains in the tropical environment.

As an example, the analysis of the antioxidant content of wild blueberries that grow in tropical areas of Mexico and Central and South America concludes that these fruits have many more beneficial antioxidants than the blueberries—already renowned as super fruits—grown the United States. These extreme super fruits from the tropics could provide more protection against heart disease, cancer, and other conditions, the report [532] suggests. The same applies to the biological diversity in tropical seas and oceans, where the bacterial forms of life predominate. Due to the large variety and mass of bacteria in the ocean, they have a noticeable effect on the exchange of hydrogen dioxide between the ocean and the atmosphere and, consequently, on the global biochemical cycle between the atmosphere and the ocean, which is an important part of the Earth's metabolism. What is especially important is the state of the ecosystems of the deep regions of the global ocean. The results of a study [184] suggest that the conservation of deep-sea biodiversity can be crucial for the sustainability of the functions of the largest ecosystem on the planet. Deep-sea ecosystems' functioning involves several processes, which can be summarized as the production, consumption, and transfer of organic matter to higher levels of the food chain, the decomposition of organic matter, and the regeneration of nutrients. This ecosystem is crucial as it supports the largest "biomass" of living things on Earth and the global ecosystem. Researchers found that a higher biodiversity supports exponentially higher rates of ecosystem processes and an increased ability of the ecosystem to exploit the available energy in the form of food sources with which those processes are performed. This is especially important for tropical sea regions. We can definitely see a link to tropical ancestors in the origins of virtually all maritime organisms. Now too we see the tropics' impact on the state of species in other regions of the world, maintaining biological diversity through countless links in the ecosystem. More than half of the animal and plant world, and an even greater proportion of all varieties of microbes and viruses, inhabit the relatively small tropical territories.

If the ecology of large areas of the tropics is destroyed, the destruction of the diversity and evolution of life forms across the whole planet will follow. There are more evolutionarily ancient forms of organisms and more genetic diversity within a species for all genetic lineages in the tropics than anywhere else. Disruption of the ecological balance in the tropics leads to the loss of much more genetic diversity than we would expect from the loss of any given species of living organism in the mid-latitude band of Earth. Bearing this in mind, the task of maintaining the tropic ecosystem is clearly a global one, if humanity does not want to experience a multitude of problems in the near future. The biological diversity of tropical life forms is truly a genetic treasure gifted by nature to man and, without any exaggeration, the basis for the survival of civilization. To recreate the disrupted ecological systems of the tropical forests is a task that could never be fully accomplished. As has been shown in one study [57], even after a 100 years have elapsed since a tropical forest was regrown, the diversity of species and genetic diversity of the recreated ecosystem in this second forest remains at a noticeably lower level—several times lower, in fact. The influence of man in general leads to a more homogeneous biotope. Homogenizing an ecosystem leads to its simplification and loss of biodiversity. In a study [183] based on the evolutionary approach, it was shown that the biodiversity of a system reaches its maximum level only given a maximally heterogeneous environment with internal biological links.

The results of a study [455] on the evolution and ecology of the Amazonian basin suggest that the incredible and fragile biodiversity of species in the Amazon took more than 50 million years to develop. The patterns of life diversity are explained predominantly by the timing of colonization of the region. If the Amazon rainforests are destroyed and the species are driven to extinction by human activities, it may take millions of years for this incredible level of biodiversity to return.

Our planet is colder and moister than it might have been due to the impact of tropical plants. One study [95] demonstrated the impact that the physiology of flowering plants has on regulating the climate in the wet jungles around the Amazon basin. The density of leaf veins is much higher in flowering plants than in plants that don't have flowers. This is important, as it assists the absorption of $CO_2$ for photosynthesis and water loss. The plant cannot obtain carbon dioxide without losing water. However, the high density of veins ensures that water is transported from the soil to the atmosphere, from where it returns highly efficiently in the form of rain. The water cycle's pace is determined by the speed of the process of water transfer between the soil and the atmosphere. This transfer would be much slower without flowering plants. Studies on the processes of transfer, carried out using the most advanced computer technology, showed that the absence of flowers in the Amazon would have led to a 40-percent decrease in the overall volume of dew in North America. Needless to say, the climate and the state of the biosphere there and in other parts of the planet would be completely different. This is another weighty reason to pay special care to preserving the biological diversity of tropical forests.

The other side of the coin is that humanity's actions are destroying the plant and animal world of the tropics and of the whole planet. Life on Earth is being extinguished as fast as ever, despite a commitment from world leaders to bring a reduction in the rate of destruction, a UN study has found. Mankind's ecological footprint has continued to grow rapidly since 2002, when 190 countries, including Britain, agreed that by 2010 they would have slowed the rate at which animals and plants were being driven to extinction. Only token efforts have been made to preserve species by designating more national parks and sanctuaries, according to the study. Animal populations have fallen by 31 percent since 1970, living corals by 38 percent, and mangroves and sea grasses by 19 percent. The annual rates of loss have shown no improvement since 2002, according to the study of more than forty international monitoring systems. A fifth of the vertebrates on the planet are on the brink of extinction. This conclusion was stated at the United Nation summit in Nagoya, Japan, by specialists working to determine which animals should be entered into the Red Book after analyzing data about 25,000 species. The researchers concluded that the worst situation at the present time is with amphibians, 41 percent of the species of which are about to disappear. The biodiversity of birds is holding up the best—only 13 percent of such creatures have been entered in the Red Book, according to BBC News. Scientists called the current extinction the sixth largest the planet has seen (the fifth took place around 65 million years ago and led to the end of dinosaurs).

However, the latest findings of some scientists show that the danger of certain species becoming extinct is hundreds of times greater than was previously thought. Brett Melbourne, a professor of ecology at the University of Colorado, states the current models are excessively optimistic, as they fail to take into account several important parameters for measuring the state of an animal population. Several species have died out since the 2010 biodiversity target was agreed at the 2002 summit in Johannesburg. Within months, the last two wild Hawaiian crows disappeared, and the last Polynesian tree snail died that year. The only known St. Helena olive tree died in December 2003 and the Kihansi spray toad died out in the wild in Tanzania in 2005. The main causes of species loss were all linked to human activities, including habitat destruction, hunting, the introduction of alien predators, the spreading of disease, and climate change. The study also found a rapid increase in the number of endangered species. The International Union for Conservation of nature (IUCN) reported recently that 21 percent of all known mammals, 30 percent of amphibians, 12 percent of birds and 70 percent of plants were under threat. The IUCN calculated that the rate of species loss was up to 1,000 times greater than the natural background rate. In addition, the loss of biodiversity—from beneficial bacteria to charismatic mammals—directly threatens human health. That's the conclusion of a work [291] published in the journal *nature* by scientists who study biodiversity and infectious diseases. Species losses in ecosystems such as forests and fields result in increases in pathogens—disease-causing organisms—the researchers found. The animals, plants, and microbes most likely to disappear as biodiversity is lost are often those that buffer infectious disease transmission. Those that remain tend to be species that magnify the transmission of infectious diseases like West Nile virus, Lyme disease, and Hantavirus. "We knew of specific cases in which declines in biodiversity increase the incidence of disease," says F. Keesing, an ecologist at Bard College in Annandale, New York, and the lead author of the paper. "But we've learned that the pattern is much more general: biodiversity loss tends to increase pathogen transmission across a wide range of infectious disease systems." In the case of Lyme disease, says co-author R. Ostfeld of the Cary Institute of Ecosystem Studies in Millbrook, New York, "Strongly buffering species like the opossum are lost when forests are fragmented, but white-footed mice thrive. The mice increase numbers of both the blacklegged tick vector and the pathogen that causes Lyme disease." Scientists don't yet know, Ostfeld says, why the most resilient species—"the last ones standing when biodiversity is lost"—are the ones that also amplify pathogens. But preserving natural habitats, the authors argue, is the best way to prevent this effect.

Global biodiversity has declined at an unprecedented pace since the 1950s. Current extinction rates are estimated at 100 to 1,000 times higher than in past epochs, and are projected to increase at least a thousand times more in the next fifty years. Expanding human populations can increase contact with novel pathogens through activities such as land-clearing for agriculture and hunting for wildlife [292]. Perhaps this is one of the aspects of that "revenge" against mankind carried out by the cybernetic device that, according to my speculation earlier (Section 3.3), is the general biocoenosis of the planet. S. Butchart, the lead author of the UN study published in *Science*, said, "Our synthesis provides overwhelming evidence that governments have failed to deliver on the commitments they made and the target has not been met. We found that the natural world is continuing to be destroyed as fast as ever. Governments have made some efforts, such as designating national parks, but the responses have been woefully inadequate and the gulf between the threats to biodiversity and government actions is growing ever wider." The population of tropical species is decreasing quickly, and humanity is consuming 50 percent more resources than the Earth can create, according to a World Wide Fund for nature (WWF) report released in October 2010. The "WWF Index," which assesses the state of the populations of almost 8,000 of more than 2,500 species of animals, plants, and fungi, shows that since 1970 the tropical species have suffered the most: their population has decreased by 60 percent in less than 40 years. It is clear enough that Protista and Monera suffer to the same, if not to a greater, extent. By 2050, the human population will have increased to 9 billion. The WWF report underlines the blow the uncontrolled use of resources and energy will deliver to the production of food—i.e., the same questions as we dealt with at the start of this book. The biological diversity of, for example, tropical forests have relevance not only for climate and ecology, but also indirectly to a sphere as important for mankind as pharmacology. I have already mentioned that more than 50 percent of modern medicines and a much larger proportion of those medicines that actually do some good are made from substances extracted from the plant, bacterial, and animal worlds of the Amazon basin. I am sure that this proportion will only grow.

Nonetheless, superfoods are also produced by nature in other regions of the planet, where the conditions exist for the concentration of energy and substances. These include the Namibian savannah, the forests of equatorial Africa, and the mountainous territories in the Tibetan plateau. There are also superfoods to be found in regions of uniquely high biological exchange, for example in upper reaches of spawning Pacific salmon in the North American continent, where a powerful mass transfer of biological material from the sea onto land takes place. This gives a spur to intensive metabolic processes over the whole regional ecosystem, from large animals to soil bacteria. Similar processes are taking place in the taiga forests of certain regions of Siberia and Gorny Altai.

The process of the mass transfer of biologically active substances can also result in localized accumulation in certain semi-deserts, for example those of Namibia and Tibet, due to the specific qualities of the ecosystem and food chains there. Similar processes also take place in the taiga regions of Siberia and Gorny Altai. Under the conditions in southern Africa or in high mountains, even rare vegetation can concentrate the animal world around it and through this symbiosis of living forms turn solar energy into energy for biological systems in various ways. The examples of the excellent superfoods that exist and grow in the bushmen's areas of habitation and on the Tibetan hills confirm this. Overall, virgin nature provides the conditions for the production of uniquely healthy plants that can provide a certain number of people with excellent nutrition with all the essential elements. The concentration of the bioenergetic potential of the surrounding nature is essential to producing plants with the qualities of superfoods even in the uniquely favorable conditions of the tropical forests. That is, if plants similar to each other grow with high density anywhere in this region, they do not, as a rule, possess the qualities that superfoods do.

Indeed, a high plant density impoverishes the soil when it comes to the specific substances consumed by these plants. At the same time, many substances in the soil are left ignored by the plants, as they are not needed in the metabolism of growth and existence. Due to the symbiotic growth of different plant varieties together, they can maximally utilize the biochemical potential of the soil and other elements of the environment. Different plants compete less with each other for the various chemical compounds in the soil. And at the same time, after transforming the chemical compounds in the soil into other, bio-organic ones, they provide each other with these either directly or through complicated chains of transformations with involvement of animal and human metabolisms. The synergy of such relationships is of benefit to all the individual living organisms and to the ecosystem as a whole. It is purely because of this that 50–60 percent of modern medicine is produced from raw plant material obtained from the Amazon forests.

Some additional factors promoting the anthropic qualities of superfoods are the processes of natural fermentation that happen alongside the growth, ripening, and storage of plants and their fruit. Humanity, during its biological evolution, consumed mass amounts of many plants and fruits not only in a fresh state, but also by picking them up after they had ripened or after harvesting and storing them for a relatively long time. In this case, the processes of fermentation carried out by the enzymes in the fruits or plants or by bacteria that entered them after they fell or after they ripened on the branch, or after storage and primitive processing, have an extremely important impact on the anthropic qualities of the finished products. The human organism is evolutionarily adapted to the consumption of food fermented to some extent just as it is to that of fresh food. During the time of hunter-gatherers, when men hunted, women and children evidently collected the grasses, roots, and fruits around them—everything that could be consumed as food. What they collected could not be stored for long, but our intrepid ancestors noticed that plants that were dried out by the sun or wind, whilst they did not have the same succulence as those fresh off the branch, were sweeter and kept longer. So the moment that man, after gathering plants and fruits, dried them out and fermented them naturally can be called the birth of the modern food industry. Happily, in those times the processing of food was not very extensive, stopping well short, for example, of creating chips for a supermarket. Otherwise, I am sure that the very existence of mankind and our emergence as a species would have been very much in doubt.

The processes of fermentation are also important for man as anatomically and physiologically we are neither purely carnivores (meat-eaters) nor are we 100-percent vegetarians. Our digestive tract is adapted to the slow digestion and absorption of the elements of food. The intensive digestion of raw meat requires a shorter intestine and digestive tract overall, as all carnivores have, to allow the quick transit of food through the organism. This is because the raw meat consumed by carnivores breaks up and secretes toxic substances rather quickly, so it is desirable for the process of digestion to take place over a relatively short time frame. Moreover, anatomically we do not possess the specially developed teeth and jaws typical of carnivores. In addition, our intestinal microflora differs significantly from the carnivore micro flora, although it also differs from the micro flora in animals that consume only plant food. As such, it would appear that our ancestors consumed meat and plants in a certain state of fermentation and in a slightly cooked state after the invention of fire.

Yet we also lack the enzymes necessary to the extensive processing of plant food and the extraction of sufficient energy from it. This indeterminate state of man is what determines what the anthropic diet is—a little of everything. From the start, man has needed the maximum variety of raw or slightly fermented animal and plant food. He cannot digest large volumes of plant products to maintain the material and energy balance of the organism. Also the process of fermentation in the human digestive tract intensifies if it has a large variety of components of animal and plant origins, which in turn improves and quickens its digestion. In principle the fermentation of food is, in many ways, the transformation of its biochemical contents in the same direction as produced by its cooking over fire. However, during fermentation under natural conditions, the processes of transforming proteins and other components are relatively slow, using a lot less energy than if fire is used. In this case, the low energy breaks up and transforms the molecules of bio-polymers. If fire is used, high-energy reactions cause the destruction of firm intra-molecular links and consequently the creation of stable compounds with active radicals—substances with a high capacity to acidify the bio-polymer compounds of the organism and cells. As is known, these substances have a great impact on the life and metabolism of cells and the organism as a whole. This acidification can lead to the premature aging of individual cells and organs and, finally, to the premature aging of the organism as a whole. Our ancestors did not cook their food as intensively as people are taught in modern culinary schools. In this way, it is entirely clear that in terms of anthropy, only plants and animals that have lived and been naturally fermented in wild nature, under conditions of a natural ecological balance (understood as widely as possible), can be counted as superfoods.

It is not the object of this work to describe in detail all superfoods available on the planet. This has been accomplished in a number of works by other authors. For example, David Wolf described in his book Superfoods [96] the effects of the main plant superfoods. There is also literature describing superfoods of aquatic and animal origins. But as an illustration, I will touch upon my own personal favorite among plant superfoods—the fruits of wild Cacao trees. From a description of the qualities of raw cacao, you will be easily able to understand all the varied effects of superfoods on man and on the system of anthropic diet advocated in this work.

## 6.3   The hidden charm of cacao

What is cacao and what is its role in the human dietary ration? To begin with, there are several categorically different sorts of cacao that are used as food.

The first and most widespread type of cacao beans (with a yearly harvest of about 3 million tons) is cultivated plantation cacao, which grows on plantations in South America, Africa, and Southeast Asia. These cacao beans are grown to be delivered into the market on a mass scale, and then used in mass-produced chocolate, butter, and hot-chocolate powders. It is notable for its cheapness and for the fact that enormous amounts of herbicides, pesticides, fertilizer, and radiological processing of the beans are used during their production. During the processing needed to turn them into chocolate, the beans are subjected to high temperatures (higher than the ones used to produce cacao-butter and powders) and leaching.

The second type is organic cacao—cacao beans that are also grown on plantations but without fertilizer. In all other ways, they are processed in the same way as the first type. This cacao is somewhat better than the first type. However, if it underwent the barbaric processes used to produce ordinary chocolate, then it will cease to have the full extent of its usual beneficial qualities. Moreover, growing the beans on a plantation under direct solar rays worsens their quality. If it has been processed using the cold method, without the use of high temperatures, organic cacao can have some part to play in an anthropic diet.

The third type is genuine cacao, grown on wild Cacao trees in actual tropical forests, without the use of fertilizer or herbicides. This cacao has the full spectrum of healthy qualities that make cacao an almost magical product with the power to heal when consumed by people. Such cacao is processed using the cold approach. As a result, its "live" qualities remain, both in the product and in its derivatives: cacao-powders, cacao-butter, and chocolate.

It is thought that mankind has been consuming cacao since the Maya learned to cook this bitter, foaming beverage. The Aztecs and Maya claimed divine origins for cacao. That is why Karl Von Linn called the Cacao tree Theobroma cacao; translated from the Latin, Theobroma means "god's food." Cacao as a beverage appeared in Europe in the XVII century. The Maya and Aztec civilizations regarded cacao as a magical cure, something seen as a blatant exaggeration in Europe, where cacao was seen and is still seen, both as a beverage and in chocolate form, as a very mildly tonic substance with a marvelous taste.

The use of the fruits of the Cacao tree in the production of ordinary chocolate involves their processing under sufficiently high temperatures and with alkalis, which alkalinizes the cacao-powder and, all taken together, completely destroys the curative qualities of the divine fruit. Raw cacao, regularly consumed by American Indians, is a product uniquely rich in minerals: magnesium, iron, chromium, zinc, calcium. If these minerals are physically-chemically present in raw cacao, studies have shown that they are absorbed by human and animal organisms much more fully and quickly than the same minerals in other products and, needless to say, a lot (hundreds of times over) more effectively than if they are in DSs. Based on the anthropic principle, we can say that the exceptional qualities of these minerals in cacao fruits are caused, most likely, by the fact that the evolution of Cacao trees and men took place under the same conditions of warm, rich in organic material, biologically diverse tropical forests. What is especially important about the mineral contents of cacao is not only that magnesium and other elements are contained in it in easily absorbable, biologically available form, but also that in terms of the amounts contained, cacao is the leader in the plant and animal worlds, especially when it comes to magnesium. For example, the plant that comes second to cacao in magnesium contents, the fruits of the Acai palm tree, has five or more times less magnesium in any given volume. Berries such as cultivated blackberries, which are incessantly recommended in dietology as a wonderful source of magnesium to fight heart disease, have ten times less magnesium than cacao; wild blackberries have eight times less; forest blackberries six times less. Magnesium is the most important element for the human heart. According to research, around 90 percent of people over forty years old in the developed countries have a deficit of magnesium, which leads to the virtually pandemic growth in the numbers of heart disease in all countries. Norman Holenberg's studies have been proven experimentally [124] on the Kuna clans, who constantly consume 5-6 small cups or 40-50 grams of raw cacao seeds per day, and as a result have much better heart health, and virtually no cancer or diabetes. Raw cacao is not a panacea, as is usually claimed in dishonest DS advertising, but it is a product essential for most people who do not have another source of anthropic food. To have a healthy heart at any age, all you have to do is consume raw cacao in some form or another for a long period of time and regularly [125].

Another vitalizing element in cacao is chrome. Its content in cacao is not very high but highly bioavailable for humans. The most important biological role of the micro-element chrome is to regulate carbohydrate exchange and the glucose levels in the blood, as chrome is a component in the low-molecular organic complex—the "glucose tolerance factor." It regulates the penetrability of cell membranes for glucose, the processes of its use by cells and its culling, working alongside insulin. It is speculated that these cells form a complex that regulates the level of glucose in the blood. Chrome increases the sensitivity of cell receptor tissues to insulin, facilitating their interaction and reducing the organism's need for insulin. It can increase the impact of insulin in all the metabolic processes regulated by this hormone. That is why chrome is essential for those suffering from sugar diabetes (most of all for type 2), as its level in their blood is reduced. Moreover, an extreme deficit of this micro-element can make the metabolism susceptible to this illness, and more than 90 percent of Americans, for example, have a deficit of chrome. The level of chrome in women goes down during pregnancy and after childbirth. This chrome deficit can be explained by "pregnancy diabetes."

A deficit of chrome in the organism, apart from increasing the level of insulin in the blood, leads to an increase in triglycerides and cholesterol in the blood plasma and ultimately to sclerosis of the arteries. The impact of chrome on lipid exchange is also connected to its role in regulating the functioning of insulin. As such, cacao consumption is important for the long-term treatment of sugar diabetes and the illnesses that go with it. As diabetes is what's called a disease of civilization, the consumption of raw cacao can unquestionably prevent its development or, given a genetic predisposition, significantly ameliorate the symptoms and course of the disease.

Moreover, experiments on animals have shown that a deficit of chrome leads to stunted growth, disrupts the nerve system, and reduces the fertility of spermatozoids. We have to underline that the abuse of sugar increases the need for chrome and at the same time its loss in urine. The consumption of raw cacao can genuinely protect the organism from a lack of chrome: 100 grams of cacao contains the daily norm, 50 micro-grams, of biologically available chrome.

Turning now to iron, cacao is again a leader among the world's plants in terms of its contents in a form absorbable by humans. The latter qualification is important, as it underlines the fact that there is quite a lot of iron in various products. However, the human organism virtually does not absorb iron in the form in which it is contained in most products, what we call the phase form. Iron deficiency is the most common nutrient deficiency in both developing and non-developing nations. Traditional treatments include iron supplements and increased meat consumption. Both of these approaches have proven to have significant limitations, however [610]. By contrast, it is found in a form accessible to humans in cacao beans and consequently the consumption of cacao is essential to maintaining its level in the organism. Iron absorption from cacao ferritin is more efficient and gives the intestinal cells more control. According to one study [609], during digestion ferritin is not converted from its large, mineral complex, which contains a thousand iron atoms, to individual iron atoms like those found in many iron dietary supplements. The iron found in meat and non-meat iron DSs enters the intestine from food one iron atom at a time. Each entry step requires the intestinal cells to use up energy. When the intestine takes in a single molecule of ferritin, however, it gets a thousand atoms inside that one ferritin molecule, making iron absorption that much more efficient.

According to data released by the World Health Organization (WHO), every fifth person on the planet suffers from an iron deficit, and 30 percent of the population of the developed countries suffer from iron deficiency anemia. There can be as much as 120 mg of iron in a kilogram of raw cacao, which is lower only than the iron content in blood serum and cow liver. Women suffer the most frequently from a lack of iron, both during youth and during menopause. A constant feeling of weakness, being quick to tire, depression, headaches, dry skin, brittle nails, hair problems—these are the symptoms of an iron deficit. But cosmetic problems are not the end of it. There are a number of studies showing a link between iron deficit and the frequency of oncologic illnesses. It appears to me that the regular consumption of raw cacao is absolutely essential for women of forty years and older. Modern medicine compensates a lack of iron by prescribing iron DSs. However, some researchers have found that no more than a few percent of the iron in such supplements plays any part in exchange between cells and the internal metabolism of cells. The rest of it, given prolonged use, accumulates in the women's tissues, as it cannot fully participate in biochemical reactions and be ejected out of the organism. The extent of harm for the organism from this process is not yet known, but given that this isn't natural for the human organism, we can be sure that a negative effect does exist and shows itself in the years after such pharmaceutical "therapy." Long-term treatment of iron deficit illnesses in women can consist of consuming raw cacao, the iron in which easily enriches the body's and blood's cells in the necessary amounts, while excess amounts are simply not absorbed by the organism and are entirely ejected from it.

Apart from this, there is research to suggest that even iron and other minerals that have accumulated in tissue in an inactive form can be ejected from the organism through the consumption of raw cacao. In this case, the bio-organic substances, the large macro-molecules of the bio-polymers, can enter "weak" valency interactions with the iron compounds and remove them from the organism as a result of such an overall compound's role in the inter-cell metabolism [129].

Another important mineral in raw cacao is zinc. A deficit of zinc shows itself through wounds healing slowly, hair loss, worsening memory, and concentration. The processes of spermatogenesis are complicated. Diabetes and peptic ulcer can develop. Raw cacao is a wonderful product when it comes to the long-term treatment of diabetes. A number of studies [128, 129, 130] have shown the positive impact raw cacao has on treating diabetes. The consumption of raw cacao over ten or more years significantly reduces the risk of suffering from diabetes [124]. Alongside this, the data laid out in several studies [131, 132] are interesting. These compared the impact of cacao rich and poor in flavonoids on the cardiovascular system of sufferers from type 2 sugar diabetes. Participants each drank a cup of ground cacao, containing either 321 mg or, in the impoverished cacao, 25 mg of flavonoids, three times per day for a month. The patients who drank cacao with high contents of flavonoids saw significant improvement in the state of their arteries. Patients who drank cacao with low contents of flavonoids had no noticeable improvement [131, 132].

An interesting study on the effects of cacao rich in minerals and flavonoids [133] demonstrated the noticeable positive effect of cacao consumed in Kuna doses on the state of the cardiovascular system in diabetes sufferers and healthy elderly people. But the ways in which the components of raw cacao effect the organism do not end at enriching it with minerals. A large part in the curative powers of cacao is played by its protein, amino-acid, peptide, and fat contents. The impact of these components, of which more than three hundred have been currently discovered, is being intensively studied.

A large number of studies, as has already been noted, have shown the absolute uselessness, and in some cases the actual harm, of artificial antioxidants that have been extracted, and even more so for ones that were synthesized. At the same time, scientists underline the certain beneficial role that antioxidants play in the processes of preventing the appearance of defective DNA, RNA molecules, and cancerous cells. But this beneficial effect is limited to antioxidants in the natural form of a fermented fruit that has not been exposed to unusual temperatures or chemical and other processes.

The modern view of oncologic illnesses is that, as has been confirmed in studies, cancerous cells might be constantly appearing in our organisms. However, up to a certain age, the healthy organism fights against them itself through the most complex mechanisms of the immune system. Micro-biological studies in vitro have shown that the extract from cacao stops the work of the carcinogenic agent in cultivated cells of the human liver. One reason for this is the increased concentration of enzymes and ferments that neutralize the toxicity of harmful substances.

Cacao is the planet's richest source of polyphenols, flavonoids, thanks to which it has its taste and color. The flavonoids of plant food in general and of cacao in particular possess a fantastically high biological activity [135]. While I have expressed certain doubts about antioxidants in this book, I do not deny their obvious value in their natural form. It is essential to note that in antioxidant activity, stability, and biological accessibility, the flavonoids in raw cacao exceed vitamin E, recommended by many defenders of DSs as one of the most important antioxidants, tens of times over. Natural antioxidants normalize the functioning of the organism's cells; prevent the formation of vein thrombosis; decrease the level of cholesterol in the blood; have an anti-allergic, anti-inflammatory, anti-radioactive effect; and widen arteries. They also fortify the walls of blood-carrying arteries and prevent the development of heart and liver disease and of diabetes [136, 137].

The flavonoids in raw cacao also fight tumors in various ways. As American scientists have confirmed, substances found in raw cacao beans partially delay the growth and accelerate the destruction of cancerous tumors in vitro [138]. Their experiment shows that these substances inhibit ribonucleotide-reductase and prevent the synthesis of mammal DNA, directly blocking the multiplication of tumorous cells. Another way they combat tumors is through the competition of raw cacao flavonoids with estradiol for space on the receptors, which reduces the stimulation estradiol has on cells of the hormone-dependent tumors [139].

Research done by Japanese scientists has further shown the impact of cacao on the immune system and on preventing cancer. Experiments have shown that polyphenols contained in cacao improve immunity. During other research, it was established under laboratory conditions that the immunity regulating effects of cacao-polyphenols does not end at their antioxidant impact. The micro-biological research of NRC in vitro has shown that cacao-polyphenols prevent the functioning of carcinogenic agents in cultivated cells of the human liver. One reason for this is the increased concentration of ferments as a result of the impact of raw cacao, which neutralize the toxicity of harmful substances [140].

Research in this direction continues, but we can already say that the polyphenols in raw cacao are a unique phenomenon. The most sophisticated review of medical research published between 1966 and 2005 on the effect of cacao flavonoids on the cardiovascular state of the human organism highlights the existence of a statistically relevant, significantly positive impact of raw cacao on the risk of premature death from diseases of the heart and arteries [143]. In addition, it has been found that cacao, evidently thanks to its flavonoids, can reduce the risk of osteoporosis, which is often caused by a drop in the level of estrogen in women during menopause. Cacao also has a positive effect on the skin, stimulating the synthesis of collagen and preventing the appearance of inter-linking chemical bonds. This in turns helps to prevent undesirable changes in the viscoelastic properties of the intracellular and intercellular matrixes, and to maintain the natural proportions between the processes of macro and nano diffusion of signal proteins. In the study [144], this effect is explained by the estrogenic impact of cacao flavonoids. Consuming cacao for a month improves the color and health of skin from within much better than any cosmetic product. According to the classification of independent expert organizations, cacao-butter has a better impact on the health of the skin than any other product [145].

Doctors at the European Dermatological Center in London advise their clients to consume black chocolate with high contents of cacao. According to research they carried out recently, the chocolate slows skin aging and protects it from the ultraviolet rays of the sun. Science suggests that the intensive impact of ultraviolet radiation leads to skin cells losing their natural elasticity and can lead to cancerous illnesses. However, the active chemical substances contained in cacao activate the defensive functions of the skin, and this leads to its rejuvenation. British scientists claim that chocolate has the greatest effect on facial skin [146].

It has been proven that substances contained in cacao make low-density lipoproteins exceptionally stable in the face of acidification in vitro, and so make capillaries firmer. As the acidification of low-density lipoproteins is the main reason for the development of arteriosclerosis, the polyphenols of raw cacao, by preventing this acidification, significantly reduce the risk of cardiovascular diseases. Raw cacao is the regulator of the exchange of fats in the organism. They prevent the weakening of the heart muscle, due both to their mineral components, such as magnesium, and to organic biopolymer compounds, flavonoids. A large number of scientific studies on raw cacao over the last seven to eight years, reviewed in a study [147], demonstrate its remarkable ability to recreate and preserve the human organism in the face of many illnesses.

Raw cacao also contains a whole range of bioflavonoids (quercetin, hesperidin, and many others), that are not synthesized in the human organism, but the introduction of which into the organism has an exclusively beneficial effect [148, 149]. That it contains substances not synthesized in the human organism but vital for it is yet more proof of the anthropy of cacao as a food and the need to consume it, in whatever form, in your daily diet. According to many studies, cacao flavonoids also have a positive impact on the cardiovascular system and the brain [147]. The benefit of flavonoids for the human brain comes from the flavonoids' direct impact on the cardiovascular system. In essence, these substances stimulate an increase in the blood flow to the brain, specifically to those regions of the brain responsible for speed of reaction and recall. Studies carried out allow us to assert that the impact of the whole complex of cacao flavonoids is strengthened by the effect of theobromine, anandamide, and feniltilamina (see below). Studies recently conducted by British scientists show that the theobromine of raw cacao makes it a more effective cough medicine than traditional medicine that uses codeine. In addition, these components don't have any of the side effects that codeine-based cough medicines often have.

Because raw cacao is extremely beneficial for mental processes, the presence of cacao beans in the daily ration guarantees a sharp increase in brain activity. From this point of view, raw cacao is very beneficial for students to help them learn. The whole world over, pupils feel sleepy while are getting ready for school in the morning. A cup of cacao and a few seeds to take with them to school and snack on during break time will improve the child's ability to learn for a short period of time.

The impact of another wonderful polyphenol in cacao, epicatechin, is that it returns elasticity to arteries, so much so that these become as elastic as arteries significantly younger than their years. Elastic blood arteries are a guarantee against early hypertonia, infarcts, and insults [151]. Epicatechin, among some other substances, has long made scientists interested in cacao. Members of the Kuna clan in Panama almost never suffer from hypertonia, or indeed from stenocardia. But as soon as they move to large cities and stop drinking their traditional beverage, their arteries become the same as any other urban dweller's. As well as in cacao beans, epicatechin is also contained in dry wine, grapes, soya, and certain other plants. But raw cacao beans are the leaders in terms of epicatechin contents [152]. According to data on the effects of cacao in Panama collected by Norman Hollenberg, a professor at Harvard Medical School, and published in the journal *Chemistry and Industry*, the risk of developing insults, heart deficiency, cancer, and diabetes is significantly low given regular consumption of raw cacao. Dietary experts note that even though the data collected by Hollenberg is purely empirical, they are sufficiently striking to call into question the very concept of a "vitamin." Currently there are thirteen scientifically recognized vitamins, and epicatechin is not among their number, as it does not correspond to the traditional understanding of a vitamin as a substance important for the normal functioning and growth of cells, a deficiency of which leads to illness. Epicatechin helps the organism combat pro-oxidant, active free radicals. Research results [231] present hard evidence that the epicatechin found in cacao beans may be a potent dietary source for defense against peroxynitrite, a highly damaging form of free radical. However, the irony is that the producers of commercial cacao and chocolate extract epicatechin from its contents due to the bitter taste of this flavonoid. We can only marvel at how the chase after the consumer's undiscriminating tastes and the sway of profit can destroy the truly vitalizing qualities of cacao. And cacao contains not just epicatechin, but also many other polyphenols that are unquestionably beneficial to the organism, even though the mechanisms for the activity of these are so far unknown to scientists. The results of their impact can only be seen in the individual effect on a human organism of raw cacao, the most powerful superfood known to modern science. These data are cited in the studies of the famous cacao specialist David Wolf, author of the *Naked Chocolate*, and confirmed by scientists from major universities in the United States, the Netherlands, and Italy.

Many studies on animals to determine the effects of cacao on the organism have demonstrated that the consumption in the Kuna doses adjusted to the size of the animal prevents atherosclerotic disease and cancer tumors more effectively than many pharmaceutical methods [153]. Moreover, based on the results of these experiments, we can suggest with a great degree of certainty that cacao must ease the symptoms caused by the chemical and radiation therapy used during the treatment of cancer.

To genuinely improve a human immune system, antioxidants have to be absorbed well by the organism. Research carried out by Nestle have proven the practical value of cacao polyphenols in a set of experiments during which samples from the plasma of healthy people were observed after they had drunk a cacao beverage. The brains of the subjects were scanned with a magnetic resonance tomography, and it was found that for sixty minutes after the cacao was consumed, blood reached the brain more intensively and increased brain activity. As a result, it was determined that the concentration of polyphenols in the blood increases significantly, reaching its peak in two to three hours' time. Laboratory experiments also showed that the polyphenols found in cacao, after their shells are removed, can regulate the immune system. During other research, also under laboratory conditions, it was confirmed that the immunity regulating impacts of cacao-polyphenols are not limited to their antioxidant effect. Work in this direction continues but we can already speak about how unique a phenomenon is the immune-modulating power of raw cacao thanks to the polyphenols contained in it [154]. German and Italian researchers [155] have reached the conclusion that cacao is better at reducing blood pressure than many medicines. At the same time, it does this in a gentle manner without the side effects of vascular dystonia and so on. A group of 470 subjects consumed cacao in the form of beans, ground cacao, and powders over the course of two weeks. This led to a decrease of their blood pressure to the same extent as in a similar group that was consuming blood pressure medicine [156]. Consuming raw cacao without interruption for more than five years significantly reduces the risk of insults [157].

Another extremely healthy amino acid in cacao is L-arginine. This amino acid is irreplaceable in the functioning of the heart, arteries, and the nervous system. It strengthens the carcass of muscle and sinew, has a positive effect on spermatogenesis, and plays a part in the detoxification processes of the liver. But only L-arginine to be found naturally in natural food is beneficial. Arginine is widely used in artificial form in sports nutrition. However, used like this, this amino acid is more likely to do harm than good to the organism. Research shows that the introduction of inhibitors of the synthesis of nitric oxide (NO) in the form of artificial L-arginine into healthy volunteers is accompanied by a significant increase in peripheral vascular resistance. The excess formation of NO from L-arginine introduced into the organism through the use of DSs plays a negative role and is one of the main causes of the most serious hypertonia [158]. And L-arginine is not the only such component of its type that cacao has to offer. More and more, new bio-organic compounds are being discovered in cacao. For example, recently a substance named Cocohill was found, which assists the growth of skin cells and the healing of wounds, removes wrinkles, and reduces the risk of gastric ulcer.

One of the remarkable qualities of cacao is that it increases brain activity after it has been consumed for a day, and at the same time ensures a quiet night's sleep. This is caused by the theobromine contained in cacao, a close "relative" of caffeine. This substance has a balanced awakening or stimulating effect, improving circulation and increasing physical activity. Scientists know of instances when cacao was of genuine help in combating stress or depressive states. Yet it has been noted in many observations that the consumption of raw cacao in whatever form assures a quiet and pleasant sleep. Unlike caffeine, which is a direct stimulant on the cell level, just as narcotic substances are, theobromine is a precursor for the creation of endogenous proteins and "happiness" hormones that are present in our organisms in youth—i.e., it has a soft but stable and prolonged effect on brain activity. No addiction of the sort that accompanies the lengthy use of caffeine happens. At the same time the effect is prolonged—i.e., does not go away after the product is no longer being consumed. Most researchers connect this effect not only with theobromine itself, but with the overall normalization of hormonal exchange under the complex impact of all the vitalizing components of cacao, which continues to exist by inertia after the product is no longer being consumed. It has been established that this inertia goes on for about two months after a year's consumption, and reaches ten years if cacao in Kuna doses is consumed for more than five years. The existence of this inertia period makes some scientists speak about the curative qualities of raw cacao.

A large content of an amino acid, tryptophan, which also helps relaxation and sleep, improves the conductivity of neurons, and has an anti-depressive effect, has been found in cacao. Cacao also contains significant quantities of anandamide, a substance that improves the transfer of nerve impulses in neurons many times over. Anandamide is partially produced endogenously in the human organism. Enzymes contained in cacao slow the breakup of endogenous anandamide in the organism and, taking into account the addition of exogenous anandamide from raw cacao into the blood, maintain a high level of it in the nervous system for a lengthy period after the cacao is consumed. This explains the special clarity of mind, quickness of reaction, and improved memory that you experience after consuming cacao.

Apart from this, some researchers [159] have noticed that in vitro cacao has the effect of slowing the growth of brain tissue filling, or glia. The amount of glia in the brain increases with age. If cacao stops the growth of glia cells, then it effectively slows the brain's aging and the probability of the appearance of brain tumors, which almost always appear in glia tissue in the brain. At the same time, the degradation of the viscoelastic properties of glia is slowed. Evidently, this effect is connected to the observed fact that in regions where cacao has been traditionally consumed, cases of senility and Alzheimer's are absent even in extremely old people.

Another interesting and important quality of raw cacao is increased sex drive [160]. Studies have noted an increase in male and female libido, frequency of sex, and degree of satisfaction during sex [161]. It has been noted, based on a question-naire of long-term couples, that their erotic desire for each other improved significantly as a result of consuming raw cacao. This relates to increased libido, but also to objective improvement in physical appearance, skin, and hair. The correct hormonal balance makes the eyes shine, makes the person happier, increases the intensity with which pheromones are produced, and consequently improves the way the individual appears sensually to the opposite sex. Apart from this, cacao contains the strongest aphrodisiacs, arginine and feniltilamin, in large quantities. As is well known, the latter is extensively produced by the brain when its owner is in love. Raw cacao is a uniquely powerful natural source of feniltilamin. When consuming cacao, a person feels better and has an improved mood thanks to feniltilamin. At the same time, this is not an agitated state, such as what happens after caffeine consumption, but rather the state of happiness and creativity typical of those in love. In addition, feniltilamin and the other components of raw cacao improve the ability to bear physical, cardio-vascular burdens and helps sportsmen achieve their best results.

Many researchers into this product note weight loss during cacao consumption [18]. This is explained by a certain dualism of the action of raw cacao on the human organism. On one side, consuming it leads to quicker exchange processes on the inter- and intra-cellular levels, stronger functioning by the endocrine system, increased levels of growth hormones, and improvement in the balance of fats in the exchange processes. Cacao also effectively prevents the breaking up and absorption of fats by creating prostaglandins in the organism during its consumption. On the other side, cacao can suppress appetite. But it doesn't work the same way as many plant products, for example of Thai medicine: by suppressing the relevant receptors. Raw cacao has its effect by improving the mineral and protein-hormonal balance to such an extent that the organism feels full of energy after consuming a smaller amount of food, by using the available reserves of fat in the exchange processes, and through a high level of hormones and the speed of metabolism.

Balanced levels of hormones are crucial in accelerating the organism's metabolism to the natural level and involving so-called "white" fat—which accumulates in areas under the skin as a result of over-eating and changes in the hormonal balance—in the processes of exchange. Some data presented in a study [18] about the effect of cacao on the metabolic processes indicate that the constantly normal level of hormones you have from consuming cacao masks the reduction in the food you have consumed at any given moment and prevents the genetic defense mechanisms of fat accumulation in the cells from activating. The fat already available is simply used by the organism in the exchange processes. Moreover, the rate of "burning" fat is equal to the rate in a non-overweight organism, which is two to eight times faster than the rate of "burning" fat in initially overweight people leading a normal lifestyle.

Cacao-butter, contained in the beans of raw organic cacao at levels of around 28–44 percent, is one of the most anthropic plant fats for man. Entering the organism, it can dissolve the fat reserves already present in cells and eject them from the organism. At the same time, it is also an excellent transport for fat-soluble vitamins and unenriched acids into the organism's cells, and is an excellent regulator of the balance of fats. In addition, cacao-butter has many polyunsaturated fatty acids that are uniquely beneficial for the heart and all other organs. Due to the firm, dense structure of cacao beans, these acids remain in their original state when the cacao is dissolved to make cacao-butter, do not acidify, and are completely absorbed by the organism, where they assist the health of the heart muscle. Also important is that the fats in raw cacao do not aid the production of "bad" cholesterol and have the ability to "dissolve" (involve in the metabolic processes) the fats that have already accumulated in fat cells.

Cacao-butter is made of three- and two-acid triglycerides. The fatty acid contents are made of such components as oleic acid (up to 43 percent), stearic acid (up to 34 percent), lauric acid, palmitic acid (up to 25 percent), linoleic acid, etc. These cacao-butter acids are what are called polyunsaturated fatty acids and are one of the few sources of the very rare, but extremely healthy, vitamin F (how healthy and essential this vitamin is was spoken of as far back as the 1940s, when a vitamin F deficiency syndrome was found that consisted of brittle, dry, and reddened skin). The uniqueness of precisely these fatty acids is that, on one hand, the organism cannot synthesize or derive them traditionally from food, but on the other, they play an extremely important role in many metabolic processes of skin cells and the tissue under the skin, and are contained in large, bio-active amounts in raw cacao. So for example, linoleic acid is essential to creating the right epidermal barrier. Without it, the barrier will literally break up into chunks (it is because of this that a deficit of linoleic acid makes the skin dry and brittle). However, extracts of this acid in the form of DSs are virtually useless, as they lose their active qualities during extraction. But if they are consumed in cacao and alongside other fatty acids and bio-organic complexes, they have their full and best effect. These same acids are precursors for the production of hormonal substances that control inflammatory reactions—prostaglandins. It is well known that raw cacao-butter-containing linoleic and oleic acids have an anti-inflammatory effect, which, when applied externally, assists the quick and effective cure of various skin diseases. The use of extra virgin cacao-butter in a cold state is actively entering the cosmetic procedures used in exclusive cosmetic salons the world over.

But cacao and its fats are also very effective when used internally. Fatty acids and all the active components of cacao taken as a whole, upon entering the area beneath the skin through the bloodstream, improve the hormonal balance and activate the cell and inter-cell metabolism, and so normalize lipid exchange, efficiently restore the epidermis, and have an excellent effect on the skin that is better than any known creams. Palmitic acid in cacao has an enormous impact on activating the synthesis of native collagen, elastin, glycosaminoglycans, and hyaluronic acids in the organism. This in turn leads to renewal of the cells in the tissue under the skin and the substance between cells in the dermis. As such, after sufficiently prolonged cacao consumption (starting a month after consumption begins), the skin becomes velvety and well-toned. At the same time, as the skin is restored due to positive internal changes within the organism, the effect is maintained for a long time. Stearic acid, which has excellent smoothing and lubricating qualities, increases the skin's stability in the face of external negative factors, improves its elasticity, and makes it velvety. It is one of the main fatty acids in human tissue.

As we can see, the acids contained in extra virgin cacao-butter are vitally important to the organism! Some of the symptoms of a deficit of unsaturated fatty acids are: dry or brittle skin; a tendency to develop eczemas; dull nails or hair; and hair loss, up to actual baldness. This is why the consumption of raw cacao both internally and externally, in the form of face and body creams based on cacao-butter mixed with ground cacao, is of excellent assistance against excessively dry skin, especially skin that suffers from a lack of nutrition, atopic eczema, the effects of aging on the skin. If a deficit of irreplaceable fatty acids leads to such negative phenomena, then nourishing the skin, externally or internally, with these substances, through the regular consumption of raw cacao and use of cosmetic procedures based on its products, will unquestionably improve the state of this dry, brittle, and irritation-prone skin. I must also note that the products contained in cacao-butter have a significant positive effect not only in solving problems of brittleness and irritations, but are also useful in stopping other skin disturbances, such as youthful acne and so on.

Thanks to the presence in cacao of a large number, in total quantity and in variety, of oligopeptides—short peptide chains formed of several amino acids—it has a positive effect on extending the life span of the organism in general. The results of an experiment to see the effects of raw cacao, carried out on volunteers in the United States and Canada, showed a noticeable therapeutic effect on those of mature age (sixty-five to seventy-five years) and old age (seventy-six to eighty-five years). Their bones became thicker, their liver function improved, and their overall tendency to get ill and their mortality went down, compared to a control group. Since 1998, scientists from Cornell University have been studying this represented group of eight thousand Americans to prove that consuming cacao and products from it extends the lifespan [162].

According to some theories of aging, disruption of the balance of polypeptides and other proteins in the organism's cells is the reason for the majority of the changes associated with aging. This applies to all the stages of osteoporosis, and changes in brain, liver, heart, and lung tissue. The most graphic of these changes is the appearance of wrinkles due to the collagen under the skin forming chemical compounds. The oligopeptides in cacao are sort of spare parts to recreate polypeptides and proteins, replacing defective parts of high molecular protein matrixes and so restoring their viscoelastic qualities. In the human metabolism, oligopeptides with a length of chain between two to three amino acids and ten amino acids are produced by the organism. However, after age forty to fifty, they are produced at a rate twenty to forty times lower than they were at age twenty. The rich variety and high contents of oligopeptides in cacao compensates for the deficit of natural oligopeptides in the organism and helps to maintain the viscoelastic state of cell and tissue matrixes. Incidentally, according to recent research the level of oligopeptides production in populations characterized by a large number of long lifespans, the level of people living in the mountainous regions of the Caucasus and Eskimos in North Canada is significantly higher (ten to twenty times) at the same age than in people who live under different conditions. This may be connected to a possible defensive reaction in the organism against the higher level of solar radiation experienced on mountains, as well as cosmic radiation, solar wind, and galactic background radiation, experienced near the magnetic poles.

All of these studies are only at the initial stage of investigating the processes of peptide exchange in the human organism, but we can already state that raw cacao is a completely indispensable product that prolongs human life with lengthy and regular consumption. The shell and cell structure of cacao, as has been shown, also have the ability to consume the salts of heavy metals. This trait of cacao is important in protecting the organism from the pollution of modern industry and the urban environment. Passing through the digestive tract, the shell of raw cacao maintains its texture in the stomach and intestine and interacts effectively with residues on the surfaces of organs, drawing out of them the heavy metals that have a tendency to accumulate in the fecal deposits of the intestine. At the same time, the shell, in interacting with the food mass in the stomach, cleans it of heavy metals and prevents them from entering other areas of the organism.

It is also important to note that raw cacao and the chocolate produced from it through the "cold" method is many times less allergic than ordinary cacao or chocolate. It is thought that one in five hundred suffers an allergic reaction to 30-40 grams of raw cacao, while one in twenty suffers it to ordinary chocolate. However, I have not seen a single allergic reaction in several years of using raw cacao and recommending it to thousands of people. The excellent impact of raw cacao, and chocolate made from it using the cold method with the addition of ground cacao shell, on such conditions as incorrect bacteria, from which 70 percent of urban dwellers suffer to some extent, has been noted. Experiments carried out by Ecuadorian scientists have shown that cacao consumption helps to restore an intestinal bacteria micro flora ruined by antibiotics over a short period of time.

Cacao-bean shell that has not been heated or had pesticides inflicted on it contains dry bacteria and, as it isn't digested in the stomach, these bacteria reach the intestine safe and sound. There they assist the recreation of a balance of intestinal bacteria micro flora anthropically suited to man.

The intestinal micro flora has long been viewed as more than a part of the digestive system. This is because, in addition to breaking up substances that are immune to our own ferments, good bacteria also "train" our immune system, prevent the multiplication of harmful micro-organisms, and synthesize certain vitamins for us. Certain Brazilian and Argentinian mass media publications assert that cacao consumption, unlike chocolate, increases the level of an insulin-like hormone that is a factor in growth. This is especially important for men as a high level of this hormone decreases the risk of prostate complaints and cancer after age fifty. A level of this hormone more than double the average reduces the risk of malign adenoma of the prostate in men sixty-five to seventy-four years old. At the same time, as has been noted many times in medical literature, the consumption of any specialized DS, including one based on cacao, does not lead to an increase in this hormone.

All this data confirms the unique qualities of raw organic or wild cacao, dubbed "the fruit of the future," and the vital importance of consuming it as part of your diet. One of the most wonderful qualities of cacao is that all the healthy substances in it are in such a "phase" state—i.e., have such a series of interactions with the surrounding molecules in the cluster biopolymer structure of the substances in cacao-beans—that they are biologically accessible for the metabolic processes of the human organism. This is not the case for all plants. For example, beta-carotin in carrots or iron in apples is in a state of low biological accessibility, even though the level of these elements is high.

I could write similar sections about many other superfoods. However, that is not the main aim of this book. My main aim is to lay out a new conceptual framework for nutrition, which can be useful in helping to deal with the nutritional situation faced with by a large part of humanity.

## 6.4   DS and vitamins—a conceptual disaster

There is currently a huge industry producing   dietary supplements (DS) and vitamins. DSs are used both as direct dietary supplements and in cosmetic products. The theory and practice of using DSs is based on using methods of chemical or physical extraction, with the application of heat, pressure, and solvents, through which some substance is derived from a biological product in concentrated form.

These derivatives are then used as active food additives and as medicine to treat or prevent this or that illness.

More or less the same applies to the natural form of vitamins. However, artificially synthesized vitamins and additives that mimic natural compounds also exist and are widely used in nutraceutical practice. When it comes to DSs and vitamins, primitive understandings of how the metabolism of the human organism functions allows their defenders to state that the beneficial substances found in some live organism will have a curative effect on humans when taken in high concentrations. Their effectiveness, it is claimed, comes from the fact that concentrated "healthy" substances will have a much better impact on the organism than consuming the animal or plant they are derived from. Moreover, dishonest advertising and pseudo-scientific articles announce these concentrates as panaceas against heart, oncologic, and other diseases.

If some dose of vitamin is good, more must be better, right? Wrong! Many studies confirm that vitamins play a crucial role in cell metabolism, but too much or too little of them has a complex negative effect on our bodies. This is particularly important as processed food, drinks, and creams enriched by a combination of vitamins, and dietary supplements containing added vitamins often make an overdose more likely than ever before. A good example is the study [473] that illuminates the value and potential harm of vitamin A use in cosmetic creams and nutritional supplements. Overuse of this vitamin could cause significant deregulation of energy production, impacting cell growth and cell apoptosis. Although the importance of vitamin A to human nutrition and fetal development is well-known, it has been unclear why vitamin A deficiencies and overdoses cause such widespread and profound harm to our organs, until now. The results of the study [473] explained why these effects occur and showed that retinol, the key component of vitamin A, is essential for the metabolic fitness of mitochondria and acts as a nutritional sensor for the creation of energy in cells. When there is too much or too little vitamin A, mitochondria do not function properly. This is why using too much processed food or DS enriched with vitamin A could lead to negative, even fatal, consequences. But we have to note that natural vitamin A from natural food is absolutely safe given a normal level of consumption of the food.

If we want to understand the effect of the substances in our environment on our organisms, then we have to abandon simplistic linear, or to be more precise, insufficiently non-linear approaches to finding their impact on the organism. Currently the theory of the impact of vitamins, DSs, and medicine is following the same path as was followed by classic physics before the start of the quantum revolution at the beginning of the twentieth century. According to the classic approach, objects can be divided into molecules or atoms, and atoms into the nucleus and electrons. This approach gave a satisfactory result when describing the simplest chemical reactions.

However, as we know, reality turned out to be much more complicated, and in this day, physicists are aware of many other quantum particles and interactions, that cannot in principle be described under the classic approach.

When it comes to studying the effect of various allegedly biologically active supplements and vitamins, the situation in biology is similar to the situation in physics when the classic approach predominated. The analogy to the classic approach is the reductionist approach in biology, which predominates currently in the theory and practice of nutrition. Molecular biology accumulates information about individual reactions and processes, which leaves us as yet uncomprehending of the process as a whole. On the basis of this imprecise and often incorrect interpretive information, and using relatively primitive models, conclusions are drawn, and then extrapolated onto the human organism. In day-to-day life, this manifests itself as new panaceas being found, which can allegedly increase the human lifespan. This includes all the theories about antioxidants, and about large doses of vitamin C, and so on.

It is true that many living plants, micro-organisms, and sea animals are healthy when consumed by people fresh. However, the supporters of reductionism do not take into account the fact that the human biochemical processes, the contents and state of chemical substances in fresh food, and the interaction of the two are so complex and divided into so many hierarchical levels that it is virtually impossible to control the metabolism effectively through the primitive concentration of whatever ingredients of fresh food and their use as dietary supplements. A good example of this is medicines, the absolute majority of which have serious side effects. This means that when we try to give some specific biochemical command signal to the organism, we have simultaneously given a variety of other signals that are absolutely undesirable and can worsen the state of the organism in the long term. In other words, whether to use medicine or not is always a compromise between the end balance of a medicine's positive and its negative qualities. It's good if the negative qualities given a certain time or dose are minimized. However, with many medicines it is doubtful if the long-term negative consequences can ever be minimized. In this case, while temporarily softening some symptom of an illness, the medicine worsens the overall state of the organism in the long term.

So for example, in the largest and longest trial investigating palmetto extract, widely used as a component of many dietary supplements (DS) for prostate health, North American scientists at eleven clinical sites tested up to three times the standard manufacturer's dose and found that this extract did not improve lower urinary tract symptoms in men with prostate enlargement [561]. There were no measurable benefits with increasing doses of the supplement in comparison to a placebo. This means that those DSs based on saw palmetto extracts, which under different brands are traded over the counter worldwide, are apparently not doing anything measurably good beyond what is called the placebo effect.

Because even very high doses of saw palmetto make absolutely no difference, people should not spend money on this herbal supplement in order to reduce the symptoms of an enlarged prostate, as it clearly does not work any better than a placebo. I do not know about the effect of natural supplementation on prostate conditions by the fruit of the saw palmetto tree. Probably natural fruit is good for it, but that needs to be checked in specially designed randomized trials.

This also applies to all other DSs, vitamins, and other such substances sold as tablets and powders. In the best-case scenario, these are useless additives. By now, however, their negative impact on people who use them regularly has been noted many times. For example, the widespread advertising drive of the 1980s and 1990s to popularize synthetic or extracted antioxidants has discredited the very idea that antioxidants can be of benefit in the eyes of the public. As shown in independent studies of hundreds of thousands of the fans of various DSs, the use of vitamins and antioxidants in the form of tablets, juices, and other concentrates has no impact on prolonging or improving the quality of life, or of preventing disease. What is more, these studies showed an increase in the frequency of oncologic diseases among active supporters of the idea of a longer life through artificial vitamins, antioxidants, and DSs. For example, there is evidence that artificial beta-carotin and vitamin E products increase the risk of death from cancer—the same beta-carotin and vitamin E, (or selenium products, coenzymes, etc.) that are in the best cases useless in combating cardiovascular illnesses, in cancer therapy, and in preventing insults. Research [445,446] shows that only the natural form of vitamin E can trigger production of a protein in the brain that clears toxins from nerve cells, preventing those cells from dying after a stroke. This process is one of three mechanisms identified so far that this form of vitamin E uses to protect brain cells after a stroke, meaning that this natural substance might be more potent than drugs targeting single mechanisms for preventing stroke damage, according to Ohio State University scientists who have studied the nutrient for more than a decade [446]. "This is one of the first studies to provide evidence that a safe natural nutrient–a vitamin–can alter microRNA biology to produce a favorable disease outcome," said C. Sen, professor in Ohio State's Department of Surgery and author of the study. "Only a natural nutritional product is simultaneously acting on multiple targets to help prevent stroke-induced brain damage."

According to data collected by the British government, artificial vitamin A increases the risk of fractures and of birth defects for pregnant women. According to Prof. Nava Dekel of the Biological Regulation Department, we still don't have a complete understanding of how antioxidants act in our bodies. New research [319] by Dekel and her team, recently published in the *Proceedings of the National Academy of Sciences* (*PNAS*), has revealed a possible unexpected side effect of antioxidants: They might cause fertility problems in females. More than that, research [320] has shown that neural stem cells, the cells that give rise to neurons, maintain high levels of reactive oxygen species (ROS)—also known as free radicals—which are mainly ions or small molecules produced in the human body's metabolic processes that are essential in regulating normal self-renewal and differentiation. Oxidative stress, however, is caused by overproduction or under-removal of these ROSs. Excessive concentration of free radicals in cells and tissue is involved in a number of disorders, including atherosclerosis, neural degenerative disease, inflammation, and cancer. Antioxidants in the simplified theoretical approach (reductionism) are thought to mop up these free radicals, reduce oxidative stress, and prevent disease. But in reality things are much more complicated. Oxygen-consuming organisms obtain their energy through cellular respiration, which is the transformation of carbohydrates and oxygen into carbon dioxide and water. This process also produces ROS that must be decomposed immediately, as they would otherwise cause damage to cells. [546]. Scientists from the Max Planck Institute for Molecular Genetics in Berlin [547] have discovered an endogenous mechanism, with the help of which cells can coordinate respiration and the degradation of ROS. Thus, the healthy cells prepare their metabolism for ROS before they even arise and therefore prevent damage to the cell. But keeping the cells in a healthy state is their main task and to fulfill this task we need to do much more than just eat antioxidants.

There is also contradictory information about the effect of vitamin D on man. The contradiction stems mainly from the form of vitamin used. If vitamin D means synthetic and extracted substances in the form of industrially manufactured tablets, then high doses can lead to vitamin D intoxication. At the same time, as is well known, if vitamin D is produced in the organism as a result of sunbathing, then there is no intoxication effect even given much higher doses. The vitamin also not does generally intoxicities the body if it is derived by the digestion process from the natural products of live nature. Vitamin D is much more effective if it is produced endogenously in the organism or enters it from products where it is produced the same way as in man—i.e., is in a biologically accessible form. For example, the shell of cacao-beans contain vitamin D if the cacao-beans were dried in the sun and fermented, while cacao-beans dried in the dark lack vitamin D even after fermentation. This indicates that the shell of the bean as it comes fresh from the pod does not contain vitamin D, and yet after exposure to sunlight during drying, vitamin D is present in small amounts in unfermented shells and in very large amounts in fermented shells. Presumably a precursor of vitamin D is present in the shell and gets converted into vitamin D by the action of both the active UV rays of the sun and fermentation. While it is too early to make definitive statements about the risks associated with routine high doses of vitamin D from DSs, past cases such as synthesized hormone replacement therapy or high doses of beta carotene or vitamin E remind us that some therapies which seemed to show promise for preventing diseases through DSs ultimately did not work out and even caused harm. Among most medical professionals there is a belief—almost a dogma—that high vitamin D levels are associated with a decreased risk of developing multiple sclerosis (MS). But the first randomized, controlled trial using DSs with high-dose vitamin D in MS patients did not find any added benefit of vitamin D supplementation [593].

A twin study [447] showed that eating a diet high in natural vitamin D or its precursors, as well as some other natural nutrients, might help reduce the risk of mascular degeneration. Age-related macular degeneration is highly heritable, with genetic factors determining up to 71 percent of the disease's severity, as shown in a previous study of this twin registry by this same research team [448]. By examining identical twins with the same genes but whose disease was at different stages, researchers were able to identify environmental and behavioral factors that may contribute to the severity of the disease. Having a relatively healthy diet could save you from this disease even if you have a genetic susceptibility to macular degeneration. This study shows that even small changes in the diet lead to wonderful results. But we cannot assume that all those twins who did not suffer from this disease had a strictly anthropic diet. Most likely, they did not. So just imagine how many diseases would be eliminated with a genuinely anthropic diet!

Despite common beliefs to the contrary, a study [208] of more than 75,000 adults found that taking supplemental multivitamins, vitamin C and E and folate, do not decrease the risk of lung cancer. And vitamin supplements do not protect against lung cancer, according to a study [209] of more than 77,000 vitamin users. In fact, some supplements may even increase the risk of developing it. Six-year vitamin E supplementation increased tuberculosis risk by 72 percent in male smokers who had high dietary vitamin C intake, but vitamin E had no effect on those who had low dietary vitamin C intake, according to a study [210] published in the *British Journal of Nutrition*. Moreover, two studies [598, 599] suggest that artificial DSs for patients after lung injury do not improve outcomes and may actually be very harmful.

A study [342] looked at the effect of vitamin E on the risk of pneumonia in a large randomized trial. Based on the results of this study, we can conclude that the impact of vitamin E from artificial sources and extracts has a pronounced non-linear character dependent on the external circumstances experienced by the organism. According to this study, vitamin E supplementation may decrease or increase, or may have no effect, on the risk of pneumonia. It is only in laboratory studies that vitamin E has influenced the immune system and protected against viral and bacterial infections. In reality, the importance of vitamin E as an artificial DS in human infections is not known. But it is well known that vitamin E as a DS could increase pneumonia or tuberculosis risk by up to 80 percent in some part of the population. In addition, men who took as DS around 400 international units of vitamin E daily had more prostate cancers compared to men who took a placebo. In a trial that included about 35,000 men, those who were randomized to receive daily supplementation with vitamin E had a significantly increased risk of prostate cancer, according to an updated review of data from the Selenium and Vitamin E Cancer Prevention Trial (SELECT) [596].

According to a recent study [379], people who eat a diet containing a common nutrient found in animal products (such as eggs, liver and other meats, cheese and other dairy products, fish, and shellfish) are not predisposed to cardiovascular disease solely based on their genetic make-up, but rather on how the micro-organisms that live in our digestive tracts metabolize a specific lipid—phosphatidyl choline (also called lecithin). Lecithin and its metabolite, choline, are also found in many commercial baked goods, dietary supplements, and even children's vitamins. "When two people both eat a similar diet but one gets heart disease and the other doesn't, we currently think the cardiac disease develops because of their genetic differences; but the studies show that is only a part of the equation," said Stanley Hazen, M.D., Ph.D., from Lerner Research Institute and senior author of the study. "Actually, differences in gut flora metabolism of the diet from one person to another appear to have a big effect on whether one develops heart disease. Gut flora is a filter for our largest environmental exposure—what we eat. Another remarkable finding is that choline—a natural semi-essential vitamin—when taken in excess, promoted atherosclerotic heart disease. Over the past few years, we have seen a huge increase in the addition of choline into multi-vitamins—even in those marketed to our children—yet it is this same substance that our study shows the gut flora can convert into something that has a direct, negative impact on heart disease risk by forming an atherosclerosis-causing by-product" [379]. According to this article, in studies of more than 1,800 patients altogether, blood levels of three metabolites of lecithin were shown to strongly predict risk for cardiovascular disease: choline (a B-complex vitamin); trimethylamine N-oxide (TMAO), a product that requires gut flora to be produced and is derived from the choline group of the lipid); and betaine (a metabolite of choline).

Instead of using vitamins and DSs in your diet, healthy and sufficient amounts of choline, betaine, and TMAO can be found in many wild fruits, vegetables, and fish. Despite the evidence that these three metabolites are dangerous pro-atherosclerotic compounds, they are still commonly marketed by producers of vitamins and DSs as direct-to-consumer supplements, supposedly offering increased brain health, weight loss, and/or muscle growth. These compounds also are commonly used as feed additives for cattle, poultry, or fish because they may make muscle grow faster. It should not surprise anybody that meat from such livestock may have dangerously high levels of these compounds. In terms of the risk versus benefits of these commonly used supplements, they have very doubtful efficiency for human health. It is not rocket science to design for yourself a relatively healthy anthropic diet for preventing heart disease – just try to avoid artificial vitamins and most dietary supplements.

The shining star among DSs is coenzyme Q10, an enzyme that in natural form is involved in energy production and that also acts as an antioxidant. But there does not exist any scientifically confirmed evidence that treatment with coenzyme Q10 as extracted DS has a statistically significant influence on any health problem. Another example [380]: elevated plasma total homocysteine (an amino acid created by the body, usually as a byproduct of eating meat) has been suggested as a potentially modifiable risk factor for coronary heart disease, stroke, and other occlusive vascular conditions. High rates of cardiovascular disease in children with homocystinuria—a rare genetic condition causing extreme elevations in homocysteine levels—led researchers to hypothesize that moderate increases in blood homocysteine levels may increase cardiovascular disease risk in the general population. As is the common practice of modern reductionism, doctors started supplementation with B vitamins, and in particular folic acid, because they lower blood homocysteine levels and allegedly reduce cardiovascular disease risk among individuals with homocystinuria. Several large clinical trials conducted in patients without the condition have been inconclusive. "Consequently, a collaboration between their investigators was established in 2004 to conduct a meta-analysis based on individual participant data from all large randomized trials of folic acid-based B-vitamin supplementation intended to lower plasma homocysteine levels for the prevention of cardiovascular disease," the authors of study [239] writes. The meta-analysis of eight trials among the 37,485 participants showed no significant difference between folic acid (folate) and placebo groups in the number of patients experiencing major coronary events. But from the 37,485 participants, 9,326 had a major vascular event during the treatment period, 3,010 developed cancer, and 5,125 died.

At the same time, a deficit of vitamin B9 or folic acid is quite dangerous for the organism and is to a significant extent the result of a relatively recent "dietary" evolution of humanity from hunter-gatherers to farmers. We, as mostly the descendants of farmers, have lost the ability to produce an enzyme that is responsible for the effective acetylation and metabolism of folic acid in the organism. Fast-acetylation alleles capable of the efficient metabolism of folates are genetically present in hunter-gatherer populations. At the same time, the slower acetylates are very common in farmer-descended populations. It appears that the drop in folate intake associated with a shift to a grain-rich diet evolutionarily favored the alleles that suppress the use of folates. Because this is a very subtle change to gene function or expression, it could be corrected via diet. The only viable way to epigenetically restore our ability to metabolize folates is to use the natural form of folic acid, normally obtained from wild grown leafy greens, in our everyday diet. Most likely, dietary supplements or vitamins containing artificial forms of folic acid and farmed leaves from the grocery shop, with very little content of bioactive folic acid, would not be very helpful. But folic acid supplements could even be harmful. Recent Norwegian research [338] suggests that there may be a connection between high levels of folic acid in pregnant mothers and the risk of the development of asthma in their children.

Another example is the treatment of postmenopausal women by modern medicines and DSs derived from soyaa. Doctors usually prescribe estrogen therapy for women. Estrogen therapy, with or without progesterone, prevents most postmenopausal symptoms. However, the Women's Health Initiative findings suggest that because "the overall risks outweigh the benefits, most menopausal women now decline estrogen therapy, increasingly seeking other alternatives." [501].

A second alternative suggested by charlatans from the DS industry is DSs based on soyaa-derived substances: soya isoflavone concentrate is extracted from the seeds. But despite what a huge commercial advertising campaign claims, soya isoflavone tablets do not appear to be associated with a reduction in bone loss or menopausal symptoms in women within the first five years of menopause, according to a report [500] published in *Archives of Internal Medicine*. Study participants received a soya isoflavone dose in tablets equivalent to approximately two times the highest intake through food sources in a typical Asian diet to ensure they received an effective dose. Overall it does not appear that soya concentrate dietary supplementation will play any role in osteoporosis prevention, although it is a well-known fact that diet with natural soyaa could help to ease menopausal syndrome and significantly postpone it. The results of this research clearly show that efforts should be directed away from the hope of a single-tablet therapy for menopausal symptoms and toward using anthropic diet treatment in the pre-menopausal period for as long as possible. In the modern Western world, women are usually experiencing menopausal syndrome at an age between forty-five and sixty (average fifty-two). But the lower boundary under normal conditions should be at least at fifty-eight to sixty. Hormone imbalance and menopausal symptoms at a young age are very much linked to the nutritional quality of the diet and exposure to toxins through the diet. When women begin to miss ovulating, they are experiencing menopausal symptoms such as weight gain and turning their hope, as a last resort, to the DS industry, which very profitably provides them with false hopes and a lot of "scientific" mumbo-jumbo.

As such, using vitamins in the artificial concentrated form of tablets, extracts, or injections is not always safe for your health. But this is never the case when consuming vitamins as part of natural food. Olympic medical authorities warned recently [559] of the dangers of commercially available supplements. It is now well established that many commercial DSs contain dangerous compounds that can cause an athlete to fail a doping test and thus could also harm users from the general public. Generally the presence of these compounds is not declared on the product label. The potential for such low levels of contamination in a dietary supplement to result in adverse health effects raises significant concerns about the manufacturing practice of dietary supplements intended for consumption by the unaware public. As is shown in Chapter 4 of this book, the low levels of contamination do not guaranty the safety of products. Sometimes it might even make them more dangerous.

Of course, in addition to vitamins, our genome plays a huge role in our body's health. The state of the genome (and the accumulation in it of alterations) is, as indicated above, the key factor that determines how the organism undergoes the aging process. DNA contains all of the genetic instructions that make us who we are, and maintaining the integrity of our DNA over the course of a lifetime is a critical part of the aging process. Scientists [365] have discovered how DNA maintenance is regulated. The finding theoretically opens the door to treatments that may enhance the body's natural protection of genetic information. Keeping DNA intact longer into the later years of life could help eliminate cancers and neurodegenerative diseases, such as Alzheimer's, and other aging-related diseases. Researchers [365] have looked into the acetylation process that regulates the maintenance of DNA. They have discovered that the intensity of the acetylation process determines the degree of fidelity of both DNA replication and repair. Acetylation directs which proteins take which route in the multiple pathways available, favoring the protection of DNA that creates proteins by sending them down the more accurate course. Small epigenetic changes in favor of the expressions of fast-acetylation gene alleles via diet alteration could make the process of acetylation sufficiently intensive in farmer-descended populations. This means that an anthropic diet, that would cause an alteration in this acetylation-based regulation, might change the average onset of cancers or neurological diseases to beyond the current human lifespan through the protection of our DNA. As these processes are extremely complex, the current stage of pharmacology is virtually incapable of selecting the correct medicine, artificial or extracted vitamins, or DSs. However, one way to regulate these processes is to consume a maximally anthropic diet, which, thanks to the many evolved mechanisms of interaction between food and the organism, has a high chance of regulating the processes in the correct way.

Many of the studies on the benefit of vitamins are done at the behest of corporate sponsors. The influence of the international corporations that produce this garbage truly has neither internal nor external bounds. While scientists in the EU and the United States have achieved a somewhat wide dissemination of the results that show the uselessness or harmfulness of vitamins and DSs among the press and the medical community, in developing countries, such as Russia, Brazil, Argentina, China, India, and so on, these results are little known to the wider public. The corrupted milieu of medical bureaucrats and local "scientists" still recommend daily poly-vitamin use to the public. For example, the Nutrition Institute of the Russian Academy of Medical Sciences has dispensed official approval for a huge number of bio-supplements and poly-vitamins conveyor-belt style, keeping the results of global studies hidden from the wider public. Given the revolting system of the pharmaceutical business in Russia, the official circles of medical science and bureaucrats are virtually carrying out "DS-vitamin genocide" of the gullible parts of the population by allowing the prescription-less purchase of most medicines.

While people generally think that "all natural" equals "safe," processed botanical preparations such as plant-based food dietary supplements (DS) may contain concentrated compounds, like alkenylbenzenes estragole, methyleugenol, safrole or-asarone, which are potentially dangerous to human health. High doses or long exposure to these compounds can cause cancer [583]. In another example, according to one study [584], high doses of phytochemicals, including flavonoids from natural plants in the unnatural form of commercial DSs, could be very unhealthy if consumed in high doses as industrially processed dietary supplements. The report cites specific examples of toxic effects, including reports of liver, kidney, and intestinal toxicity related to consumption of high doses of green-tea-based DSs. The risk of such toxicity may be greater in individuals taking certain medications, or with genetic traits that increase the bioavailability of phytochemicals, the researchers said. So a DS called genistein (from plant extracted "phytoestrogen" present in many DSs) can undermine breast cancer treatment, researchers [585] report. These DSs have complex biological activities that are different from the natural plants on which they are based and are not fully understood.

About half of the adults in the United States consume dietary supplements regularly. About a hundred million men and women in the United States alone are taking those DSs to relieve some health-related conditions. However, these hopes are likely in vain. In reality these DS might stimulate the growth of cancer and exacerbate other diseases. Public demand for DSs containing large amounts of plant-derived substances that fuels the huge multi-billion-dollar DS production industry has outpaced scientific knowledge on the actual health benefits, best dosages, and risks of those processed phytochemicals, antioxidants, enzymes, etc. The complex effects of the DSs that contain phytoestrogens and many other substances in reality now are a subject of an ongoing experiment on humans in which the outcome is not only unknown, but certainly presents an immediate danger at least to some vulnerable part of population.

At the same time, scientific studies demonstrate the unquestionable benefit of "real" antioxidants and vitamins given at certain concentrations for the vital functioning of the organism. This may seem like a contradiction of all I have been saying, but in reality it is not. It is true that antioxidants and vitamins are beneficial and necessary for the human organism and those of other higher animals in their "live" (biologically accessible and bio-active) form. This is the form in which they exist in nature and in which they are designed to be consumed by every step in the food chain, humans included. The same applies to all other food components natural to humans. This is the basis for the anthropic principle of diet—only those plants and lower animals are maximally healthy for man that have undergone their evolutionary path under the same circumstances as our ancestors—i.e., been a part of the overall strange attractor of the metabolism of man and his environment. Speaking of the presence of antioxidants and free radicals in the body, I must note that numerous studies have shown that a strong but controlled production of oxygen radicals by the immune system is important for subduing illness, and a change in the balance between antioxidants and active radicals can be very damaging for the health of cells and the body.

During man's evolution, his inner metabolism conformed evolutionarily to the overall metabolism of his environment, which was his source of food. At the same time, primates, much like other representatives of the animal world, had their niche, which included consuming plants. Of course, these plants were beneficial for man to varying degrees. Clearly, simply by virtue of the simplest mathematics of such a variation, there were plants that were virtually useless or actually harmful to man, but at the same time there were plants that had extraordinary qualities in terms of their benefit to the organism when consumed. A well-known fact, which I've discussed above, confirms the anthropic principle of diet: 50-60 percent of medicines in the world are still produced from the plants of tropical forests, mostly from the Amazon basin.

Incidentally this too can be explained using the flicker-noise theory. If we consider all possible live products and their active components as potential food products for a species, then from the perspective of the flicker-noise theory during the process of the evolution of a species' metabolism, the peak of the spectral distribution of flicker-noise against individual sources of food, such as plants or animals and their components, shifts to the "lower frequencies" in terms of the extent to which they participate in the specie's metabolism (see Section 2.6). This change is fortified by the reverse interaction of the species' metabolism with its environment. In simper terms, this means that while a species' metabolism evolves in whatever biological environment, it attempts to involve as many different plants and animals in its diet as possible. Clearly, if all else is even, a species that achieves this to a greater extent will have more survivability.

However, we have to remember that this process has an evolutionary character. Adaptation to new sources of food might not take place during the life of a single generation. That is, the majority of organisms of this species will be burdened by the new sources of food, while a minority will through evolution discover the ability to use a new source or new sources of food. How this occurs can be seen by viewing living organisms as strange attractors of the metabolism. As a result of the chaotic reactions with new types of substances from new sources of food, a multitude of new trajectories of reactions in the phase space of the strange attractor of the metabolism appear, which lead to the appearance of new stable phase states of the strange attractor. As already noted, this is an evolutionary process, during which certain mutations in gene type assure the evolution of the attractor to a new state. This can lead to large changes in the appearance of specimens of the type, for example a change in size and so on. Equally clearly, this process takes place in nature alongside other changes in the environment, such as in temperature, intensity of solar radiation, atmospheric contents, etc., which I do not touch upon in this work.

The response of the strange attractor of the metabolism to changes in the content of substances entering it with food is an evolutionary mechanism that cannot be circumscribed by an individual. In this case, a mechanism of culling (negative) selection is in effect, which is characterized by the extinction of specimens living through the old metabolic processes and the flourishing of new specimens that are adapted to the new conditions through a modified metabolism. At the present moment, due to the dramatic changes that have been taking place in the contents of modern human diet, a large part, if not all, of the modern human population is at the stage of biological hardship—of selection through culling. There is, I think, simply not enough time for positive selection—i.e., for the greater appearance of specimens with the necessary characteristics—as the changes in the environment and diet are taking place too quickly in comparison to the time typically needed for the mechanisms of positive selection to come into effect. It is likely that the neo-evolutionary development of mankind will in some way counter these changes in the future, either through improvement in diet quality, or through changes in the Homo Sapiens species itself in relation to eating as a method to acquire energy and substances from the environment. I find it difficult to even suppose what sort of changes might happen to man as a species, but that this is the next direction in the evolution of our species I have no doubt.

Biochemists have long known that "live" chemical compounds act differently than the same compounds after extraction or as synthetic "copies." This is due to a large number of factors, such as the order of atoms in macro-molecules and their physical structure, the presence of other molecules, the hetero-phase structure of the environment, etc.—what in this book we call the metabolic phase state. The presence of the molecules of chemical substances in a disanthropic phase state—i.e., fully or partially uninvolved in the metabolism or with the tendency to change the natural trajectories of the cascades of metabolic reactions—leads to the accumulation of disanthropic compounds in cells and the space between cells in human organs, with their removal from the organism impossible and its gradual poisoning by the molecules of these substances inevitable.

The negative effect of the imperfect molecules of the synthetic substances created in large quantities during the industrial chemical synthesis of vitamins is that they often play the role of anti-vitamins, due to their chemical similarity to the molecules of vitamins. Thanks to the similarity of molecules, end caps, or other molecular structures, they take the place of vitamin molecules in the structures of biochemical compounds during the chemical reactions in the organism's metabolism, which distorts the mechanisms of reaction flow and the order of the transformation of compounds on the inter- and intracellular level. As a result the frequency of the cancerous growth of cells, their death, and the disruption of their function increase. This is the ultimate first cause of many illnesses.

The existence of the various chemical forms of vitamins known as vitamers has long been known to science. During the process of chemical synthesis, vitamers and anti-vitamin molecules are always created alongside vitamins. Under the conditions of mass production, the separation of anti-vitamins and other side products of chemical reactions is a difficult and practically (economically) impossible task. These pharmaceutical corporations don't make much of an effort either, as increasing profits is their sole aim. Also, during the processes of chemical and heat-based extraction of the molecules that appear in them out of natural sources, the vitamins are transformed into anti-vitamins and/or less beneficial vitamins. And as I have already said, the extraction of natural molecules out of bio-organic clusters, in which they are in a live state, in most cases completely eliminates their expected beneficial effect on an actual human metabolism.

Apart from this, the current state of scientific knowledge about the interaction of natural vitamins in the metabolic processes is quite rudimentary. A work [19] cites many contradictory assessments of the role of vitamin C in the metabolism. The usual mechanism for the effect of vitamin C is described as follows: the molecules of the vitamin, when interacting with free radicals, acidify and turn them into harmless molecules. Then the acidified molecules of the vitamin return to their non-acidified state through the action of a specific enzyme. Everything said above about vitamin C is entirely true, but not the whole truth. It is true that the reactions of the molecules of vitamin C follow this path, but the results of the reactions vary widely, depending on the molecular content of the environment where the reactions take place. Just as it can prevent illness, vitamin C can in some cases cause illness or even death of the organism. It is interesting that, with the exception of higher primates, virtually all organisms, plant or animals, synthesize vitamin C independently in their own organism. At some point in the process of evolution, we lost the gene responsible for the final stage of the synthesis of this vitamin in our organism. We cannot say that this is definitely for the worse. If we lost it during the process of evolution, then this loss brought us more good than harm overall.

Despite the widespread point of view about the importance of only the antioxidant function of vitamin C, this vitamin is needed in our organism for a wide range of reactions. For example, it is vitally important in reactions involving the enzymes that synthesize collagen. Collagen structures make up 25 percent of the mass of the proteins in our organism. In the absence or deficit of vitamin C in the metabolic processes that synthesize collagen, their filaments form incorrectly. Scurvy is the visible result of this. In addition, a deficit of vitamin C leads to the synthesis insufficient levels of Carnitine in the organism. But this substance is extremely important for "burning" fat and for its role in the inner metabolism of cells, in particular in mitochondrates, and for many other reactions in the organism. Among these, one of the most important, from the point of view of food absorption, is the transformation of trivalent iron, which is usually present in food and impossible to dilute, into the divalent biologically active form that can be diluted and therefore easily absorbed by the organism. As is well known, iron is an extremely important element in mammal blood. All this also applies to the metabolism of copper in the organism. But odd as it may seem, aside from its antioxidant effect, vitamin C also has a pro-oxidant one [19], and this threatens the organism with all the problems usually associated with the effect of free radicals. The organism defends itself from the pro-oxidant impact of vitamin C, but being broken down so quickly, the vitamin C does not have time to accumulate both in the body as a whole, and in individual cells, environments, and organs. This balance can only be maintained if natural vitamin C from anthropic food enters the organism. Synthetic or extracted vitamin C, especially in high concentrations, is extremely harmful whatever its many defenders might say, and carries the threat of a deformed balance of concentrations in the metabolic reactions. Everything said above about vitamin C is equally true of other vitamins and supplements. Their use can cause serious imbalances in the organism, with far-reaching consequences.

The activity of the molecules of vitamins and biological components of plants or their fruits does not exist in isolation from their "live" phase state. It is well known, for example, that the three-dimensional structure of proteins, amino-acids, and peptides inside a live cell is cardinally different from their structure once they have been removed from the cell. It is also known that their three-dimensional structures have the major impact on their chemical interaction in cells. Even if identical in their atomic contents and order, molecules with different three-dimensional structures (stereometry and conformation) act completely differently in chemical reactions. That is, the biologically active components with beneficial qualities are only healthy in combination with the complex of other substances in a plant's cells—i.e., in their specific metabolic phase state in live nature.

Under the conditions of actual intracellular and intercellular metabolism, a uniquely varied and complex set of biochemical processes take place, which at the current time have been little studied. The difficulty of studying and understanding the mechanisms of these processes is connected to our insufficient knowledge of the details of how live organisms function on the molecular level, given the variety, complexity, and ambiguity of these processes. For example, biochemistry increasingly understands that quantum mechanisms of the informational interaction between the molecules of a substance play a significant role in the processes of molecular transfer in a live organism. The three-dimensional structure of the molecules plays the decisive role. For example, the extreme importance of the order of amino-acids and the spatial structure of proteins and peptides is illustrated by the existence of prion proteins.

Prion proteins are contained in the membranes of nerve cells. In some cases, a normal protein can change the spatial structure by "re-arranging" an amino-acid chain into another order. Such an altered protein can, upon contact with a normal protein, turn it into a new "alternative" prion through information exchange on the quantum level. The uniqueness of this process lies in the fact that material transfer between the molecules does not take place, and that only the quantum set of electron clouds of molecules outside of the recipient, along with its three-dimensional structure, changes under the influence of an alternative prion. Moreover, prion proteins can infect humans, and in many ways behave as viruses, with the ability to multiply in the organism.

The transformation of prion proteins can be seen on the level of the organism in damage to brain tissue, in which a multitude of tears appear; as a result, the brain starts to resemble a sponge. For this reason, illnesses caused by prions are known as spongiform encephalopathies. In people these are known as Creutzfeldt-Jakob disease and Kuru. This example also demonstrates the extreme importance of the details of the structure and state of molecules, and their interaction with the molecules around them, on the character of the flow of reactions in the organism. Only a tiny proportion of the interactions and processes involved in biochemical transformations in our organisms is known to modern science. For example, the biochemical processes of intermediate compounds, which appear during the transformation of substances after their entry into the organism and before their exit, are still little studied. The time span of these intermediate compounds varies from seconds to hours. At the same time, they interact with the surrounding molecules of various substances that are byproducts of the compounds of our food. The mechanism and result of their reactions depends on many characteristics of molecular environment, including the mineral, organic, and stereometric contents of the biopolymer molecules.

As such, the consumption of synthetic vitamins is unquestionably dangerous and harmful. But even vitamins obtained through extraction from natural plants and animals do not possess those qualities that their consumers expect in an absolute majority of cases. This is caused by:

1.  The peculiarities of the processes of extraction, which include the use of active chemical reagents and environments that change (destroy) the structure of molecules and their environment.

2.  The processes of the acidification and mechanical destruction of the molecules of vitamins and bio-organic substances, for example during drying, chafing, and other technological processes.
3.  Changes in the phase state of the molecules of vitamins in the final substance (a pill) compared to the live form, which eliminate or reduce anthropic qualities.

However, if synthetic vitamins are dangerous to consume, then ones industrially manufactured or made by handicraft from natural sources by fraudulent companies are in most cases simply useless—which is unquestionably a step in the right direction, albeit a minuscule one, especially compared to the economic expenditures made by the consumer. For example, one study [230] demonstrates the complete uselessness of the vitamin D extracted from both its natural live environment in plants and its synthetic variant.

The evolution of primates and their predecessors in the evolutionary chain took place starting in the sea environment and ending in the environment of tropical forests. This is confirmed by the fact that man's closest relatives in the animal world, monkeys, still inhabit almost exclusively an environment of tropical forests. These forests differ from almost all other landscapes on Earth by the variety of their plant world, micro-organisms, animals, and insects, and by the intensity of the solar radiation, a year-round source of energy for biochemical transformations. For example, the soil of the Amazon jungles contains thousands of varieties of micro-organisms, which cannot be found in the soil of mid-latitude regions and the majority of which are unknown to modern science. The life activity of these micro-organisms is more intense, which, finally, increases the intensity of the metabolism of the soil in the Amazon jungles hundreds of thousands of times over. The amount of an enormous variety of bio-organic substances, which are used by plants for nutrition, play a part in the metabolism of plant cells and give their parts—roots, stems, leaves, bark, and fruits—their curative, vitalizing effect, that is contained in the Amazon basin soil is greater by the same hundreds of thousands of times over. The combination of all these mentioned features ensure the rich micro-element and bio-organic contents of that flora and fauna and the diversity of metabolic processes. As such, we can with predict with great certainty that the plants growing in the forests of the tropical regions are the healthiest and most evolutionarily suitable for man.

Many plants of tropical origin are used as the raw material for DSs. However, despite expectations, the process of crude extraction creates extracts that not only do not increase the impact of the healthy qualities of the plants used as raw materials, but are if anything either entirely useless or are good only for short-term consumption as a form of medicine. All the money in the world spent on advertising and promoting these extracts as various sorts of "panaceas" cannot change this. Nonetheless, as I have noted, more than half of all medicines worldwide are produced from plants that grow in the tropical forests. However, medicine produced from the sub-strata of live plants, even if it does help in treating some illness, has a number of negative side effects alongside the benefits. In the majority of cases, these side effects are the result of the active components of a curative plant being extracted from its natural environment, where its side effects can be compensated by the bio-organic surroundings, and/or because the side effects appear due to the absence of this environment from their anthropic phase state. By and large, such medicines can only be used in the short term. Otherwise, their positive effect will pale in comparison to the harm their side effects cause.

It is not only that the three-dimensional biopolymer and peptide structures of live material are extremely important for the metabolic processes, but also that the physical-chemical interactions in these three-dimensional structures are crucial for the impact of minerals and vitamins. If a substance is extracted from a live natural object, not only the three-dimensional structures of proteins, peptides, amino acids, and their polymer chains are destroyed, but also the physical-chemical interaction of minerals and vitamins with these structures, which leads to the loss or manifold decrease of their effect. In most cases, the side effects of a medicine are toxic effects, caused by the need to increase the dose of the medicine due to its low effectiveness, and/or by the accumulation in the organism's tissue of elements that play no part, or little part, or play their part poorly in the metabolism. As a result the organism is poisoned to some extent by medicine or DSs. The metabolism in live cells, which is determined by quantum interaction and non-linear inner structures, is so complex and little studied by modern science that the creation of genuinely effective DSs through a reductionist, primitive linear approach is impossible.

The mechanistic, linear approach to the study of nature has long been abandoned in other areas of science, but it leads a strange life in the "science" of reductionist justifications for the use of vitamins and bio-additives. In my opinion this is connected to the fact that the main aim of the production of DSs is to maximize profits. There is nothing wrong with this aim if we do not lose sight of human ethics. But the vitamin and DS industry has gone down the path of extracting super-profits with minimal investment, taking advantage of the uninformed public to cloud the public consciousness with primitive pseudo-scientific approaches and wild advertising claims. This is a vein mined both by large, multi-national pharmaceutical corporations that produce "miracle" vitamin mixtures, and similar and relatively small firms that have joined the gold rush by producing "natural" bio-additives against all ailments. Both of these flourish through the irresponsible and shameless advertising of, at best, useless substances. Mass products of minute production costs and dubious quality are advertised as panaceas.

Medicine is an expensive thing. One reason of no little importance is that, in addition to the immediate costs of research, developing the technology, and production, medicinal substances have to be certified. In most countries this requires large-scale and expensive clinical trials to prove the existence of the claimed benefit and to show any side effects. In addition, there is the real risk that the trials will demonstrate some undesired and powerful side effect, and the medicine will be banned.

Any business is faced with certain temptations: what if they could sell their product with the same gigantic margins as enjoyed by medicine, with the public's awe before this "medical advance," but without the trouble of trials and research? And, what is also important, without getting into trouble with the law when their "miraculous" substance does nothing to help the consumer? There is a simple solution: they can call their creation not a medicine, but a biologically active supplement. No need to prove the effect! This, then, is how the government keeps us safe. Unlike medicine, which has to cure in some way, DSs just have to not be poisonous. That is, the recommendation of any certifying organ proudly emblazoned on the packaging means only that the product is not poisonous but in no way is a guarantee of its curative qualities. Sadly, as already noted, in a series of cases, certain DSs have nonetheless proved poisonous after prolonged use.

In recent years, pharmacological spam on the Internet has become a dangerous weapon in the hands of dishonest advertisers for hundreds or thousands of different panaceas. Weight loss remedies have become a large part of Internet charlatans' output. Globally, people who consider themselves overweight or obese (very often wrongly) waste billions of dollars, pounds, euros, and so on every year on food products that "imply" that they aid weight loss, but are totally ineffective, says a nutritional expert on the *British Medical Journal* website [240]. In the United States alone, people spend around 40 billion dollars a year on such products. The people have been fooled and commercially exploited by the DS industry and Internet marketing.

Spammers have the most love for extracts from Hoodia Gordoni. This cactus has already been used by Bushmen to sate hunger. Its worldwide popularity was caused by the mighty marketing machine set into gear by Pfizer. But their work was in vain: the desert plant is poorly suited to cultivation on an industrial scale, while the concentrates used in DSs were found to have negative side effects that rendered them unacceptable for human use. In addition, the direct weight loss impact of products based on Hoodia was found to be negligible—on the level of a placebo. Nonetheless, wild demand for the plant led it to the brink of extinction, while the market was flooded by a tide of foods containing dried or ground Hoodia. The ban placed by many countries on this "natural appetite suppressant" finished the work of exiling this dubious product into the world of spam.

At the same time, no side effects are observed when Hoodia is consumed in a live state either by Bushmen or by Europeans undergoing a course of treatment in

Namibia. Hoodia genuinely does suppress the appetite wonderfully and allows one to eschew a significant proportion of the daily ration. The side effects of DSs based on Hoodia are easy to explain—the process of drying, grounding, and subsequent monstrous acidification destroys all the qualities it possessed in its live state, leaving behind only, to differing extents, the toxic products of the breakup of the molecules of the organic compounds that had been contained in the live Hoodia.

The economic model of pharmacological enterprises and corporations is built on the exploitation of any exclusivity, either in the form of patented methods of production, or of extremely complex chemical and technological processes, or of gigantic expenditure on marketing, including advertising. The overall margin on vitamins and DSs varies from hundreds to many thousands percent. For example, a major corporation in the field of drug production will make several billions of dollars from each drug brought to market just during the period of patent protection. After this period ends, second-tier companies get involved to produce "generics," which also bring in billions of dollars. Research on the methods of developing new bio-products in the United States has shown that this is an enormous, well-organized industry that generates enormous profits for all participants at each stage of developing a product.

The process begins with the study of the mooted product in venture firms. The average expenditure on the synthesis and testing of a product, including clinical trials, varies from 20 to 80 million US dollars. If the trials are completed successfully and the US Food and Drug Administration grants its approval, the capitalization of the venture firm goes up to a billion dollars at the very least. A large pharmaceutical company buys this venture firm for a billion dollars or more and in turn extracts 5 billion dollars or more, depending on the product, from its production and marketing. At the same, time I should note that the sum of a billion dollars only applies to relatively rare medicines, which treat an ailment suffered by no more than a few thousand per year of the 300 million living in the United States. We can imagine how much profit they gain from medicines for more widespread illnesses.

The middle companies operating on the DS and mineral markets are even more shameless. These obtain and extract everything imaginable from all known curative plants and fruits, sea and land creatures, and sell this to the broad public in the form of some sort of panaceas for the majority of illnesses. Just as there's no reliable, independent proof of the existence of little green men, there's no independent evidence for the benefit of an absolute majority of DSs. All the evidence comes from interested parties—UFOlogists in the first case, and producers of supplements and associated researchers in the second.

Certain parts of curative plants if correctly prepared and stored, as an example, might have a limited beneficial effect, but they are not DSs proper, as no one usually ascribes them magical properties. Why then do these corporations not develop the market for genuinely beneficial products? This is easy enough to explain: they cannot obtain exclusive rights to any part of the business and consequently ratchet up prices and their margins. As a result, despite the obvious benefit of such products, they are of no interest to them. In addition, it has been observed that the use of certain DSs, multi-vitamins, and mineral complexes can lead to hypervitaminosis, especially in children. The latest results of this research show that this is caused by the multi-vitamin tablets containing high concentrations of various "vitamins" in not-entirely-absorbable forms. According to some research, the unabsorbed part of the multi-vitamin complex simply poisons the organism, accumulating in individual organs and disrupting their functioning. This problem never arises as a result of the consumption of highly anthropic live products because an excess amount of some mineral or organic substance from anthropic food is easily removed from the organism. In this way, the research of many members of the scientific community testifies that the consumption of DSs and multi-vitamins is not without danger to your health and can lead to the development of a variety of illnesses. Breaking the anthropic principle of nutrition in favor of sheep-like credulity toward the advertising of DS producers can cause direct harm to your health.

The ignorance of many "scientists" put forward by commercial pharmacology and medicine is astounding too. They think that the essence of any food or plant is a collection of certain chemical elements. In this, they forget the simple fact that, chemically, pork mince is very similar to a pig; and if we add to it a number of sub-products, bones, and the animal's skin it would become entirely identical. Yet still the pork mince wouldn't oink or wag its tail. The collection of all the elements that make up a pig is only actually a pig if all its components find the inter-dependent state of a live animal. The same applies to plants. The synergy of live nature produces in them substances that play a part in the phase trajectories of the metabolism only in the live state or in a state of natural fermentation with highly anthropic qualities for man. Thanks to the mutual evolution of humans, animals, and plants, a number of the latter two have become suitable for human consumption, but only in the anthropic state—i.e., in the state of greatest appropriateness for consumption. This does not mean that God or nature created them according to some plan. All it means is that during the process of evolution, the metabolism of our forebears and of man himself adapted, through mutual interaction, to the type of metabolism that consumes the metabolic forebears of the mineral and biochemical components we need to consume now. A simple example of this is the indispensability of amino acids from those types of plants that man has been in a symbiotic evolutionary relationship with.

Maintaining the anthropic state of plants poses a great challenge in terms of its processing and storage, which are incompatible with primitive methods of drying (including vacuum drying), extraction, and thermal processing. The failure to develop and practice the use of such technologies in the commercial processing of plants into DSs and vitamins is the main reason for the uselessness or harm to the human organism caused by their commercial versions.

Studies about the influence of natural substances in cancer prevention are conducted on the basis of the conversion of live food with known anti-cancer properties to the form of pills or extracts. The vast majority of these studies lead to disappointing results—the effect of pills and extracts in real medical testing in the best case is absent, or in some cases that are referred to in this book, give negative results. The main problem in such an approach is that the conversion from live food to the pills, in fact, means that scientists are denying the difference in state of chemical substances in a broad sense in living organisms and in the extracted dead substances. However, like a life itself is unique in respect to inanimate nature, so natural food substances in the natural state are equally unique in respect to the extracts and pills. This uniqueness lies in the fact that the living natural substances in the anthropic food are providing maximum longevity and the highest possible state of health for a particular individual. In addition, for people who have been exposed over the long term to disanthropic nutrition, anthropic foods can restore metabolic processes and the expression of the genome to an evolutionarily determined state, in whole or in part. Ignoring this uniqueness of living or anthropically fermented food in relation to DSs or vitamin pills explains the worthlessness and futility of the latter for human health.

## 6.5  Are the commercial superfoods really that super?

While the products of live nature in a natural state, so-called superfoods, are truly a gift from God to humanity and an essential part of an anthropic diet, many of their commercial versions, to be found on the shelves of health food stores, are useless at best. One of the main problems of a healthy diet is how to deliver the anthropic qualities of food harvested from a wild state or produced in a tightly controlled, organic system of agriculture to the consumer. Many of the superfoods sold in commercial chains or Internet stores are genuinely made from superfoods. But they do not possess even a fraction of their qualities. Why does this happen?

The reason is that the initial superfoods undergo technological processing, which includes their drying, grounding, steaming, extraction, and so on. Depending on commercial considerations, these products of processed superfoods are sold as tablets, powders, dried leaves, infusions, and so on. It cannot be denied that some of these products are delivered with their beneficial qualities intact, but these are a minority. Moreover, even this part of commercially available superfoods is often kept under inappropriate conditions, while for an absolute majority the question of prolonged storage has not even been studied. The regulators pay this no attention, being content that, in their opinion, commercially available superfoods do not cause great harm, while the costs of detailed investigations are prohibitive.

The drying and grounding of superfoods, just as with any other product, exposes them to a level of oxygen thousands and tens of thousands times as great as during the natural storage of these products of live nature. During the grounding of a product, its particles become thousands of times smaller than is normal for that product. At the same time, the surface area of the product grows thousands of times over, which results in a thousand-fold increase of the intensity of the processes of interaction with the oxygen in the air, which is a very strong acidifier. In addition, the intensity of acidification increases because, in a ground state, the product lacks a protective outer layer, which has evolutionarily formed in the fruits, leaves, and roots of a majority of plants.

The acidified elements of a processed biological product are a set of compounds which are entirely different in their chemical properties and have a completely different set of biological effects on the organism. During the metabolic processes, they play a part in entirely different reactions and generate unexpected cascades of reactions. What is more, a large variety of so-called pro-oxidants—free radicals, or molecules with highly acidifying qualities—appear in a product in such a state. The presence of such molecules presents a fairly high threat to the human organism. High levels of oxygen combined with the other free radicals produced by excess oxygen could lead to the formation of the most toxic free radicals, such as, for example, peroxynitrite. Upon reaching a certain level in a cell, peroxynitrite becomes even deadly. A potent cytotoxic agent, peroxynitrite can damage a variety of biomolecules such as proteins, lipids, and DNA, and is considered one of the major pathological causes of several diseases. Since most biological tissues of foods produce both nitric oxide and superoxide, oxygen that is enhancing the formation of these precursors is enhancing the formation of peroxynitrite. The oxidative activity of peroxynitrite involves direct reactions with cellular constituents such as antioxidants, the production of reactive species capable of initiating lipid oxidation, and catalyst-dependent nitration reactions that may in addition redirect the reactivity of peroxynitrite to favor protein oxidations. Thus, peroxynitrite reactions in food have very deleterious effects, leading to a decline of the anthropic functional properties of that food. To prevent the peroxynitrite from undergoing damaging activity in a body, it is better to use phenolic rich food, such as cacao for instance. In addition, as shown in study [229], certain spices have protective capacities against peroxynitrite-mediated molecular damage. At the same time, we have to bear in mind that an acidified product might contain not only peroxynitrite, but also tens of thousands of types of similar free radicals with differing degrees of activity. By the way, if you see a shop label informing you of the presence of whatever antioxidant in the product, don't just assume that the producer has decided to bless you with healthy substances. Often the addition of antioxidants is done only to lengthen the expiration date and has nothing to do with making the product healthier for you.

When we say something about products ground to a dust, the same applies to dried products with a porous structure. These form a majority of the market. Even without grounding, the pores of such products already contain a developed surface area that can interact with the oxygen in the air. The speed with which porous and ground products acidify is extremely quick. Within a few days, such products become not just useless but actually harmful for the organism. This is confirmed by a wealth of evidence from the use of superfoods. In the United States, 44 percent of all expenditure on non-traditional medicine has been on such products as fish oil, glucosamine, and Echinacea. No one is denying the value of fish oil or Echinacea when consumed in their natural form, from fresh fish or vegetables. But the results of research have shown the complete ineffectiveness of commercial forms of Echinacea, which loses all its healthy qualities after being dried, ground up into a flour, and presented as an extract. Extraction also changes the healthy qualities of the natural forms of superfoods unrecognizably.

Speaking of fish oil in the commercial capsule form, there is published research [230] to suggest that it can lead to the development of certain types of cancer. While polyunsaturated fatty acids such as omega-3 fatty acids and omega-6 fatty acids are essential nutrients for everything from brain function to cell function, the human body is unable to make them. But at the same time, these fatty acids are the most vulnerable components in human cells and in food because of their high sensitivity to oxidation caused by highly reactive oxygen molecules. When polyunsaturated fatty acids are oxidized, they form a new compound with the oxygen in the bloodstream that impairs the function of the cell membrane. In addition, these fatty acids are also used in the body to produce an array of fatty acid-derived hormones that are very important for cells. Under the influence of advertising, people buy dietary supplements such as capsules of fish oil, omega-3 fatty acids, or flaxseed oil to get these nutrients. However, extracted DS forms of polyunsaturated acids are very different from the same acids in a food which contains them: fish, seeds, etc. In natural food form, these acids are not as susceptible to oxidation as in DS form. And scientists who conduct aging research know that the one critical characteristic of people who live very long lives is not taking DSs but eating the right food. In this particular case, it is fatty fresh wild sea fish and seafood which are rich in polyunsaturated fatty acids.

In this way, even the products of live nature with well-documented healthy qualities lose these completely after the processes of drying or extraction. Fish oil has its maximal healthy effect on the human organism when it enters it from live or weakly processed fish. In some cases, the healthy qualities of the oil and the fish remain at least partially, if these were processed correctly and if these were made on the basis of a high-quality oil; extra virgin olive oil is one example among several.

Many herbal remedies available over the counter in pharmacies and health food shops lack any properties beneficial for health. People generally believe that herbal medicines are somehow different to other medicines because they are "natural." But that is not the whole story. A number of popular herbal remedies sold in powder or dry form have lost their initial lively qualities through the oxidation process. Producers are missing key details in their production processes: in particular, it is of paramount importance to prevent the influence of oxygen, which does not happen in production. Many of these products are even unsafe to use because of their high content of ROS (reactive oxygen species) owing to oxidation in the time of production and shelf life. These products sell unlicensed, and consequently are not required to meet any standard of safety, and over half of these are marketed as food supplements. A new EU law recently came into force regulating the sale of traditional herbal medicines (as in the United States) such as St. John's wort and Echinacea, but not regulating the rate of oxidation, which is crucial for their safe use and final benefits in a product for the customers. The majority of over-the-counter herbal products did not contain any of the key production information required for evaluation their vitality. Customers have a right to reliable and comprehensive information when buying herbal medicines—information which tells them whether or not they are produced under proper conditions.

Apart from oxygen, or more precisely in combination with it, heating the superfoods up to high temperatures has a part to play in destroying its healthy qualities. For example, 70–90 percent of the vitalizing qualities of such a king of the superfoods world as raw cacao disappear already at the first stage of its industrial processing—the removal of the bean skins under temperatures of 130–150 degrees Celsius. Any such method of processing is deadly for a majority of the products of live nature.

One of the most subtle processes of the interaction between food and the organism is the transformation of food biopolymers during the metabolic process in the digestive system. This interaction has evolved such that the food biopolymers deconstruct into monomers and oligomers in the digestive system, which can be used in the metabolic processes and the construction of the biopolymer structures of the organism. If the function of the digestive system is distorted, especially in the intestine, as a result for example of the disruption of the diversity of the microbiome, the processes of the deconstruction of biopolymers are also disrupted. What you get is oligomer or monomer compounds with qualities unsuitable for the organism. These compounds cannot build the viscoelastic matrixes of the organism or build them in distorted forms.

The other important source of the creation of irregular compounds from the products of the deconstruction of biopolymer structures is intensive industrial or culinary processing. The effect of active chemical agents, oxygen, and heat deconstruct the biopolymer molecules of the initial food in a way other than how they are deconstructed under normal conditions of low-intensity processing, for example during natural fermentation or in the intestine. As a result you get scraps of molecules in the form of free radicals or other compounds unsuitable for the organism. The presence of active radicals or debris from the biopolymers of food has a negative impact on the metabolic processes in the organism because these substances are not natural for it. In addition, on entering the organism's metabolism, they distort the trajectories of the cascades of chemical reactions and the construction of polymer matrixes, and thereby change many of the traits of cells and tissues (see Chapter 3). This in turn leads to changes in many functions of the cells, the aging of tissue and other ailments.

# 7
# The Light at the End of the Tunnel

## 7.1  The theory and practice of anthropic diets

In a recent academic review [184], a researcher at the University of Minnesota's School of Public Health has concluded that food as a whole, as opposed to specific nutrients in DS form, may be crucial to having a healthy diet. This notion is contrary to popular practice in the food industry and government, where marketers and regulators tend to focus on total fat, carbohydrate, and protein, and on specific vitamins and added DSs in food products, not the food items as a whole. "We are confusing ourselves and the public by talking so much about nutrients when we should be talking about foods," said David Jacobs, Ph.D., the principal investigator and Mayo Professor of Public Health at the University of Minnesota. "Consumers get the idea that diet and health can be understood in terms of isolated nutrients. It's not the best approach, and it might be wrong." Jacobs, with coauthor Prof. Linda Tapsell of the University of Wollongong in Australia, argues that people should shift the focus toward the benefits of entire food products and food patterns in order to better understand nutrition in regard to a healthy human body.

These researchers focus on the concept of food synergy—the idea that more information about the impact on human health can be obtained by looking at foods as a whole rather than at a single food component (such as vitamin C, or calcium added to a container of orange juice). Jacobs and Tapsell provide several examples in which the single-nutrient approach to nutrition has not proved to benefit health: Long-term randomized clinical trials, considered the gold standard for making judgments about nutritional treatment and health, have failed to show benefit or have suggested harm for cardiovascular events as a result of isolated supplements of beta-carotene and B-vitamins. A similar large experiment also did not show benefit in total fat reduction. In contrast, myriad observations have been made of improved long-term health as a result of foods and food patterns that incorporate these same nutrients naturally occurring in food. An understanding of the interactions between food components in both single foods and whole diets opens up new areas of thinking that appear to have greater application to the contemporary population's health issues, particularly those related to chronic lifestyle disease, Jacobs said. "It is this new understanding that reminds us emphatically of the central position of food in the nutrition-health interface, which begs for much more whole-food-based research, and encourages us in both research and dietary advice to, 'think food first,'" Tapsell said.

Researchers in Sweden have found [242] out what effect multiple, rather than just single, foods with anti-inflammatory effects have on individuals. The results of a combined diet study show that bad cholesterol was reduced by 33 percent, blood lipids by 14 percent, blood pressure by 8 percent and a risk marker for blood clots by 26 percent. A marker of inflammation in the body was also greatly reduced, while memory and cognitive function were improved. It is obvious enough that even in the best-case scenario the drugs or DS products with health claims affect only one or maybe a couple of metabolic pathways. This leads to deformation of the phase space of the metabolism as a whole. But with a combination of the right food with a wide spectrum of anthropic elements interacting with each other in the human metabolism you can, in a simple and striking way, affect many metabolic pathways simultaneously.

If we look at the absorption of food in the context of the human organism as a strange attractor for the organism, all these ideas find their fundamental confirmation. The approach of modern dietology is based on the recommendation of some food component or other, often ignoring the fact that these components are to be found in different degrees of human biological accessibility in the different products and DSs. A normal anthropic diet contains the widest spectrum of organic and mineral formations, which in the environment of the human metabolism reciprocally interact within different parts of the phase space of the metabolism. It is precisely the reciprocal interaction of the metabolites in food that produces the best impact on the overall condition of the organism's metabolism.

Anthropological studies have shown that humans' hunter-gatherer omnivore diets have not changed significantly over evolutionary time and tend to share certain stable characteristics for last one hundred to two hundred thousand years. The human race maintained relatively steady diets and only slightly changed their feeding strategies over the territory they occupied on Earth. Humans could eat seafood, meat, insects, and land and aquatic plants, which were relatively abundant in any location of their habitats. Such a diet enabled our ancestors to survive for long periods of ice age, droughts, and so on. Humans were clearly the main beneficiaries of their diverse eating habits and have since forty to fifty thousand years ago spread to nearly every corner of the world. But several thousand years ago, when humans were still at the hunter-gatherer level of development, humanity fed itself with several thousand different sorts of plant. Each equatorial region had a variety of plants, well known to the hunters and used as a regular foodstuff that numbered no less than a hundred different sorts. Since the Neolithic revolution in agriculture, the number of different naturally occurring plants and fruits in the human diet has gone down hundreds of times over. At present, ten main cultures, including the grain and bean ones, supplemented by two to three types of root-crops, make up 80 percent of all food and animal feed produced worldwide. Several cereal cultures provide 50 percent of the calories in the modern world's diet [390]. If we take into account that the diets of a large part of cattle have a high content of the same cereal cultures, then the proportion of the modern human diet that comes from this narrow range of cereal cultures might even be a lot higher.

Bearing in mind the degradation of food as a result of many centuries of selection and modern methods of processing (see Section 6.1), these cereals don't have even a small part of the value of their wild ancestors. The plant sources of nutrition to be found in nature, which have also disappeared from the human diet, had provided the human metabolism with a countless number of anthropic food ingredients, to which man in each region was evolutionarily adapted. The same applies to the part of the diet consisting of animal meat. Only a small percentage of wild animals were domesticated. Similarly to the narrow spectrum of plants that have been cultivated, this has deprived our metabolism of the intake of genetically and metabolically varied food.

I must also note that, since the earliest times, people have unconsciously added anthropic elements to their ration as needed. One extremely exotic example of this is geophagy—the eating of earth! But according to many researchers, geophagy is not simply a consequence of food shortage. The first written account of human geophagy comes from Hippocrates more than two thousand years ago [464]. Since then, the eating of earth has been reported in almost every country. In my view the most probable explanation for human geophagy is that it has some special nutritional value. For example, many reports indicate that geophagy is often associated with anemia, but several studies have shown that cravings for earth continue even after people are given iron as dietary supplements (DS) [465]. This in no way disproves the theory of nutritional value, but is only evidence that artificial iron DSs do not contain that form of iron that the organism requires. People crave dirt because it provides natural minerals, such as iron, zinc, and calcium in highly anthropic states. And according to data [464], people often eat earth during episodes of pathogenic gastrointestinal stress. It's unlikely the intestinal problems are caused by the dirt itself because people usually boil the clay before eating it. It means that earth probably has an anthropic protective effect, working as a shield against ingested parasites and pathogens, and/or as an absorber of food toxins.

It is obvious that, in the time that has elapsed since the Neolithic revolution, the human organism has to some extent attempted to adjust to the changes. However, to change the processes created by millions of years of evolution over such a short period of time, especially on the genetic level, has proven beyond nature—especially as the intensity of deviations from the anthropic in the processes of eating have accelerated rapidly over the last one to two hundred years. In last few decades, as a result of the chemical pollution of our environment and genetic manipulation, these deviations have taken on a practically explosive character. These many incompatibilities with our biological evolution and present evolved state are the cause of all the degenerative diseases or, in other words, the instability of the strange attractor of the phase space of the metabolism. It is precisely this instability that often leads to the breakdown of the most complex evolved structures of our organism at all levels of their hierarchy. A memorable example of this is the degeneration of the evolutionarily appropriate state of our cells during cancer, which are noted in Chapter 3. The changes to the human genome that took place since the start of the Neolithic revolution have been caused for the most part not by changes in our diet, but by the wide spread of epidemic diseases. We cannot deny that some individual changes in the genome were related to diet, for example in genes connected to the absorption of milk, alcohol, and certain seaweeds and so on. However, the depth of genomic changes connected to diseases that have appeared as a result of changes in the conditions of human life greatly exceeds the depth of changes connected to diet. The transition to settled agriculture and animal domestication has led to the evolution of micro-organisms and viruses characteristic of domesticated cattle into micro-organisms carrying diseases fatal to humans [390]. What is more, the spectrum of these diseases is not exhausted by the smallpox, plague, cholera, influenza, and other diseases known to us today, but also contains many infectious diseases that have not reached us today for various reasons. One of the main reasons is the presence in the human population of genetic alleles strongly immune to these diseases.

The deterioration in the conditions of food consumption as a result of the Neolithic revolution and the development of agricultural society took place slowly and without the dramatic changes that might have produced an evolutionary answer comparable to the answer to decimating epidemics. These changes in food consumption mostly caused epigenetic changes in the genome.

There is a known fact that the human brain has slowly shrunk over the last twenty to thirty thousand years. The reasons for such changes of the main organ of the human body are unknown. It is possible that one of the most important reasons for the brain shrinkage may be changes in human diet, especially since the beginning of the Neolithic revolution around thirteen thousand years ago. And there is some indirect evidence. People with diets high in natural vitamins and some fatty acids, etc., are less likely to have the brain shrinkage associated with age or Alzheimer's disease than people whose diets are not high in those nutrients, according to a study published by the American Academy of Neurology [628]. The gradual depletion of diversity of the diet associated with the glacial period and the beginning of agriculture also led to food consumption low in anthropic nutrients, which in turn caused the brain shrinkage. Of course, this has been influenced also by the development of culture, language, and writing during that period, since the collective intelligence and the exchange of information in society does not stimulate the intensification of brain work. However, the lack of anthropic food for the population could be the trigger for this process. I think that for such a relatively short period of time, changes in brain size had only epigenetic causes. This means that our genome still carries genes that determine brain size slightly larger than now. Consequently, long-anthropic childhood diet may cause reversible changes in the epigenome and probably make our children a little wiser, though I must admit that this process may take two to three generations. But in any case, the anthropic diet is not only an investment in your own health, longevity, and cognitive abilities, but also in the health and intelligence of your children and grandchildren.

Thus, man still remains a creature biologically optimized for the anthropic diet, which is the diet closest to that of hunter-gatherers. The reversibility of epigenetic changes in the genome allows us to return the metabolism to its normal state with the help of an anthropic diet. The anthropic diet is not as confusing and complicated as one might think. It›s about choosing foods that provide your body with the energy and wide spectrum of natural nutrients—biologically available for digestion and metabolically interacting with each other in the body—that it needs to perform. The focus of the diet should be not on only the number of nutrients but mainly on the extent of natural presence and biological availability of them in a food. For example, new research [343] shows that men and women who regularly eat berries, preferably wild or organic, may have a lower risk of developing Parkinson's disease, while men may further lower their risk by regularly eating organic apples, oranges, cacao, and other sources rich in dietary components called flavonoids. Another recent study [344] also confirms that good diets fight bad Alzheimer's genes: diets high in fish oil from natural sources have a beneficial effect in patients at genetic risk. This means that anthropic food in a diet could counteract the negative effects of Alzheimer's disease-related genes in the population.

The abnormal diet that has formed over the last few thousand years, and especially over the last few centuries, with its reliance on a high carbohydrate intake, is taken in most modern theories of dietology as some sort of an ideal. But in spite of the universal recommendation of modern-day nutritionists, humans did not evolve on a high-starch carbohydrate diet. The human diet for more than 2–3 million years was based on lean meat and seafood supplemented mainly by non-starch or low-starch fresh or/and fermented fruits, leaves, and vegetables.

And despite common beliefs, the results of a study [318] by the Harvard School of Public Health showed that a low-carbohydrate score was not associated with a risk of coronary heart disease. There was no evidence that the relationship was modified as a result of physical activity levels, body-mass index, or the presence or absence of hypertension, diabetes, or hypercholesterolemia. The findings in the study [318] did suggest, however, an association between low-carb diets high in vegetable sources of fat and protein and a low risk of heart disease . According to the authors, "neither a low-fat dietary pattern nor a typical low-carbohydrate dietary pattern is ideal with regards to risk of coronary disease; both have similar risks. However, if a diet moderately lower in carbohydrates is followed, with a focus on vegetable sources of fat and protein, there may be a benefit for heart disease." The authors found that, when vegetable sources of fat and protein were chosen instead of animal sources, the low-carbohydrate-diet score was associated with a 30-percent lower risk of coronary heart disease.

The results of another study [388] show that curbing starchy carbohydrates like potatoes is more effective than cutting calories for individuals who want to quickly reduce the amount of fat in their liver. The disease, linked to high levels of triglycerides in the liver, affects about a third of Americans. High levels of triglycerides can lead to inflammation, cirrhosis, and liver cancer.

The participants in the study [388] assigned to the low-carb diet limited their carbohydrate intake to less than 20 grams a day—the equivalent of a small banana or a half-cup of egg noodles—for the first seven days. For the final seven days, they switched to meals that matched their individual food preferences, carbohydrate intake, and energy needs. Those assigned to the low-calorie diet continued their regular diet. Researchers restricted the total number of calories to roughly 1,200 per day for the female participants and 1,500 per day for the males. After fourteen days, researchers had analyzed the amount of liver fat in each individual. They found that the study participants on the low-carb diet lost more liver fat. In addition, as shown in recent findings [580], an intermittent, low-carbohydrate diet was superior to a standard, daily calorie-restricted diet for reducing weight and lowering blood levels of insulin, a cancer-promoting hormone.

In my view, these studies demonstrate with complete clarity the idea that only following a low-carbohydrate anthropic diet, and not any intermittently fashionable movements such as the Atkins diet and other popular diets, brings health benefits. We must also note that natural superfoods of wild origin serve as the best source for such an anthropic diet.

For example, if we look some basic recommendations of modern dieticians, about the health benefits of whole grains, we will find that they are false. On the contrary, medical studies of populations with high levels of unleavened whole grain breads in their diet show that vitamin D deficiency is widespread. The biological availability of all nutrients in whole grain remains very low. Most whole grains and related products are devoid of certain vitamins. Even if they have some vitamins and minerals, all of them are in a state poorly absorbable by human body.

As mentioned in J. Diamond's work [390], humanity failed to domesticate a lot of food plants because they were inconvenient to grow. The nutritional value of domesticated plants for human health, apart from getting maximum calories from them, was generally neglected. Yes, cereals and pulses are a good source of energy and proteins. High-carbohydrate cereals now account for half of all human energy intake. But while they provide energy and proteins for a balanced diet, cereals and pulses do not provide essential anthropic elements of food. The quality of such a diet can guaranty survival for humans but not the possibility of a long life and absence of degenerative diseases.

As another example, we have the recommendations to consume products or DSs rich in calcium. A study published online in the *British Medical Journal* [186] found that these products commonly taken by older people for osteoporosis are associated with an increased risk of a heart attack. According to a commentary in the *Journal of the American Society Nephrology* (JASN), negative health effects linked to taking too much calcium are on the rise. The incidence of the so-called milk-alkali or calcium-alkali syndrome is growing in large part because of widespread use of calcium and vitamin D supplements [187]. Recent research [391]—a seven-year trial of over 36,000 women—demonstrated that calcium DSs increase the risk of cardiovascular events and are useless in preventing osteoporosis.

If we turn to such a currently fashionable DS as resveratrol, a superstar for modern nutritionists, we will find that according to a study [188], its use leads in certain cases to negative consequences in the form of the accelerated growth of cancerous tumors. Concentration, timing, state, and mutual interaction are crucial when it comes to any dietary compounds. There are many factors involved in the efficiency of resveratrol that are linked to its state (natural or extract), whole diet, and gene expression. For example, because of the huge number of claims about the protective qualities of resveratrol, researchers studied [400] whether resveratrol could protect against radiation injuries. They found that resveratrol protected cells from radiation in flasks but absolutely did not protect mice (stand-ins for humans in the laboratory) from radiation damage.

A University of Florida review of research finds that resveratrol from red wine, grapes, and other fruits cannot prevent aging [454]. Despite numerous clinical studies on the tonic effects on animals of resveratrol as an extracted dietary supplement, there is little evidence that it benefits human health. "We're all looking for an anti-aging cure in a pill, but it doesn't exist," said H. Hausenblas, one of the researchers involved in the study [454]. Resveratrol probably has considerable potential to improve health and prevent chronic disease in humans, but only in natural form and in combination with other correlating factors such as metabolism, food fermentation, the chemical interplay of molecules derived in a body from food and food metabolites, genetics, age, and many other factors that play a role.

There is no end to similar examples of the failure of modern nutritionology. To take another, there is the completely idiotic campaign to persuade people to drink more water. How many times have we heard the myth that drinking eight glasses of water a day is good for our health? But how many of us have gone along with this idea and felt better? The recommendation to drink six to eight glasses of water a day to prevent dehydration "is not only nonsense, but is thoroughly debunked nonsense," argues GP Margaret McCartney in a recent online *British Medical Journal* (BMJ) [480]. There is currently no clear evidence of benefit from drinking increased amounts of water, she says, yet the "we-don't-drink-enough-water" myth has endless advocates, including the UK NHS. Organizations, often with vested interests, reinforce this message, she says. For example, Hydration for Health (created by French food giant Danone, makers of bottled waters including Volvic and Evian) recommends 1.5 to 2 liters of water daily as "the simplest and healthiest hydration advice you can give." It also claims that "even mild dehydration plays a role in the development of various diseases." Untangling the evidence presented by Danone "results in weak and biased selection of evidence," McCartney argues. Danone says we need "informed choices," but their own evidence does not support their call to action.

The effects of the components of food are not limited to their immediate effects at the early stages of digestion. In reality the arrival of the constituent parts of the spectrum of anthropic elements of food results in multiple mutually interconnected cascades of metabolic reactions. The components of food have their effect also at the level of their metabolites in the enormous number of hetero-phase structures of an organism. At the moment only the phenomenological approach, based on a rich experience of the use of superfoods in nutrition, allows us to correct the diet of modern man in developed countries in the direction of greater anthropy.

One of the questions that concerns modern dietologists and engenders endless debates is the question of what we are from the point of view of healthy nutrition: carnivores, herbivores, or omnivores? The defenders of an herbivore diet rely for the most part on the herbivore evolution of man's antecedents up until the epoch that began approximately 3–4 million years ago. Starting from this period, man's antecedents, having begun to master stone work-tools and improve their social organization, slowly turned from being the herbivore prey for large predators into the strongest and most dangerous predator on the planet. This established the start of the carnivore style of life and food consumption of hominids. However, hominids did not entirely lose the ability to fulfill their need for mineral and bio-organic compounds, and partially their energy needs too, through vegetable food. The carnivore component of our evolution went on for a relatively long period and, undoubtedly, had an impact on our metabolic character. Man needs a certain amount of animal protein, the source of which are terrestrial and underwater organisms. But at the same time, plant food and plant protein are absolutely essential for him. Man's omnivore nature, in combination with other factors, has ensured his survival during periods of unfavorable climatic changes and given him an advantage in competition. It is precisely the fact that modern man's antecedents, and he himself during the early period of his existence, easily included in their nutritional ration practically any products of living nature which ensured their evolutionary advantage and led to the development of the brain and the growth of cognitive capacities. Genetic changes in our digestive system have allowed us to handle such high-energy food as protein or fat well. People who follow a vegan lifestyle—strict vegetarians who try to eat no meat or animal products of any kind—may increase their risk of developing blood clots and atherosclerosis or "hardening of the arteries," which are conditions that can lead to heart attacks and stroke. That is the conclusion of a review [345] of dozens of articles published on the biochemistry of vegetarianism during the past thirty years.

I think that man's intermediate status as an omnivorous creature is indicated by his intermediate length of digestive tract compared to the rest of the animal world. In carnivores the tract is approximately three times longer than the length of the body. In herbivores it is seven to ten times. So on this most telling of parameters we are below the herbivore level and are relatively close to pure carnivores. All other comparisons, such as the possession by carnivores of claws and horns and their absence in humans, do not play an important part. For our somewhat rational predecessors, it was no great effort to compensate for the lack of horns and claws with fire and labor tools. A more important distinction between us and carnivores, I think, is our different concentration of acid in the stomach. But this too means that we are well adapted to the digestion of a relatively small amount of animal protein in combination with the enzymes, juices, and cells of plant food.

Out of the components of the modern diet, the high-starch carbohydrate part is relatively the least characteristic of man. Over the whole history of man's evolution during the last several million years, we have never had in our diet a large amount of carbohydrate food of such types as cereals, corn, potatoes, and certainly not in the mashed, baked, or boiled state. The main sources of carbohydrates were non-starchy vegetables: fruits and plants. Doubtless early men did consume various grains and roots, which had a significant starchy component, but all this was more a reserve diet or a small part of the regular diet. The carbohydrate part of the diet, if the carbohydrates are drawn from fruit, berries, certain vegetables, and greens, is an essential part of our diet, and one which man is evolutionarily adapted to. But the consumption of carbohydrates from grains and products such as potatoes as a significant element of the diet, which has formed since the days of the Neolithic revolution, is a positive from the point of view of social evolution, but exclusively harmful from the point of view of people's health. Practically all the achievements of modern medicine are directed toward the battle against the negative consequences of the predominance of starchy carbohydrates in mass nutrition. With the modern development of medicine and overall comfort of human life, if we had stuck to the high-anthropic diet of hunter-gatherers, then we would have never heard of the majority of modern degenerative illnesses of the metabolism and a life of less than 100 years would be seen as an early end. Plants contain a huge amount of cell structures essential for man, which are regarded by the majority of modern dietologists only as a ballast for improve peristalsis. In actual fact, the starch of plant food is the richest source of a multitude of anthropic compounds involved in the enrichment of food in the intestine and later in a countless multitude of cascade metabolic reactions in the body. The cells of plants contain the precursors of many compounds vitally important for the organism to carry out its functions. It is preferable to consume it without thermal processing or with the minimum of it.

At the current time, there are several conceptions of diet relatively close to the anthropic diet: the paleo diet, functional food, organic food, raw food, and its well-known sibling vegetarianism. All have their own limitations.

The **paleo diet** is based on the idea that we should eat the same way as our hunter-gatherer ancestors. A sensible idea, but hard to implement, as at present it is difficult to find or grow a fairly large number of products that would have the same nutritional value. Even if that might be a useful approach, which ancestors should we follow? Our early ancestors, 2.5–3 million years ago, followed diets almost exclusively of plant-foods. Beginning at two hundred thousand years ago or so, many hunter-gatherers consumed meat and seafood or fish as a main part of their diet. Certainly we have to avoid the diet adopted most recently (ten thousand years ago or less) based on starches like rice, corn, wheat, potatoes, etc., because in terms of the time scale of evolution, ten thousand years is only a brief moment.

Even the methods of practicing such a diet that have been preserved, for example the extensive fermentation of an animal's carcass in the soil practiced by Eskimos, is unacceptable for modern man for several reasons. Modern man is vulnerable to the pathogens that are to be found in such food. Eskimos and certain other inhabitants of the planet Earth who live in its distanced ecosystems, little touched by human civilization, have still retained their natural resistance to certain pathogens, but most importantly they have a number of ferments and a bacterial environment in their digestive system that are absent in the majority of other populations. The other reason is that this process is hard to put into practice technologically outside the environment inhabited by the Northern peoples. The last reason is aesthetic and cultural preferences in cuisine. When the characteristics of a food cross the boundary of what you find acceptable by these lights, no force on earth will make you eat it.

As a matter of fact, Eskimos and other peoples who have lived until relatively recently in sufficient isolation also suffer from the introduction of the Western diet into their lifestyles. Passing over the question of infectious diseases that they have no immunity to, the Western style of eating brought a metabolic shock to the lives of these people, which is apparent in the appearance of a multitude of metabolic illnesses and of early deaths. A particular effect has been the spread of alcoholism, caused by the absence in the metabolic systems of these people of the ferments to break down alcohol. And it isn't just alcohol, but also many other staples of the modern junk diet, for example too much sugar and carbohydrates, that have an extremely strong and negative effect on their metabolic state. Thousands of years are needed to change the fermentation bio-chemical peculiarities of a population. As such, we cannot use the paleo diet of our ancestors as a guide to what we should eat.

**Organic food** also has its limitations. First, there are very few purely organically grown products on the market. The commercialization of this sphere of agriculture has led to fairly lax standards for monitoring the organic nature of these products. So, for example, in many countries a product can be labeled organic if no chemical fertilizers have been used on its soil for the last three years. However, it has been established that hundreds of pollutants caused by the use of agricultural chemicals stay in the soil and affect what is grown in it for thirty years or more.

Second, even produce that has been grown organically on medium-sized farms does not have high anthropic qualities due to the lack of biodiversity in the soil, caused by the multitude of reasons mentioned earlier (see Section 6.2). Organic foods have been often favored by customers because of the nutritional value and health benefits they are supposed to have. Consumers who want to consume food stuffs that have been grown in the absence of pesticides and other chemicals also consume organic foods extensively. But a recent study conducted at the University of Sydney has shown that just because organic foods have been grown without pesticides and herbicides, it doesn't necessarily follow that they have the implied nutritive value [193]. There is no convincing evidence that eating more fruit and vegetables can reduce chances of developing cancer [385], in particular if you are eating fruit and vegetables from supermarkets or similar sources.

Third, animals reared without the use of hormonal or other additives are fed in most cases on feed grown on ordinary fields. It is not enough to feed animals grass and other natural feeds. It is important that these feeds do not themselves contain chemical pollutants. In addition, even when producing organic meat, farmers often use, officially and unofficially, antibiotics and other medicines, and also genetically modified feed. Judging an organic product requires you to carefully and thoroughly study the conditions under which it was produced. In addition, organic food is for the most part used as a marketing device, with no connection to the actual methods of growing plants and rearing animals. However, despite all this, the consumption of genuinely organic products can be sufficiently beneficial in achieving an anthropic diet when it comes to reducing the level of chemical pollution in our food. Yet that is still insufficient to create a truly anthropic diet and can be seen only as the first step towards this goal.

**Raw food** is also a fairly controversial direction in the modern diet. Around two million years have passed since the invention of fire. Over this period, biological evolution has significantly altered our digestive system and the whole metabolism of modern man in the direction of adjusting to the use of fire in preparing food. Certain modern populariers of the various cocktails and mousses made of grass base their allegedly beneficial qualities on the claim that more than 98 percent of genomes in man and chimpanzees are identical. And plant food predominates in the diet of chimpanzees. The naivety of such an approach is completely obvious.

First, the consumption of highly biologically accessible protein food of animal origins that has been prepared with fire made us human and ensures the functioning of the brain at a level exceeding the intellectual activity of a chimpanzee, although it is true that raw consumption is highly beneficial when it comes to many seafoods and fish. Second, as was shown earlier, the functioning of the metabolism is determined not only by the presence of these or those genes but also by their expression. In the case of chimpanzees and humans, it is clear that the expression of the genes differs significantly. Third, even such a small difference between the human genome and that of a chimpanzee is critical when it comes to styles of food consumption. The diet of a chimpanzee is in no way acceptable for a human. Fourth, the creation of various cocktails and mousses out of greenery is itself disanthropic for our diet as human food consumption involves the masticating food and enriching it with certain enzymes from the saliva in certain proportions. The same applies to the entrance of liquid greenery with a high amount of juice into the stomach. Our stomachs, much like the stomachs of chimpanzees, are adapted to partially grinded green mass with the slow secretion of juice. Liquid greenery with immediately available juice to a significant extent perverts the digestive process in the stomach, and then in the intestine.

At the end of the day, almost all differences between humans and chimpanzees can be traced to diet [386]. As shown in a study [457] of primates, the relationships of the gut microbial communities matched that of their host. In other words, chimpanzees and humans digested food differently due to the different microbial populations within their guts, and these gut microbes have been tracking the evolution of their hosts for millions of years. The types of gut bacteria that populate the guts of primates depend on the species of the host as well as, but to a much lesser extent, where the host lives and what it eats. For all great apes and humans, a host's species, rather than its diet, has the greatest effect on its gut bacteria diversity. As such we cannot rely, as many fashionable "dietologists" do, on the genetic similarity of humans and, say, chimpanzees to give recommendations about the "correct way to eat" based on observation of the latter.

This does not mean that I am calling on you to stop including greenery in your diet. The plant component of nutrition is the cornerstone of the theory of anthropic diet, but only when consuming plant food the anthropic, natural way. The evolved condition of the human metabolism is such that even small deviations from the natural state of food can lead to large changes in the phase space of the metabolism. For example, consuming two to three glasses of freshly squeezed orange juice, the equivalent in mass to two to three large oranges, significantly alters the functioning of the stomach and intestine. When these oranges are consumed naturally they are fermented in the mouth and then slowly flow into the stomach. If you drink a glass or two of orange juice, you get the "shock" impact on the stomach of a pure product that has not gone through the necessary stages of enzyme enrichment. In addition, the stomach and intestine are not powerful enough (they lack the necessary bacteria and enzymes) to effectively absorb such a shock quantity of juice. Moreover, there is evidence that two glasses of freshly squeezed orange juice a day increase the risk of gout. In the opinion of specialists, the reason for this effect is the accumulation of uric acid in the joints under the influence of the bioactive fructose in fresh juice. Orange juice is the most benign example and has minimal negative effects. But many varieties of leaves and stems in their dispersed state in a liquid cocktail made from certain plants can produce a much greater shock to the digestive system and metabolism of an organism. As a result, the entire characteristics of a plant will not be realized or the juice will actually cause the organism harm, instead of the advertised benefit.

Compounds from row plants have a significant beneficial effect on human health because of their high content of polyflavanols and other substances such as phytoestrogens. Phytoestrogens have been the subject of intense scientific debates in past years. The results of several studies of cells as well as epidemiological findings suggest that they have a cancer protective effect. Another observation that may be interpreted in this direction is that Asian women are less frequently affected by breast cancer. Their soya-rich diet contains large amounts of another type of phytoestrogens, isoflavones. We can find an even higher content of isoflavones in row cacao beans and in cacao butter. The most important type of phytoestrogens in our diet is lignans, which are contained in seeds, particularly flaxseeds and with other types of phytoestrogens in cacao. In the gut, these substances are turned into enterolactone, which is absorbed by the mucous tissue [549].

**Functional foods** are foods that claim to provide health benefits beyond basic nutrition due to certain physiologically active components, which may or may not have been manipulated or modified to enhance their bioactivity. Apologists of these foods say that they may help prevent disease, reduce the risk of developing disease, or enhance health. As an example, by feeding their hens a modified diet, some farms have increased the amount of omega-3 in the eggs they sell. These eggs are considered to be functional food because their higher omega-3 content could improve the health of consumers whose diets are deficient in that fatty acid. And there is no end to similar modifications presented as functional products. But what proof is there that consuming, over a long period, omega-3 in excess of what there should be in the eggs is beneficial for human health? This is a glaring demonstration of the reductionist approach: if omega-3 is good, then we should enrich absolutely everything that we can with it.

First, is it good for the chicken that its organism is producing extra omega-3? Clearly, to feed a chicken on supplements that lead to this result is to change the chicken's natural metabolism. Do we know in what direction? That the metabolism of a chicken is virtually as complex as the human metabolism leads me to another question: has even a single healthy person worldwide been made healthier by being fed a collection of artificial food products? Obviously, the answer is no! The inevitable change in the phase space of the metabolism will lead to illness.

If these questions are answered honestly, then it becomes clear that, under the guise of functional food, we are being offered the eggs of a diseased chicken. At the same time, the long-term effects of consuming such "healthy" eggs are unknown and have not been investigated by anyone. While polyunsaturated fatty acids such as omega-3 fatty acids and omega-6 fatty acids are essential nutrients for everything from brain function to cell function, the human body is unable to make them. But these fatty acids are also the most vulnerable components in human cells because of their high sensitivity to oxidation caused by highly reactive oxygen molecules in the body. When a polyunsaturated fatty acid is oxidized, it joins with the oxygen in the blood stream to form a new compound that impairs the function of the cell membrane. In addition, these fatty acids are also used in the body to produce an array of fatty acid-derived hormones that are very important for cell functions. Under the influence of advertising, people buy dietary supplements such as capsules of fish oil, omega-3 fatty acids, or flaxseed oil to get these nutrients. However, extracted DS forms of polyunsaturated acids are very different from the same acids in foods containing them: fish, seafood, seeds, etc. In natural food form, these acids do not have such high susceptibility to oxidation. And scientists who conduct aging research know that the one critical characteristic of people who live very long lives is not taking DSs, but eating the right food. In this particular case, it is fatty wild sea fish and seafood, which are rich in polyunsaturated fatty acids. Omega-3, omega-6, and all other fatty acids are beneficial to our organism only if we get them not from DSs, but from natural, live sources, which themselves have healthy, natural metabolisms as a result of a balanced, anthropic diet.

The dreams of scientists to teach tomatoes to produce antibiotics, or apples or pigs to produce other medicines, and so on, run into one not insignificant obstacle: these will then be different sorts of tomatoes, apples, or pigs that man has not evolutionarily learned how to consume. Yes, it is possible that for certain products no harm will be immediately apparent, but we, our children, and grandchildren are not guinea pigs. We cannot, if we find out in twenty years' time the critical harm done by such products, console ourselves that a negative outcome is as important for science as the hypothetical positive one would have been. I'm sure you will agree, my readers, that personally for you, that is too minute a reward for risking your own health and that of your children. Much as with vitamins and DSs, business interest in functional foods is spurred on by the prospect of possible commercial gain by creating a new niche in food retail. This too is the spur for the science-esque advertising of functional foods. Another example of functional food is margarine fortified with plant sterols. We scarcely need to mention the well-known disanthropic qualities of margarine that can't be changed through any "fortification" process by any supplements. All other functional foods have approximately the same value.

In contrast to these various approaches to diet, the anthropic diet advocated in this work is based on the principle that the result of food consumption should be the enrichment of the organism with substances evolutionarily equivalent to food in its anthropic state, with the aim of the maximal restoration of the phase space of metabolic reactions to the genetically correct state. This complicated task is achieved by the formulation of specific diets based on the need to enrich the organism with natural produce that has high biological value and a wide variety of organic compounds and minerals in a biologically accessible form. The idea can be tailored individually, based on the person suffering from this or that illness, or can be designed for general categories of the population based on their age. Finding such diets is a fairly complicated task. However we can indicate the general principles and strategies used in designing such diets.

First, as the problems associated with disanthropic diets start from the earliest days of a life, and often are caused by epigenetic factors at work during the child's development inside the womb, the application of anthropic diet principles has to start with the health of the mother. A diet filled with anthropic superfoods has to be a part of a mother's nutrition in the period leading up to conception, so as to prevent possible epigenetic changes during the period of pregnancy and during the whole period of breast-feeding. Thereafter, anthropic components should always be present in the children's diets. Of course, the food itself must be adapted so that it can be fully utilized by children. It is extremely important to keep the child on an anthropic diet right until the pubertal age, until the point at which the formation of a healthy adult metabolism takes place. Deviations in a child's metabolism, or in other words, cracks in the phase space of metabolic reactions during an immature age, can become fixed through the mechanism of expression of some gene, for the child's entire life. This will consequently lead to premature aging and the appearance of degenerative diseases. For example, the excessively early onset of menstruation in girls can affect the state of their health over their entire subsequent lives. Puberty in women normally occurs between eleven and fourteen years of age. If a girl reaches a particular weight (around 45 kg or 100 lbs), the onset of puberty is triggered. The heavier the child, the earlier puberty occurs, possibly affecting the risk of later diseases. Scientists in a study [287] have discovered thirty new genes that control the age of sexual maturation in women. But the expression of many of these genes also act on body weight regulation or biological pathways related to fat metabolism. According to the authors of this study, "Several of the genes for menarche have been associated with body weight and obesity in other studies suggesting some women may have a genetic suscep-tibility to weight gain and early puberty. It is important to understand that these 'genetic factors' can be modified by changes in lifestyle. Efforts to reduce or pre-vent childhood obesity should in turn help reduce the early onset of puberty in girls." This evidences the importance of correct nutrition for girls during the ages leading up to puberty.

The metabolic system of a child during infancy and childhood is uniquely sensi-tive to changes in the composition of food. The first, immediate changes take place in an important part of an organism's metabolism—in the composition of micro-flora in the intestine. A baby's first peas are a life-changing event, at least when it comes to the microbes in its gut. A study [194] has shown that profound shifts in the number and diversity of baby gut microbes occur when he samples new foods. While the baby is breastfeeding, the bacteria in his stomach contain numerous genes useful for breaking down milk sugars. When the baby moves to a diet of solid foods, there are more bacteria with genes that influence starch digestion and so on. But the assembly of the human infant gut microbiome involves a relationship between life events and the microbiome composition and function. So the ingestion of table foods causes a sustained increase in the gut of the abundance of Bacteroides; enrichment of microbiome genes associated with carbohydrate utilization; vitamin biosynthesis, and xenobiotic degradation. Then the signal the metabolism sends as a result of food entering the intestine causes a cascade of metabolic reactions that affect the entire organism. But low-fiber Western diets deter "good bacteria" in the baby's micro-biome. The stools of African children contained almost three times as many short-chain fatty acids than that of Europeans or Americans [194]. These acids, which are generated by microbes associated with diets containing a very high proportion of vegetables, have the ability to kill harmful gut bacteria such as salmonella and help protect against inflammation. Child allergies are often the result of an excessive inflammatory response to otherwise harmless agents.

A typical example is a study [195] published in the *Journal of Nutrition & Dietetics* that found that some month-old babies had been introduced to high fat, salt, and sugar foods, despite health authorities recommending exclusive breastfeeding to six months of age. Almost one in four mothers had introduced fruit juice, biscuits, and cakes to their infants by six months of age. This is a worry because eating habits developed early in life usually continue throughout a person's lifetime—and an overweight child is much more likely to become an overweight adult. As a result, in a recent Australia-wide survey, up to 20 percent of children aged two to three years were found to be overweight or obese. What happens at home has the biggest effect on what children eat, so any effort to address children being overweight and obese must start at home.

The same situation is found everywhere in the developed world [196, 197]. Fifty-three percent of food products specifically targeted to babies and toddlers in Canadian supermarkets have an excessive proportion (more than 20 percent of calories) coming from sugar, according to a new study by the University of Calgary. Eighty-nine percent of kids' foods provide poor nutritional quality, but packaging claims and healthy images could be misleading parents. Keep in mind that confectionery, soft drinks, and bakery items were excluded from the survey. A recent study [199] from the University of Arizona criticizes the US food and beverage industry for failing to shift their marketing efforts aimed at children and parents. The report said misleading advertising continues to contribute to epidemic levels of obesity. As I already mentioned in previous chapters, food factors during childhood have a decisive impact on the risk of contracting cancer later in life [75].

Experts worldwide fear that the childhood obesity epidemic could lead to large numbers of adults developing diabetes, causing serious and lasting health complications for future generations. A study [451] published in the *International Journal of Obesity* indicates that girls whose mothers are classified as clinically obese are significantly more likely to struggle with weight problems in childhood, with a similar relationship existing between obese fathers and their sons. The findings showed that the same trend does not exist between mothers and their sons and fathers and their daughters—meaning that behavioral factors (diet pattern), rather than purely genetic factors, could be the key to unraveling the causes of the current obesity epidemic affecting children [451].

Another major international study [452] collating and analyzing worldwide data on diabetes since 1980 has found that the number of adults with the disease reached almost 350 million in 2008, more than double the number in 1980. The research reveals that the prevalence of diabetes has risen or at best remained unchanged in virtually every part of the world over the last three decades [453].

Contrary to earlier findings, excess body fat in the elderly decreases life expectancy. Research recently published in the *Journal of the American Geriatrics*

*Society* showed that men over seventy-five with a body mass index (BMI) greater than 22.3 had a 3.7-year shorter life expectancy, and women over seventy-five with a BMI greater than 27.4 had a 2.1-year shorter life expectancy than their thinner counterparts [509]. But preventing people from gaining weight to a level above the BMI level indicated is not that difficult a task, if they keep to a certain diet—not even an anthropic one—and a moderately active lifestyle. At the same time, we see this leading to a noticeably longer lifespan. Just imagine what would happen if we were to hold to even a partially anthropic diet for a sufficiently significant part of our lives. I am sure that the result would exceed all expectations. In addition, the rising level of obesity in populations around the world places an increasing burden on the people's health, on healthcare systems and on overall economies. Projecting from the data sets published in their study [530], researchers predicted the following impacts on the US by 2030:

- Obesity prevalence among men would rise from 32 percent in 2008 to approximately 50 percent and among women from 35 percent to between 45 percent and 52 percent.
- 7.8 million extra cases of diabetes.
- 6.8 million more cases of coronary heart disease and stroke.
- 539,000 additional cases of cancer.

Total medical costs associated with treatment of these diseases, which are preventable by an anthropic diet, are estimated to increase by US$48–66 billion per year.

What children eat is a fundamental cause of excess weight that many parents do not pay sufficient attention to. A study [200] of healthy children, ages seven to nine, found that fat in the liver, abdominal fat, and fat oxidation predicted insulin resistance and appear to be early markers for the metabolic syndrome via a mechanism of impaired lipid metabolism and fat oxidation. Impaired metabolic function may be due in part to pre- and post-natal factors that are modified by eating habits and current physical activity. Low or high pregnancy weight and/or birth weight, wrong diet, and low physical activity collectively create a phenotype for poor metabolic function leading to increased risk for insulin resistance in young children. New research [632] confirming the importance to mammals of the mother's diet around the time of conception and in early pregnancy was published recently online in the *FASEB Journal*. The results suggest that the mother's diet may cause an increased risk of obesity and type 2 diabetes throughout her babies' lives due to epigenetic changes in the offspring related to obesity and disease. From the health point of view, in some sense we are what our mothers ate. Researchers conducted experiments involving sheep to investigate twin pregnancies and the effects of altering nutrition around the time of conception and early pregnancy. Specifically, scientists examined the brain tissue of fetal sheep before birth and found that there were changes in the genes that control food intake and glucose levels that may lead to obesity and diabetes. They found epigenetic changes with alterations in not the genome sequences, but its structure which affects the genes' behavior in later life [633].

One of the main preconditions for maintaining a healthy metabolism in children is maintaining the correct bacterial environment in their intestines. An anthropic diet is one of the preconditions for the creation in children of a variety of intestinal micro-flora. It's no use to feed children with advertised probiotics and other garbage. The intestinal microbiome is also formed by the child's lifestyle. If the child grows up on a farm, next to a cattle yard, he is much less likely to suffer from allergies and other illnesses both in childhood and later on. But not every family has the opportunity to create such conditions for a child in, for example, an urbanized city environment. The solution consists of correcting the diet and the use of specific nutritional supplements from the natural products of virgin woodlands. The best woodlands to choose as a source are the forests around the Amazon basin, but that is not crucial. There are several regions in the world which can supply highly anthropic food produce.

Another aspect of the anthropic diet for both children and adults is the consumption of products of live nature, preferably from wild sources. Using even genuinely organic products is, as discussed, only a palliative measure. The value of organic products is in most cases that they are also free of chemical pollutants, even if they don't possess all the essential anthropic components. It is essential to reduce the starchy carbohydrate portion of the diet. Starch is indeed the first drug mankind became addicted to in its history. The consumption of fruits, grains, or roots containing a high level of starch created a blissful state of satiation. As a result, humans became habituated to it even in the Neolithic epoch.

This is especially relevant for overweight children. One of the main indicators of future metabolic syndrome in children is excess weight. As has been demonstrated in a study [201], it is more effective to correct the state of metabolism of people with excess weight by limiting the carbohydrates in their diet then by reducing the amount of fat. What is more, plant fats such as virgin olive oil and fish fat from sea fishes are an absolutely essential part of the anthropic diet for children, as indeed for adults. The diets of people with a hereditary predisposition to certain illnesses have to be especially rich in anthropic food. Looking at the example of Indians in the Unites States, we can clearly see that a hereditary predisposition to diabetes, for example, will not become a reality given correct nutrition during childhood and adult life. What is more, the correct diet helps fight diabetes and achieve a long lifespan.

Most long practicing doctors know that most diabetic patients develop severe complications, but 15–20 percent never do, even if they carry the disease for a period of twenty-five to forty years until the age of eighty-five. What is it that protects them? The answer is their diet! These patients are putting a great deal of effort into changing their diets in a more or less anthropic direction—completely excluding modern processed food and eating more fresh fish, seafood, and natural herbal food. The same applies for most other illnesses.

According to a study [481] of 25,000 obese people, dieting to lose weight is a waste of time. The study proves dieters nearly always put the weight back on. Even those who showed some success normally regained the weight within twelve months or less. Another piece of research [482] also confirms that the human body is evolutionary designed to strongly resist attempts to lose weight, but has very weak mechanisms to prevent weight gain. Incidentally, this demonstrates that it is the most ancient evolutionary mechanisms that determine the metabolism of human food absorption, which include an inbuilt resistance to weight loss. The insignificant genetic changes of the last few millennia have not touched this sphere. This is important not only in relation to excess weight, but also in relation to the most important aspect of our metabolism: how anthropic the food that it receives is. A reliable solution for the problem of excess weight and the illnesses connected to it can only be achieved under the conditions of an anthropic diet, when the organism will be nourished with anthropic substances from wild nature for a period of time. At the same time, only restoring our metabolisms to the maximal functioning evolutionarily possible for man can solve this problem. Of course, I am not here denying the importance of other aspects of our lifestyles, such as moderate physical exercise and so on. I am simply saying that diet is the most important factor by far.

And it is important to note again that for the problem of excess weight not to appear at all, a person must maintain a correct anthropic diet from the youngest age possible. Animal experiments have shown that a high-protein diet during development primes the body to react unhealthily to future food binges. A study [217] on juvenile animals, published in BioMed Central's open-access journal *Nutrition and Metabolism*, suggests that lasting changes result from altering the composition of the first food that is consumed throughout the period of growth into early adulthood. This is the first study to investigate the long-term effects of high-protein or fiber diets during development on the response to future food intake. It is evident that a long-term diet high in protein, when mismatched with a high energy challenge, has negative effects on body mass and hormones and genes involved in glucose and lipid metabolism. However, a fiber-enriched diet in juvenile animals may provide some protection from high-fat and high-sugar diets in later life. That is, preserving a metabolism from childhood onwards allows the disanthropic aspects of a diet to be fought more easily and their damage minimized in later life. This is important as the diet of a young adult is sometimes outside his control for objective reasons, such as military service and so on. In this way, cultivating the correct metabolism in children acts here as a future "shield" for their health from unfavorable dietary conditions later in life.

The pollution of children's food during the process of production presents another threat to children's health which starts from the youngest age. A study [203] published in the journal BMC Pediatrics by a team at Keele University in Staffordshire, led by Dr. Chris Exley with Shelle-Ann M Burrell, demonstrated the vulnerability of infants to early exposure to aluminum, serves to highlight an urgent need to reduce the aluminum content of infant formulas to as low a level as is practically possible. Infant formulas are integral to the nutritional requirements of preterm and term infants. While it has been known for decades that infant formulas are contaminated with significant amounts of aluminum, there is little evidence that manufacturers consider this to be a health issue. Aluminum is disanthropic and is linked to human disease. There is evidence of both immediate and delayed toxicity in infants and a high possibility of this toxicity having an influence on health in later adulthood. Commercially available branded infant formulas used by literally millions of parents to feed children of up to twelve months of age and older are still significantly contaminated with aluminum. The concentrations of aluminum in the milk formulas varied from 200 to 700 µg/L and would result in the ingestion of up to 600 µg of aluminum per day. The concentrations of aluminum in infant formulas are up to forty times higher than are present in breast milk. These concentrations are all several times higher than are allowed in drinking water. They are clearly too highly toxic for such a vulnerable group as infants. These results are horrifying; the concentrations are monstrous, placing in danger the health of future generations. Aluminum is a chemical element that is non-participated in biochemical evolution due to a complete lack of its biologically reactive forms. The human body does not have protection against the toxicity of aluminum. The only protection from aluminum for early humans was the absence of it in everyday life. But the wide use of aluminum in modern life means that we had lost this type of protection, and it is now a toxic enough part of the human metabolism. Results of the arrival of biochemically reactive aluminum in our life, especially in the children's menu, are obviously quite worrisome and potentially very dangerous. Aluminum has been shown to cause severe ecological damage in territories where it has contaminated the land, killing animals, trees, and so on. Knowingly exposing our babies to this substance is unthinkable.

More dangerous still is the fact that many other pollutants of mineral and organic nature that enter during the production of food produce find their way into the diets not only of adults, but of children too, and in the same monstrous concentrations; and the captains of industry and regulators in many countries treat this as a cause of little concern. But even that pollution in a food product, which in the opinion of regulators has a low level of concentration, can be very dangerous, as shown in Chapter 4 of this work.

Speaking specifically of adults—i.e., people doomed by decades of incorrect nutrition—the correction of such a state is achieved through several main methods. First, the organism must be detoxified. But don't think my dear reader that I am sending you to one of the many charlatan clinics that flourish in this niche throughout the world. The process of detoxification includes the gradual replacement and squeezing out of toxins to be found in the organism by anthropic substances or their metabolites. Believe me when I say that anyone consuming a modern Western diet has toxins in his organism. But what is more, the Western diet causes a dramatic overreaction of the immune system to sepsis, a condition of systemic bacterial infection. An experimental study [252] published in the open access journal *BMC Physiology,* has shown that a diet high in saturated fat, sugars, and cholesterol greatly exaggerates the inflammatory response to sepsis and that the combined effects of sepsis and diet would exacerbate hepatic inflammation. And an especially deadly mixture is the combination of sepsis and inflammation with the toxins of any nature.

But the study [520] shows that not all obesity is the same and some body fat may actually be toxic. Some obese people develop chronic diseases such as diabetes and heart disease, and others do not. The data from this research [520] suggest intrinsic defects in the critical adipose tissue cells that are relevant to an increased risk of metabolic syndrome and consequently of diabetes and cardiovascular disease. Metabolic syndrome can be reversed through anthropic diet. To reverse the process of metabolic syndrome and detoxify, with a condition made more severe by excess weight and sepsis, it is necessary to utilize certain techniques to gradually rid ourselves of excess weight; that will soften the second toxification the organism will undergo due to the toxins produced by fatty tissue, in combination with a highly anthropic diet. But it is most important for the person, after the initial detoxification and weight loss, to keep to the anthropic principles of diet for the rest of her long life.

In addition, the process of gaining weight is not an uncomplicated one. It is often thought that this is a tendency to accumulate the high-energy component, fat, that has developed in humanity as a result of the frequent lack of food provision during our evolution. But there is an alternative point of view. Eating a high-energy diet leads to the early development of obesity and metabolic syndrome, apparently through an inability to cope with the energy density of the diet. From this point of view, it looks like we have been looking at the obesity problem from the wrong angle. It appears that getting fatter may be a defense of our metabolism against the worst effects of the disanthropy of our everyday food, rather than their direct cause. The risk of getting heart disease, diabetes, coronary artery disease, and so on depends mainly on excess fat in the bloodstream. This happens when fatty tissue in the obese overload, and also in normally slim people if they are eating too much food. The reason for this is, in the first case, the release of fat into the blood stream from excessively loaded fat cells, but in the second, the inability of a lean person's metabolism to store excessive fat. In both cases we have similar results, the aforementioned host of diseases. These diseases appear as a result of the negative inflammatory response of our immune system. Of course the situation is significantly worse for overweight people because the release of many pollutants accumulated in fat tissue is occurring alongside the release of fat to the bloodstream [188]. But even in lean people, excessive fat may provoke the immune system to an inadequate response. That the accumulation of toxins in fatty tissue can lead to the development of certain diseases is confirmed, in my view, by a study at Boston University School of Medicine (BUSM) and Boston Medical Center (BMC) [428]. Researchers have shown that the type of adipose, or fat, tissue is a significant contributing factor in the development of inflammation and vascular disease in obese individuals. The research team identified that 30 percent of the obese subjects demonstrated reduced fat inflammation, less insulin resistance, and a vascular function similar to a lean person's despite severe obesity. In my opinion, the study suggests that humans prone to inflammation in association with weight gain may just have more toxic substances accumulated in adipose because of their diet habits.

And consequently these people are more susceptible to cardiovascular and metabolic disease risks. Even in cases of severe obesity, the relatively low level of harmful products accumulated in adipose tissue could prevent their emission into the body and prevent systemic inflammation. The anthropic diet could, in the long term, at least partially eliminate toxic substances accumulated in adipose and could potentially combat the development of several debilitating obesity-related disorders.

Research [425] reveals that the molecular processes linked with diet that take place during the perinatal period leaves marks on the fetus epigenome, lasting throughout the individual's lifetime. During intrauterine life and lactation, malnutrition brings about modifications involving DNA, leading to metabolic pathologies at the adult age. It could be malnutrition as a lack of food or as under-supply to the body of certain essential proteins, amino acids, and so on. Influencing such programming mechanisms through an anthropic diet is essential for the prevention of metabolic diseases in the population. The study [430] also found that exposure to a high-fat diet before birth modifies gene expression in the livers of offspring so they are more likely to overproduce glucose, which can cause early insulin resistance and diabetes. This is why experts worldwide fear that the childhood obesity epidemic, strongly linked to intrauterine life and lactation, could lead to large numbers of adults developing diabetes, causing serious and lasting health complications for future generations. A study [451] published in the *International Journal of Obesity* indicates that girls whose mothers are classified as clinically obese are significantly more likely to struggle with weight problems in childhood, with a similar relationship existing between obese fathers and their sons. The findings showed that the same

trend does not exist between mothers and their sons and fathers and their daughters—meaning that sometimes behavioral (diet pattern) and epigenetic, rather than
pure genetic, factors could be the key to unraveling the causes of the current obesity
epidemic affecting children [451]. Formerly known as "adult-onset diabetes," type
2 diabetes is increasingly seen in children at earlier and earlier ages [435A recent
study [434] published by the American Physiological Society also offers the strongest evidence yet that vulnerability to type 2 diabetes can begin in the womb, giving new insight into the mechanisms that underlie a potentially devastating disease
at the center of a worldwide epidemic [435]. According to Peter W. Nathanielsz,
senior author of the study, "We pass more biological milestones before we are born
and in the early weeks of life than at any other time. Poor maternal nutrition, which
translates to less sustenance for growing fetuses, is a stubborn problem in parts of
the U.S. and the developing world." Thus, poor (by quantity or quality) nutrition
at critical periods of development can hinder growth of essential organs such as the
pancreas, which sees a significant decrease in its ability to secrete insulin [435].
The researchers conclude that even moderate nutrient deficiencies during pregnancy
result in offspring predisposed to type 2 diabetes, particularly if they are exposed
to other risk factors in later life, such as a totally disanthropic Western diet leading
to obesity. A fetus may also receive less of some food nutrients due to teenage pregnancy, where the growing mother competes with her offspring for resources.

Obesity also has certain social consequences for the people affected. Negative
attitudes against overweight people are becoming a cultural norm around the world.
In addition, it is common to think about obese people as people who lack the willpower to stop eating. This is not necessarily true. A study [202] has discovered the
reason why some people who eat a high-fat diet remain slim, yet others pile on the
weight. They discovered that a high-fat diet caused brain cells to become insulated
from the body, rendering the cells unable to detect signals of fullness and pleasure
from food to stop eating and signals to increase energy use. This study shows that
obese people are not necessarily lacking willpower. Their brains just do not know
how full they are and how much fat they have stored, so the brain cannot send signals to the body when it's time to stop eating.

Also, food intake is associated with dopamine release. The degree of pleasure derived from eating correlates with the amount of dopamine released. Evidence from the
study [221] shows obese individuals have fewer active (those not insulated from the
body) dopamine receptors in the brain relative to lean individuals and suggest obese
individuals overeat to compensate for this reward deficit. As a result, the body's ability to lose weight is significantly reduced. And it is a good answer to the question of
why only a fraction of Americans who have ever been overweight or obese lose weight
and maintain that loss. This is a huge problem: two-thirds of the United States adult
population is overweight or obese, with a body mass index of more than 25.

One report [489] might help to explain why it's so frustratingly difficult to stick to a diet. When we don't eat, hunger-inducing neurons in the brain start eating bits of themselves [489, 490]. That act of self-cannibalism turns up a hunger signal to prompt eating. The cellular process uncovered in neurons of the brain's hypothalamus is known as autophagy (literally, "self-eating"). The authors say their findings suggest that treatments aimed at blocking autophagy may prove useful as hunger-fighting weapons in the war against obesity. Research [489] shows that lipids within the so-called agouti-related peptide (AgRP) neurons are mobilized following autophagy, generating free fatty acids. Those fatty acids in turn boost levels of AgRP, itself a hunger signal [490]. Fatty acids released into the circulation and taken up by the hypothalamus as fat stores break down between meals and may induce autophagy in those AgRP neurons. Earlier research [490] showed a similar response in the liver. High levels of fatty acids in the bloodstream, induced by a disanthropic high-fat diet, might alter the hypothalamic lipid metabolism at the epigenetic level, "setting up a vicious cycle of overfeeding" [490].

In this way, the results of these studies demonstrate that limiting the food intake is not always effective or possible on its own, and that fatness can only be cured by returning a metabolism inundated by disanthropic nutrition to normal functioning through a prolonged anthropic diet. Treatment by anthropic diet aimed at eliminating the wrong metabolic pathways can make you less hungry and burn more fat, a good way to maintain energy balance in a world where calories are cheap and plentiful [490]. This new point of view on obesity is very helpful for distancing fatness from disease and for finding a strategy of anthropic therapy to create weight loss diets without harmful effects to the body. But often, people not inclined to fatness, and so who have few fat deposits despite consuming a poor diet in the anthropic sense, think themselves invulnerable to metabolic illnesses. But according to a study [475], having a lower percentage of body fat may not always lower your risk for heart disease and diabetes.

In addition, weight loss can only be achieved if the dieter pays great attention to his or her body's metabolism. There are two hormones, leptin and ghrelin, that play opposite but important roles in food intake and weight maintenance. If either of them does not function properly, weight gain is the result. When it comes to food digestion, leptin signals the brain when it's time to eat, while ghrelin signals the brain when it's time to stop eating [313]. In a study [312] published in the *Journal of Clinical Endocrinology & Metabolism*, researchers analyzed the role of the plasma levels of hormones such as ghrelin, leptin, and insulin on weight recovery in 104 overweight people following a hypo-caloric diet. After eight weeks, the group that had regained more than 10 percent of the weight lost was found to have higher levels of leptin and lower levels of ghrelin. According to the authors [312], "Some obese or overweight patients who gain more weight following a diet could even be identified before they embark on their weight-loss therapy, just by looking at their plasma levels of these hormones." After finding an imbalance of hormones in plasma, the only method to restore leptin and ghrelin balance is by eating anthropic foods as close to nature as possible and by eliminating sugars and carbohydrates from your diet. There are no readily available drugs that can ensure these two hormones have a balance at the level they should [313].

Anthropic therapy for the metabolism uses a combination of special diets and other measures in an attempt to remove "toxins" from the body and strengthen the body's immune systems against disease. The strategic goal of anthropic therapy for adults is to return the phase space of their metabolism maximally to the natural state, and to exclude from their diet substances that lead to the unnatural crosslinking polymer matrix with a change in their body's viscoelastic properties. For a million years of evolution, early humans relied on natural foods and herbs for energy and for medicine. Only in the past ten thousand years or so, we have step-by-step forgotten our evolutionary "roots" in favor of modern food and patent medicines. While pharmaceuticals have their value, we should not forget the well-documented, healthy properties of whole foods.

But in contrast to the authors of all those countless diets that promise the fulfillment of all your dreams about your health, as a serious researcher, I can only say that the result of these procedures is limited by the state of your organism. Anthropic therapy for the metabolism is based on the belief that toxic substances in food and the environment build up in the body and create chemical imbalances that lead to changes in the metabolism in two ways:

1.  by changing chemical reactions in the phase space of the body's metabolism;
2.  via genes as commonly understood and epigenetic mutations which alter the trajectories or pathways of the cascades of chemical reactions in the body's metabolism.

These changes in the metabolism lead to degenerative diseases such as cancer, arthritis, multiple sclerosis, and diabetes. Anthropic therapy rids the body of toxins and strengthens its immune systems and resistance to diseases. In addition, when we talk about the healthy and vitalizing qualities of natural, anthropic food, we have to remember that many of our illnesses are simply the consequence of consuming disanthropic food. A natural, anthropic diet is good not so much because it helps cure illnesses, but because it prevents them in the first place. But the potential of anthropic nutrition, no matter what the sellers of natural produce claim, is quite limited. In contrast to all known types of metabolic therapy, anthropic therapy does not use any commercially available forms of DSs, enzymes, vitamins, or minerals. Only whole or processed superfoods kept in an anthropic state are used. Only if we manage to correct the disorders in the metabolism and the organism remembers its natural vital functioning, do we get a positive result. But often, if the process of degenerative changes in the organism has gone too far, the consumption of anthropic food has an unquestionably positive but limited effect.

So we can achieve a stable positive impact on children, and in most cases eliminate the unhealthy condition entirely. A study [281] published in the *New England Journal of Medicine* confirms the hypothesis that infant feeding plays a role in initiating the disease process leading to type 1 diabetes in children carrying an increased genetic disease risk, and that dietary intervention in infancy can prevent the disease process leading to type 1 diabetes.

For an adult, much depends on age, lifestyle from as early as birth, hereditary factors, and physical exercise. If her disanthropic diet has had a strong, maybe even irreversible, impact on the expression of genes in the cells of her body, then the result will depend on what part of the cascades of metabolic reactions can be restored to a normal level. That the problem of incorrect nutrition has taken on a pandemic character is demonstrated by the spread of cancer. The World Cancer Research Fund has stated in a recent report [550] that according to estimates there are about 3 million cancer cases a year globally that are linked to the wrong diet and a lack of physical activity. It is a figure that the Fund expects will rise dramatically over the current decade. People tend to think of cancer and diabetes as largely being diseases originating from chemical or radioactive pollutants. But increasing obesity rates and cancer are coming from epigenetic changes connected with eating more disanthropic processed food. It is the true source of a global health problem of great magnitude.

For example, diabetes can virtually not be cured once it has started, but ensuring that the sufferer is less reliant on injections can be done in virtually all cases. Pre-diabetic conditions can be more easily corrected, and it is possible to prevent the illness or significantly soften its impact or slow its progression. Researchers at Washington University School of Medicine [597] have restored normal blood sugar metabolism in diabetic mice using a compound the body makes naturally. They suggest that it may one day be possible for people to take the compound much like a daily vitamin as a way to treat or even prevent type 2 diabetes. I do not know when that is going to happen, but certainly boosting production of these compounds in the body at least partially is possible right now, without waiting for the often empty promises of science, by using natural high anthropic value food.

The same applies to other illnesses linked to the metabolic syndrome. Simple methods, such as complete or partial fasting, do not help to restore the disordered metabolic processes. Here only an anthropic diet can help. By accompanying food with anthropic signals to the phase space of metabolic reactions, which produce changes in the expression of genes in the cells of an organism in the direction of the natural, genetically determined state, we can achieve the gradual return of the strange attractor of an ill organism to the form the attractor takes in a healthy metabolism.

Speaking of fasting, there are many ossified myths in modern dietology. Among such myths is the extrapolation onto mammals of the evidence that fasting has a positive impact on length of life in worms. Researchers [401] studied the effect of food restriction on fat and weight loss in forty-one genetically different strains of mice. The scientists then correlated the amount of fat reduction to life span. The answer: Mice that maintained their fat actually lived longer. Those that lost fat died earlier. "Indeed, the greater the fat loss, the greater the likelihood the mice would have a negative response to dietary restriction, i.e., shortened life," said James Nelson, Ph.D., professor of physiology, the leading author of the study [401]. "This is contrary to the widely held view that loss of fat is important for the life-extending effect of dietary restriction. It turns the tables a bit" [402].

Another important question closely related to nutrition is the use of alcohol. In a 2010 article in *Time*, John Cloud cites a stunning statistic: based on data collected over twenty years, it has been found that the highest rates of mortality are among teetotalers, irrespective of whether they were drinkers before that twenty-year span or not. Heavy drinkers have a slightly lower rate, while it is moderate drinkers who have the lowest one. The statistical sample consisted of people aged fifty-five to sixty-five who had required an ambulance within the last three years. The group of 1,824 people was monitored by researchers for twenty years. This statistic requires a little explication. Even though alcoholism increases the risk of cyroid of the kidney and cancer of the mouth, esophagus and kidney, heavy drinkers live longer than teetotalers. One explanation of this paradox sometimes offered is that many of those counted as teetotalers in such studies were in reality alcoholics in the past, and managed to destroy their health before quitting alcohol. However, this explanation fails to withstand scrutiny, as former alcoholics were removed from this sample. The conclusion to be drawn from this study, which is based on genuine scientific research, is that drinking is healthy.

Moderate drinking can significantly reduce the risk of dementia and cognitive impairment, suggests an analysis of 143 studies by a team of researchers at Loyola University Chicago Stritch School of Medicine [527]. That conclusion is based on reviewing studies that included more than 365,000 participants. Indeed, moderate drinkers were 23 percent less likely to develop cognitive impairment or Alzheimer's disease and other forms of dementia. Wine, according to the study, was more beneficial than other types of alcohol. As an explanation of the benefits of moderate drinking, the authors suggest that reasonably small amounts of alcohol might, in effect, make brain cells more fit. Alcohol in small amounts stresses cells and thus trains them up to cope with major stresses in the life that could cause dementia.

Recent research [544] shows that the pattern of alcohol consumption—moderate drinking every night, or binge drinking on weekend nights only—may also be more important in determining alcohol's damage on heart health than the total amount consumed. "People need to consider not only how much alcohol they drink, but the way in which they are drinking it," said lead study author John Cullen, Ph.D., research associate professor in the Department of Surgery at the University of Rochester Medical Center. "Research shows that people have yet to be convinced of the dangers of binge drinking to their health; we're hoping our work changes that." After discounting other factors, such as smoking, that might affect their health status, the authors [545] found that women who drank 5-15g of alcohol per day (between a 1/3 and 1 drink per day) had about a 20 percent higher chance of good overall health when older compared to non-drinkers. Furthermore, women who drank alcohol regularly had a better chance of good overall health when older than occasional drinkers: compared to women who didn't drink, women who drank five to seven days a week had almost a 50 percent greater chance of good overall health when older. The authors conclude, "These data suggest that regular, moderate consumption of alcohol at midlife may be related to a modest increase in overall health status among women who survive to older ages." They add, "The 2010 US Department of Agriculture dietary guidelines note that moderate alcohol consumption of up to one drink per day for women and up to two drinks per day for men may provide health benefits in some people." Their data support this recommendation and provide evidence suggesting that light-to-moderate alcohol consumption at the levels of one to two drinks per day or slightly less at midlife may benefit overall health at older ages in American women.

Alcohol is an old (possibly from one-hundred-thousand to one-million-year-old) companion for mankind, Even primitive hunter-gatherers were perfectly able to discover that fruit, berries, roots, and grains, if covered in water and left in hot weather for one, two, or three days, could turn into a revitalizing and cheering beverage. Moderate alcohol consumption is beneficial irrespective of your genotype—whether you have the genetic predisposition to absorb alcohol or not. In addition, primitive forms of alcoholic beverages in the guise of pure wine or beer brought undoubted benefit to a man's organism due to the presence in them of many anthropic substances, including vitamins, enzymes, and, strange though it is, antibiotics. Chemical analyses of the bones of ancient Nubians show [219] that they were regularly consuming tetracycline, most likely from something like modern beer or wine. The source of antibiotics in the Nubian beer-like drink was the grain used to make the fermented gruel, containing the soil bacteria Streptomyces, which produces tetracycline. In addition, alcohol itself is a product present in the metabolism of even teetotalers. Endogenous alcohol plays a part in a multitude of the trajectories of metabolic reactions on the level of cells and in the digestive system. As such, the consumption of alcohol is important in the anthropic diet and an indispensable part of it. The health benefits of wine and the presence in it of components in their natural form are underlined by many studies. But at the same time, there is no reason to believe that various extracts from the plants, fruit, and berries used in fermentation, such as resveratrol and others, are a beneficial supplement to our nutrition in the form of tablets and other such things. Resveratrol and the other components of wine have a beneficial effect on our organism only in their natural, "live" form, and in combination with the thousands of other organic and mineral compounds in wine. The mighty advertising drive behind resveratrol as a panacea from many illnesses and from aging withstands no scrutiny. As has already been noted, the effects of resveratrol in this form are far from clear and it can lead to the accelerated growth of malign tumors.

While the physiological effects of high alcohol consumption have been established as detrimental in humans, many studies show that low to moderate alcohol consumption, equivalent to one or two glasses of wine or beer a day, results in an improvement in cardiovascular health and increased longevity. Low doses of alcohol can be beneficial not only to humans but probably also to all animals, especially under stress conditions. Scientists who study the biochemistry of aging found that minuscule amounts of alcohol can more than double the life span of a tiny worm known as *Caenorhabditis elegans*, which is used frequently as a model in aging studies [606]. And as in humans, high alcohol consumption for worms is very harmful. It looks like the mechanism of slower aging in a worm has something in common with findings that moderate alcohol consumption in humans has cardiovascular and other health benefits mentioned above.

The moderate consumption of alcohol in the form of wine, beer, and other relatively weakly alcoholic beverages is an essential aspect of a healthy, anthropic diet. However, the recommendation to use alcohol is not, of course, an invitation to abuse it. Excessive or inappropriate use of alcohol leads to increased mortality from causes linked with trauma and socially aberrant behavior, and categorically cannot be recommended for children.

How moderate alcohol consumption in humans may have health and aging benefits is unknown, partly because the process of aging is itself not fully understood. For more than thirty years, the prevailing explanation of why we get old and lose health has been tied to what is called oxidative stress. This theory postulates that when molecules like free radicals, oxygen ions, and peroxides build up in cells, they overwhelm the cells' ability to repair the damage they cause, and the cells age. An industry of "alternative" antioxidant therapies—such as vitamins and DSs with extreme doses—has sprung up as the result of this theory. However, clinical trials have not shown that these treatments have statistically significant effects [243]. A healthy, balanced diet is very important for reducing the risk of developing many diseases associated with old age, such as cancer, diabetes, and osteoporosis. There is no clear evidence that dietary antioxidants can slow or prevent aging. There is even less evidence to support the claims of any anti-aging beauty product [244].

What is more, according to scientists from the Swedish medical university Karolinska Institutet, fear of free radicals may be exaggerated. A report [355] of their study [356], published in *The Journal of Physiology*, shows that free radicals act as signal substances that cause the heart to beat with the correct force. The idea that free radicals are generally dangerous and must be counteracted is, however, a myth, according to scientists who have conducted a new study of the role that free radicals play in heart physiology. "As usual, it's a case of everything in moderation. In normal conditions, free radicals act as important signal substances, but very high levels or long-lasting increases can lead to disease," says Professor Hakan Westerblad, who led the study. When the body is subject to different types of stress, the sympathetic nervous system stimulates receptors known as beta-adrenergic receptors on the surface of heart muscle cells. This leads to several changes inside the cells, one of which is the phosphorylation of proteins. This leads to the contractions of the cells becoming stronger and the heart beating with greater force.

In the study, the Swedish scientists show that stimulation of the beta-adrenergic receptors also leads to increased production of free radicals in the mitochondria of the cells, and these then contribute to stronger contractions of the cells. When the researchers exposed the cells to antioxidants, a major part of the effect of beta-adrenergic stimulation of the heart muscle cells disappeared.

The results of an original article [356] reveal a previously unknown regulatory mechanism of force production in the heart, and may lead to a better understanding of various types of heart deficiency. "Free radicals play an important role, since they contribute to the heart being able to pump more blood in stress-filled situations," says H. Westerblad. On the other hand, persistent stress and such very active radicals as peroxides can lead to heart failure. And chronically increased levels of free radicals, even if they are not particularly active, may be part of the problem as well [355].

A study [531] found that elevating free radical levels in the hypothalamus directly or indirectly suppresses appetite in obese animals. It means that animals and people have these reactive oxygen species (ROS) as signaling molecules to stop eating. According to the authors of the study, the crucial role of free radicals in promoting satiety as well as processes associated with aging may explain why it has been difficult to develop successful medical treatment for obesity without major side effects. Oxidative stress, or an imbalance of ROS throughout the body tissues, has been linked by most researchers to an array of diseases, from cancer to heart disease and stroke. However, scientists in one study [429] and in several others have demonstrated that cells can also use ROS in a controlled way to send signals needed for normal functions.

The cells that line the gut intestine live in close contact with bacteria and normally form a barrier that keeps bacteria away from other organs. They can repair small gaps in the barrier, which breaks down in intestinal diseases, by migrating into the gaps [429]. The researchers showed in vitro and in vivo that intestinal epithelial cells produce ROS internally when in contact with bacteria, when it enters the gut with food. The ROS induced by the bacteria stimulate the formation of focal adhesions (where cells attach to the matrix that surrounds them), structures on intestinal epithelial cells that act as anchors for their movement. Antioxidants from foods that inhibit ROS prevent the bacteria from promoting wound healing. Authors of the study [429] believe that the ROS production they observed stimulates tissue maintenance and is a marker of cohabitation and adaptation, rather than pure defense. The finding of this study suggests that large amounts of antioxidants in the human gut could interfere with the ability of bacteria to promote intestinal healing.

Many people take a variety of vitamins, minerals, and phytochemicals in a pill or extract forms as supplements, and most of them are not efficacious in that form, many researchers say. For example, a recent study [595] has found that if you want some of the many health benefits associated with eating broccoli or other cruciferous vegetables, you need to eat the real thing. That's because a key phytochemical in these vegetables is poorly absorbed and has a negligible value if taken as an artificial dietary supplement (DS). People believe that optimal levels of popular supplements such as vitamins C, E, and fish oil, for instance, can be difficult to obtain through diet alone. That is wrong. Yes, of course, for getting necessary levels of anthropic substances from food, you have to correct the metabolism of your body by anthropic dieting for a long enough period. But it is the only way to solve the problem of nutritional deficiency because the use of artificial DSs is inefficient and in many cases may be harmful.

Most recently there has been what some entrepreneurs call "bio-identical" or "all-natural" hormones. What they mean by these terms varies from substances made from vegetables—such as soya or yams, which some claim have estrogen-like effects—to, more commonly, drugs that are exactly the same as hormones prescribed by endocrinologists for specific diseases. Dr. T. Perls [245] remarked, "The terms bio-identical or all-natural, particularly in the case of the drugs prescribed by endocrinologists, misleadingly convey a sense of safety to the gullible customer. Arsenic is all-natural too, and it even has some medical uses, but it is anything but safe." An assessment [245] of each of the purported anti-aging hormones has shown that in terms of anti-aging, the risks of these hormones outweigh the little or no benefit they confer. The marketing of these hormones, particularly growth hormones and anabolic steroids (anabolic steroids are variations of testosterone), for anti-aging should be stopped.

Speaking of such an aspect of dieting as the curing of illnesses with specific products, which is in most cases speculation, just as with vitamins and DSs. Be careful if you see on the Internet or on the news an announcement of this kind: "cure cancer naturally," with references to a lot of "scientific evidence." Claims like this are based on a best-case scenario and merely on what goes on in Petri dishes or in mice. There exist countless Internet sites that offer cancer-fighting foods, which might lead people to think they can ward off this disease simply by eating certain foods. There are some foods associated with a lower risk of getting cancer, but in reality cancer prevention can be achieved through anthropic dieting for a long time and of course through your lifestyle. Only if you can restore your own metabolism to its genetically determined state can you be sure you will avoid cancer and other degenerative diseases. That's not an easy task, by the way, for most of us.

Recently the new term "nutraceutical" has been much used in print and in the advertisement of certain products. Nutraceutical, a term combining the words "nutrition" and "pharmaceutical," describes a food that is supposed to provide health and medical benefits, including the prevention and treatment of disease. Nutraceuticals, according to [218], is a broad umbrella term used to describe any product derived from food sources that provides extra health benefits in addition to the basic nutritional value found in foods. Products typically claim to prevent chronic diseases, improve health, delay the aging process, and increase life expectancy.

There is minimal regulation over which products are allowed to display the nutraceutical term on their labels. Because of this, the term is often used to market products with varying uses and effectiveness. The definition of nutraceuticals and related products often depends on the source. For this reason, dieters are so involved with trying to keep up with modern diets that they are more likely than non-dieters to choose wrong foods that are promoted as the healthy choice. The medical community wants the nutraceutical term to be more clearly defined in order to distinguish between the wide varieties of products out there. There are many different types of products that fall under the category of nutraceuticals but do not have any noticeable benefit for health.

## 7.2   Smart technology for smart food

One of the main problems connected with the anthropic style of diet is the problem of delivering the products to the final consumer without losing their anthropic qualities during a presumably long period of storage. The solution to this problem lies, in my view, not in especially rapid logistical methods of transporting fresh anthropic products, which will, most likely, be too expensive for the ordinary consumer, but in developing an affordable technology for the processing of the products of live nature so as to ensure the prolonged retention of their original vitalizing anthropic qualities.

As we have already noted above in Section 6.6, modern, commercially available superfoods sourced from live nature, while possessing at first unquestionable and exceptional anthropic qualities, lose them completely during their processing. The main negative impact on their anthropic qualities comes from the process of oxidation. Products become oxidized during their processing for two main reasons, which can impact the product both individually and together. The first cause of the oxidation of products is their grinding to powdered state or drying with the formation of a developed porous surface. During this, the several fold growth of the surface that interacts with the oxygen in the air is characteristic. As a result, the rate of oxidization increases in comparison with the natural or coarsely crushed state of a product, at minimum in proportion to the increase in the surface area. Why at minimum? Because, due to the non-linear character of interaction between oxygen and the product in its natural state, caused by, for example, the appearance of protective layers made out of the oxidized flesh of fruit or leaves on their surface, the process of oxidization is slowed somewhat under natural conditions. In a highly ground state, these protective layers do not form, and as the linear size of part of a product is smaller than the typical thickness of a protective layer, the oxidation of the mass of a product in reality takes place disproportionally quickly.

Based on this, we have to accept that products such as bread are by definition absolutely disanthropic substances, which in the best case simply do not directly cause serious harm to the organism. Dried and ground fruits and berries are especially harmful due to this sort of processing, as the processes involved in their intensive oxidation cause the appearance of a multitude of active free radicals. Such dried products, made out of peaches, apricots, berries, and vegetables, have become popular on the market. They are usually produced through vacuum drying and then being ground into flour. These products' cans have detailed descriptions of their nutriological contents, which match the nutriological contents of a biologically fresh product, but with no relation to the actual contents. These sorts of powder are widely used as "natural" additives to many food products and beverages. In the long term, their use as an addition to a diet leads, in my opinion, to a significant risk of harm to the consumer that has to date been poorly assessed.

The other cause of oxidization is the use of high temperatures during processing. With both causes of a negative insecticides and pesticides effect operating concurrently, they can be synergistically strengthened. The various processes of chemical and thermal extraction also have a destructive, disanthropic effect on the quality of the original product, especially those involving the use of strong alcohol. Bearing this in mind, we can briefly formulate several technological preconditions for the preservation of the vitalizing qualities of the anthropic products of live nature:

The prevention of oxidation during processing.

A sufficiently long period of time during which the product can be stored without destroying its physical-chemical qualities or bio-organic compounds.

Foreswearing the processes of chemical and physical extraction of substances from a product.

Besides this, scientists are reporting new evidence that a centuries-old food preservation technology—high-pressure processing (HPP), which subjects food to pressure—more than doubles the levels of certain healthy natural chemicals in fruit and kills harmful bacteria, viruses, and mold in food. The force in HPP does not squash the food, which can be fresh, processed, or liquid. The pressure does change the molecular structure of the microbes in food in ways that kill bacteria, molds, and viruses. The technique is also known as "pascalization" in honor of the seventeenth-century French scientist B. Pascal, famous for research on the effects of pressure on liquids [533]. According to this review, HHP processing increased the concentration of total carotenoids in avocado or papaya by more than 50 percent. Individual members of this healthy family of chemicals increased by up to 513 percent. It appears that cells in the fruit had shifted to make more antioxidants and other chemicals in order to cope with the stress from HPP.

I am currently developing, on the basis of the conceptions laid out in this book, specific methods and technologies to process live superfoods that will ensure the preservation of their anthropic qualities. It is possible that research in this direction will be laid out in my next book, which would be published if this large and difficult task is completed and sufficient evidence has been gathered.

# Conclusion

The aim of this book is an overview of the processes of the absorption of food in the human organism. We have briefly connected this to the evolution of food consumption, and to economic and technological issues. This book is not a collection of answers for any question, but is more a guide for the public to ascertain the extent to which their diet is anthropic and to find ways to increase the anthropic character of that diet. The task was to show that there is a certain heuristic value to the unusual approach to problems of diet used in the book, relating these to the broadest conception of human and social evolution, the non-linear laws of the working of the organism, and bio-chemical, mechanical, and epigenetic interrelations. I am fully aware about the imperfect precision of many arguments from the mathematical, biological, and physical-chemical points of view. All original conceptions in this work are hypotheses, which I have attempted to substantiate by doing the relevant research in accessible literature and by using his own experience and knowledge.

Specific applications of this work, to study and develop technological processes and diets to incorporate various supplements, are at various stages of completion, and as such I will be glad to hear from anyone interested in mutual work to research these issues both in theory and in practice.

As I mentioned above many times, degenerative diseases, such as cardiovascular disease, diabetes, arthritis, and cancer represent the major global health problem of the twenty-first century and affect all age groups. This book aims to create an integrated method of dieting, using systems of dietology to prevent degenerative diseases as a whole. In modern life, loss of biodiversity, climate change, and industrialization, affects the available nutrition in a food and lead to disanthropy in everyday life. As all degenerative diseases are influenced by similar environmental, dietological and lifestyle factors, such as nutrition and exposure to chemicals, individuals often have several diseases that could be prevented or better controlled by application of the anthropic diet. Of course, this problem requires many more studies. As a first step, the integration between results from primary care, dietology, and public health data, and the creation of appropriate reliable mathematical models to provide a better understanding of the link between disease progression in populations and dieting habits would help. A better understanding of the degenerative diseases of the metabolism, through systems of dietology, would allow us a more efficient use of superfood resources and focus attention on prevention as well as control, thereby reducing the cost and burden of care to society. At the same time, we have to develop technologies for processing products, especially superfoods, with a focus on the prolonged preservation of their therapeutic qualities and commercial accessibility for the public.

Governments should work with more urgency to stop the spread among the population of degenerative diseases of the metabolism, such as heart disease, cancer, and diabetes, that threaten the health and life of people of industrialized and developing nations alike. Degenerative diseases of the metabolism account for 60–70 percent of all deaths worldwide. Trends also suggest that the major risk factor for these diseases is an incorrect diet, which leads to hypertension, high glucose levels, and obesity among the public. Metabolic disorders among the population are all on the rise, especially in developing countries.

A rapid increase in prosperity has gone hand in hand with the surge in obesity in many developed and developing countries. In the press, there is a lot of compelling evidence on the development of the diabetes and obesity epidemics worldwide in tandem with that of economic prosperity.

In addition to the health consequences, the huge costs of treatment and the negative effects on the economy take devastating tolls on the financial situations of individuals, families, and countries. Unless degenerative metabolic disorders are tackled, goals aiming at the prosperity of mankind cannot be achieved nor can economic development be sustained. The current global disanthropic lifestyle will have even more drastic consequences for those in the developed world who hold hopes for modern health care that it will never fulfill.

While almost everybody in the world is being affected by the disanthropic lifestyle and diets, those already on or below the poverty line will find themselves in even more desperate circumstances. They go without the proper anthropic food and medicine to compensate the disanthropy of their diets, putting them at greater risk of illness and early death.

Not only does this book give a new paradigm in how we think about dieting, but it hopefully gives readers a new paradigm for preventing metabolic syndrome diseases—that is, by anthropic dieting therapy.

# Literature

1. **V. Vernardsky**. Научная мысль как планетное явление, Moscow, Russia, «Наука», 1991.
2. **P. Glansdorff, I. Prigogine**. *Thermodynamics Theory of Structure, Stability and Fluctuations*. London: Wiley-Interscience. 1971.
3. http://www.enviroliteracy.org/index.php
4. http://www.un.org/apps/news/story.asp?NewsID=10007&Cr=computer&Cr1
5. http://www.eurekalert.org/pub_releases/2002-11/acs-ttp110502.php
6. http://www.prweb.com/releases/2007/09/prweb555778.htm
7. http://www.democrats.us/editorial/wise041105.shtml
8. The Guardian, 10.11.2009
9. **E. Schrödinger**. *What is life?* Macmillan, 1946.
10. **J. Diamond**. *Collapse: How Societies Choose to Fail or Succeed*. New York: Viking Books, 2005
11. http://transmission.lenin.ru/Diamond-Mistake.html
12. **M.Brooks**. *13 Things That Don't Make Sense*. London: Profile Books, 2010.
13. **M. Keschner** Шум типа 1/f. // ТИИЭР. - 1982. - Т.70, № 2.- С. 60 – 67. Russia
14. **T. Istomina et al.** Хаотическая динамика в нарушениях сердечного ритма. Russia. http://www.mks.ru/library/text/biomedpribor/98/s1t7.html
15. **Robert H. et al.** *The Theory of Island Biogeography*. Princeton University Press Centenary, 1905-2005.
16. **Robert H. Whittaker.** *Communities and Ecosystems*. Macmillan, 1975.
17. **N. Lane.** http://nick-lane.net/On%20the%20origin%20of%20barcodes.pdf
18. **F. I. Zapparov, D. F. Zapparova.** О'Какао! Спорт и Культура, Moscow 2009.
19. **N. Lane.** Oxygen, Oxford University Press Inc, N.Y., 2002.
20. **S. Kapitsa et al.** Синергетика и прогнозы будущего, http://www.iph.ras.ru/~mifs/kkm/G12.htm
21. **Yale University.** "Why Synthetic Estrogens Wreak Havoc On Reproductive Syst." *ScienceDaily* 2 April 2008.
22. **J. G. Bromer et al.** "Bisphenol-A Exposure in Utero Leads to Epigenetic Alterations." *The FASEB Journal*, 2010.
23. **Soto A.M, Sonnenschein C.** "Environmental Causes of Cancer: Endocrine Disruptors as Carcinogens." *nature Reviews Endocrinology*, May 25, 2010.
24. **Journal of the National Cancer Institute.** "Non-Cancer Deaths More Common Among Breast Cancer Survivors." *ScienceDaily*, 15 February 2008.
25. **C. J. Patel et al.** "An Environment-Wide Study on Type 2 Diabetes Mellitus." *PLoS ONE*, 2010; 5 (5).
26. **B. Leber et al.** "Proteins Centrosome Clustering in Cancer Cells." *Science Translational Medicine*, 2010; 2 (33):
27. **N.Lane.** *Power, Sex, Suicide*. New York: Oxford University Press , 2005.
28. **Visikailo et al.** Взаимодействие природных систем. http://fireball.izmiran.ru/RC/13/D/Vy.doc
29. **E. Zababahin, I. Zababahin.** Явления неограниченной кумуляции. М.: Наука, *1988*. 171 с. Russia
30. **Laboratory of Radiation Biology.** Joint Institute for Nuclear Research http://lrb.jinr.ru/index-e.html
31. **Bernini et al.** "Human Phenotypes in Metabolic Space and Time." *Journal of Proteome Research*, 2009.
32. **Federation of American Societies for Experimental Biology.** "Children Metabolize Differently." *ScienceDaily*, 28 April 2010
33. **A. C. Skinner et al.** "Multiple Markers of Inflammation and Weight Status." *Pediatrics*, March 2010.
34. **Cold Spring Harbor Laboratory.** "A Gene-Regulatory Network Governing Metabolism." *ScienceDaily*, 26 February 2008.
35. **American Society for Cell Biology.** "Major Theories Human Cellular Ageing." *ScienceDaily*, 30 Dec 2008.
36. **M. Schafer et al.** "Nrf2 Establishes a Glutathione-Mediated Gradient of UVB Cytoprotection in the Epidermis." *Genes & Development*, 2010; 24 (10)
37. **Washington State University.** "Crop Residue May Be Too Valuable For Biofuels." *ScienceDaily*, 19 July 2008.
38. **H. Blanco-Canqui.** "Energy Crops and Their Implications on Soil and Environment." *Agronomy Journal*, 2010.
39. **I. Sipahi et al.** "Angiotensin-Receptor Blockade and Risk of Cancer." *The Lancet Oncology*, 2010

40. Soucy-Faulkner et al. "Requirement of NOX2 and Reactive Oxygen Species for Efficient RIG-I-ntiviral Response through Regulation of MAVS Expression." *PLoS Pathogens*, 2010.
41. American Academy of Sleep Medicine. "Links Insomnia to Increased Risk of Death." *ScienceDaily*,15 June 2010.Web.
42. G. Nagel et al. "Effect of Diet on Asthma and Allergic Sensitisation in the International Study on Allergies and Asthma in Childhood (ISAAC) Phase Two." *Thorax*, 2010; 65 (6).
43. M.J. Feigenbaum. "Universal Behavior in Nonlinear Systems." *Los Alamos Science*. 1980, v.1, No.1, pp.4
44. M. C. Lucas I. G. Priede. "Utilization of Metabolic Scope in Relation to Feeding and Activity by Individual and Grouped Zebrafish, Brachydanio rerio (Hamilton-Buchanan)." *Journal of Fish Biology*. , V. 41 Issue 2, 175-190
45. S. Kim. Изложение докладов Симпозиума по системной биологии. Балтимор *2009* http://m-batin. livejournal.com/33321.html
46. E. S. Lysenko, R. S. Lijek, S. P. Brown, J. N. Weiser. "Within-Host Competition Drives Selection for the Capsule Virulence Determinant of Streptococcus pneumoniae." *Current Biology*, 17 June 2010.
47. University at Buffalo. "Study On Toxin That Tainted Spinach, Shiga Toxin." *ScienceDaily*, 11 December 2007.
48. A. Breitkreutz et al. "A Global Protein Kinase and Phosphatase Interaction Network in Yeast." *Science*, 2010; 328 (5981).
49. Y. Xu et al. "Leukocyte Pyruvate Kinase Expression is Reduced in Normal Human Pregnancy but Not in Pre-eclampsia." *American Journal of Reproductive Immunology*, 2010
50. L. Multigner et al. "Chlordecone Exposure and Risk of Prostate Cancer." *Journal of Clinical Oncology*. 2010.
52. Ahsan et al. "Arsenic Exposure from Drinking Water, and All-Cause and Chronic-Disease Mortalities in Bangladesh (HEALS): A Prospective Cohort Study." *Lancet*, 2010.
53. American Chemical Society. "Rice In US Contains Less-Dangerous Form Of Arsenic." *ScienceDaily*, 21 May 2008. Web.
54. S. Tay et al. "Single-cell NF-κB Dynamics Reveal Analogue Information Processing." *nature*, 2010
55. N. Kosaka et al. "MicroRNA as a New Immune-Regulatory Agent in Breast Milk." http://www.silencejournal.com/content/1/1/7
56. B. F. Voight et al. "Twelve Type 2 Diabetes Susceptibility Loci Identified Through Large-scale Association Analysis." *nature Genetics*, 2010.
57. University of Chicago Press Journals. "New Study Explores Patterns In Species Diversity And Genetic Diversity." *ScienceDaily*, 28 July 2005. http://www.sciencedaily.com/releases/2005/07/050727060325.htm .
58. Michigan State University. "Childhood Malnutrition Weaken Brain in Elderly." *ScienceDaily*, 3 July 2010. Web.
59. NYU Langone Medical Center. "UVA Radiation Damages DNA in Human Melanocyte Skin Cells and Can Lead to Melanoma." *ScienceDaily*, 2 July 2010. Web.
60. American Chemical Society. "PCBs Threaten Whale Populations For 30-60 Years." *ScienceDaily*, 10 September 2007. Web.
61. K.A. Vuori, M.Nikinmaa "M74 syndrome in Baltic salmon." http://www.ncbi.nlm.nih.gov/pubmed/17520930
62. Van Zeebroeck et al. "Transport and Signaling via the Amino Acid Binding Site of the Yeast Gap1 Amino Acid Transceptor." *nature Chemical Biology*, Dec 7, 2008
63. B. S. Metcalf et al. "Fatness Leads to Inactivity, but Inactivity Does Not Lead to Fatness: A Longitudinal Study in Children" (EarlyBird 45). *Archives of Disease in Childhood*, 2010
64. V. Konstantinidou et al. "In Vivo Nutrigenomic Effects of Virgin Olive Oil Polyphenols within the Frame of the Mediterranean Diet: A Randomized Controlled Trial." *The FASEB Journal*, 2010.
65. A. Camargo et al. "Gene Expression Changes in Mononuclear Cells from Patients with Metabolic Syndrome after Acute Intake of Phenol-rich Virgin Olive Oil." *BMC Genomics*, 2010.
66. University of Rochester Medical Center. "Obesity in Men Linked to Common Chemical Found in Plastic and Soap." *ScienceDaily*, 15 March 2007. http://www.sciencedaily.com/releases/2007/03/070314110441.htm
67. S. H. Sicherer, A. Muñoz-Furlong, J. H. Godbold, H. A. Sampson. "US prevalence of self-reported peanut, tree nut, and Sesame Allergy: 11-year Follow-up." *The Journal of Allergy and Clinical Immunology*, May 12, 2010.
68. C.Emery. "Food Allergy Prevalence Rises Dramatically." http://www.medpagetoday.com/AllergyImmunology/Allergy/17040
69. J. Jaenike, R. Unckless, S. N. Cockburn, L. M. Boelio, S. J. Perlman. "Adaptation via Symbiosis: Recent Spread of a Drosophila Defensive Symbiont." *Science*, 9 July 2010: Vol. 329. no. 5988, pp. 212 – 215.
70. J.H. Park et al. "Development of Type 2 Diabetes following Intrauterine Growth Retardation in Rats is Associated with Progressive Epigenetic Silencing of Pdx1." *Journal of Clinical Investigation*, 2008.
71. http://www.unisci.com/stories/20013/0813012.htm
72. P.C. Pickhardt C. L. Folt, C. Y. Chen, B. Klaue , J. D. Blum. "Algal Blooms Reduce the Uptake of Toxic Methylmercury in Freshwater Food Webs." http://www.pnas.org/content/99/7/4419.full 2002.
73. http://www.unisci.com/stories/20022/0425021.htm

74. http://preventdisease.com/news/articles/red_meat_gene_prostate_cancer.shtml

75. http://www.unisci.com/stories/20014/1203013.htm

76. Society for General Microbiology. "Bacteria 'Launch A Shield' To Resist Attack." *ScienceDaily,* 9 March 2009.

77. http://www.rosbalt.ru/2010/10/29/785267.html

78. American Society for Microbiology. "Bacterial Communication Encourages Chronic, Resistant Ear Infections." *ScienceDaily,* 6 July 2010. Web.

79. Wade M. Hicks, Minlee Kim, James E. Haber. "Increased Mutagenesis and Unique Mutation Signature Associated with Mitotic Gene Conversion." *Science,* Vol. 329. no. 5987, pp. 82–85.

80. M. Sarovar et al. "Quantum Entanglement in Photosynthetic Light-harvesting Complexes." *nature Physics,* 2010

81. M. H. Traka et al. "The Dietary Isothiocyanate Sulforaphane Modulates Gene Expression and Alternative Gene Splicing in a PTEN Null Preclinical Murine Model of Prostate Cancer." *Molecular Cancer,* 2010.

82. S.K. Mazmanian et al. "Pro-inflammatory T-cell responses to Gut Microbiota Promote Experimental Autoimmune Encephalomyelitis." *PNAS,* July 19, 2010.

83. M. Peters et al. "Arabinogalactan Isolated from Cowshed Dust Extract Protects Mice from Allergic Airway Inflammation and Sensitization." *The Journal of Allergy and Clinical Immunology,* 2010.

84. S. F.A. Grant et al. "Disease-associated Loci are Significantly Over-represented among Genes Bound by Transcription Factor 7-like 2 (TCF7L2) in Vivo." *Diabetologia,* July 17, 2010.

85. K. Fulzele et al. "Insulin Receptor Signaling in Regulates Postnatal Bone and Body Comp." *Cell,* 2010.06.02.

86. M. Ferron et al. "Insulin Signaling in Osteoblasts Integrates Bone and Energy Metabolism." *Cell,* 2010.06.03.

87. M. Imielinski, C. Belta. "Deep Epistasis in Human Metabolism." *Chaos: An Interdiscip. Journal of Nonlinear Science,* 2010; 20 (2).

88. American Institute of Physics. "Can Chaos Theory Help Predict Heart Attacks?" *ScienceDaily,* 23 July 2010.

89. A. Büttner-Mainik et al. "Production of Biologically Active Recombinant Human Factor H in Physcomitrella." *Plant Biotechnology Journal,* 2010.

90. USDA/Agricultural Research Service. "Better Control of Reproduction in Trout and Salmon May Be in Aquaculture's Future." *ScienceDaily,* 23 July 2010.

91. Adey et al. "Scale Microchemistry as a Tool to Investigate the Origin of Wild and Farmed Salmo salar." *Marine Ecology Progress Series,* 2009.

92. Ford JS, Myers RA. "Assessment of Salmon Aquaculture Impacts on Wild Salmonids." *PLoS Biol* 6(2) 2008.

93. University of Rhode Island. "Development of More Muscular Trout Could Boost Commercial Aquaculture." *ScienceDaily,* 11 March 2010.

94. Ghaderi et al. "Implications of the Presence of N-glycolylneuraminic Acid in Recombinant Therapeutic Glycoproteins." *nature Biotechnology,* 2010.

95. C. Kevin Boyce, Jung-Eun Lee. "An Exceptional Role for Flowering Plant Physiology in the Expansion of Tropical Rainforests and Biodiversity." *Proceedings of the Royal Society B,* June, 2010.

96. D. Wolfe. *Superfoods.* North Atlantic Books, Berkeley, California, 2009.

97. F. I. Zapparov. Влияние неньютоновских свойств жидкости на процессы конвективного теплообмена.- Диссертация...... канд. техн. наук.- Казань, Russia, 1983.

98. F. A. Garifullin, F. I. Zapparov. Конвективное движение в надкритической области для жидкости второго порядка. ИНЖЕНЕРНО-ФИЗИЧЕСКИЙ ЖУРНАЛ Том 38, 6, 1980, Moscow.

99. J. Gapiński et al. "Size and Shape of Micelles Studied by Means of SANS, PCS, and FCS." *Langmuir,* 2010; 26.

100. J. Szymański et al. "Diffusion and Viscosity in a Crowded Environment: From Nano- to Macroscale." *The Journal of Physical Chemistry B,* 2006; 110 (51).

101. P. P. Bhat et al. "Formation of Beads-on-a-String in Break-up of Viscoelastic Filaments." *nature Physics,* 2010.

102. H. Tanaka. "Viscoelastic Phase Separation." *Journal of Physics: Condensed Matter.* 2000;12:R207–R264.

103. F. J. Iborra. "Can Visco-elastic Phase Separation, Macromolecular Crowding and Colloidal Physics Explain Nuclear Organisation?" *Theor Biol Med Model.* 2007;4: 15.

104. Massachusetts Institute of Technology. "Biomedical Engineering - The Next Century." *ScienceDaily,* 21 February 2008.

105. American Institute of Physics. "Molecules Delivering Drugs as They Walk." *ScienceDaily,* 3 August 2010.

106. C. Hunte, V. Zickermann, U. Brandt. "Functional Modules and Structural Basis of Conformational Coupling in Mitochondrial Complex I." *Science,* 2010.

107. University of Florida. "Dietary Restriction Cleans Cells." *ScienceDaily,* 24 August 2007. Web.

108. Duke University Medical Center. "Why Cells Starved Of Iron Burn More Glucose." *ScienceDaily,* 10/06/2008

109. University at Buffalo. "How Dietary Iron Is Used By Cells." *ScienceDaily,* 3 July 2007. Web.

110. Yu. Romanovskyi, V. Teplov. Физические основы клеточного движения. УФН. Том 165, 5, 1995

111. British Medical Journal. "Calcium Supplements Linked to Increased Risk of Heart Attack."   http://www.eurekalert.org/pub_releases/2010-07/bmj-csl072810.php

112. G.J. Jing et al. "Aberrant Expression of Nuclear Matrix Proteins during HMBA-induced Differentiation of Gastric Cancer Cells." *World Journal of Gastroenterology*, 2010; 16 (17).

113. C.L. Sampieri, S. de la Peña, M. Ochoa-Lara, R. Zenteno-Cuevas, K. León-Córdoba. "Expression of Matrix Metalloproteinases in Human Gastric Cancer and Gastritis." *World Journal of Gastroenterology*, 2010; 16 (12)

114. M. Abdullah et al. "Helicobacter Pylori Infection and Gastropathy: A Comparison between Indonesian and Japanese Patients." *World Journal of Gastroenterology*, 2009; 15 (39)

115. S.M. Frisch, Y. Francis. "Disruption of Epithelial Cell-Matrix Interactions Induces Apoptosis." http://jcb.rupress.org/content/124/4/619.full.pdf , 02.15.1994.

116. Project On Emerging Nanotechnology. "Nanotechnology Used 500 Products." *ScienceDaily,* 23 May 2007.

117. Project on Emerging Nanotechnologies. "Nanotechnology Surges into Health and Fitness Products." *ScienceDaily,* 7 October 2007. Web

118. Project on Emerging Nanotechnologies. "New Nanotechnology Products Hitting The Market At The Rate Of 3-4 Per Week." *ScienceDaily*, 25 April 2008. Web.

119. University of Idaho. "Could Nanotechnology Make A Health Food?" *ScienceDaily*, 18 February 2009. Web.

120. J. P. Celli at al. "Helicobacter Pylori Moves Through Mucus by Reducing Mucin Viscoelasticity." http://web.mit.edu/nnf/publications/GHM133.pdf

121. G. Guigas, C. Kalla, M. Weiss. "Probing the Nanoscale Viscoelasticity of Intracellular Fluids in Living Cells." *Biophysical Journal,* Volume 93, Issue 1, 316-323, 1 July 2007.

122. Weissman et al. "Protein Aggregation and Protein Instability Govern Familial Amyotrophic Lateral Sclerosis Patient Survival." *PLoS Biology*, 2008; 6 (7).

123. http://www.davidwolfe.com/

124. http://www.dfhcc.harvard.edu/membership/member-profile/member/586/0/

125. http://www.foodnavigator.com/content/view/print/176790

126. http://www.womenshealthmatters.ca/news/news_show.cfm?number=502

127. http://www.worldcocoafoundation.org/info-center/document-research-center/Cacao_HumanNutrition.asp

128. http://ezinearticles.com/?How-Can-Cacao-Help-With-Diabetes?&id=1646895

129. http://www.ncbi.nlm.nih.gov/sites/entrez?cmd=Retrieve&db=pubmed&dopt=Abstract&list_uids=14709774

130. http://www.zdr.ru/news/2008/08/18/temnyi-shokolad-ot-diabeta-i-ozhirenija/index.html

131. http://www.rmtp.org/index.php?option=com_content&task=view&id=80&Itemid=2

132. http://www.ncbi.nlm.nih.gov/sites/entrez?Db=pubmed&Cmd=Retrieve&list_uids=15712597&dopt=abstractplus

133. http://209.85.129.132/search?q=cache:Hmp05rd36lwJ:www.davidkatzmd.com/media/DarkSide.Katz.6-6-06.ppt+Kelm+cacao&cd=6&hl=en&ct=clnk

134. http://www.ncbi.nlm.nih.gov/pubmed/15087264?dopt=Abstract ) 135.http://www.envirohealthtech.com/cacao.htm

136. http://209.85.129.132/search?q=cache:QCQDwWB0ZkIJ:www.paulengemann.com/filemanager/science%2520tab/HeartHealthy%2520Compound%2520in%2520Chocolate%2520Identified.pdf+Kuna+island+health&hl=en&ct=clnk&cd=20

137. http://www.ncbi.nlm.nih.gov/pubmed/15755830?dopt=Abstract

138. http://www.sciteclibrary.ru/rus/catalog/pages/6673.html

139. http://www.medstream.ru/news/2342.html

140. http://www.cacaolife.com/

141. http://www.antirak-center.ru/index.php?catid=35&page=287

142. http://www.worldcocoafoundation.org/info-center/document-research-center/documents/Ding2006.pdf

143. http://ezinearticles.com/?Health-Benefits-Of-Cacao-Chocolate&id=831370

144. S.Park et al. "Cell Motility and Local Viscoelasticity of Fibroblasts." *Biophysical Journal*, V. 89, Issue 6, 4330-4

145. http://www.goodguide.com/products/178436-derma-e-cocoa-butter-e-replenishing-cr

146. http://www.izvestia.ru/news/news220629

147. http://www.worldcocoafoundation.org/info-center/document-research-center/Cacao_HumanNutrition.asp

148. http://www.pubmedcentral.nih.gov/articlerender.fcgi?tool=pubmed&pubmedid=16390538

149. http://www.ajcn.org/cgi/content/full/77/6/1466

150. http://www.chocolate.org/health/epicatechin.html

151. http://www.xagena.it/news/medicinenews_net_news/254fa7691a1ca09ccdceaf216a7afc13.html

152. http://www.ncbi.nlm.nih.gov/pubmed/14640573?dopt=Citation

153. http://www.ncbi.nlm.nih.gov/pubmed/15777537?dopt=Abstract

154. http://www.naturalnews.com/022610.html

155. http://www.envirohealthtech.com/cacao.htm

156. http://archinte.ama-assn.org/cgi/content/short/166/4/411

157. http://www.ulimana.com/health-benefits-of-chocolate-raw-cacao.html

158. E. Anggard. Nitric oxide: mediator, murderer, and medicine. Lancet 1994;343:1 199-206

159. http://books.google.com/books?id=r_XsYYsmF68C

160. http://ezinearticles.com/?Eat-Chocolate-and-Enjoy-Better-Sex&id=1160933

161. http://www.holistic.com/

162. http://news.bbc.co.uk/hi/russian/sci/tech/newsid_3254000/3254261.stm

163. http://cat.inist.fr/?aModele=afficheN&cpsidt=15217290

164. T. J. Mitchison, G. T. Charras, L. Mahadevan. "Implications of a Poroelastic Cytoplasm for the Dynamics of Animal Cell Shape." *Semin. Cell Dev. Biol.* 19, 2008:215–233.

165. Y. Jamali, M. Azimi, M.R.K. Mofrad. "A Sub-Cellular Viscoelastic Model for Cell Population Mechanics." *PLoS ONE* 5(8), 2010: e12097. doi:10.1371/journal.pone.0012097

166. Corona et al. "Extreme Evolutionary Disparities Seen in Positive Selection across Seven Complex Diseases." *PLoS ONE*, 2010; 5 (8): e12236 DOI: 10.1371/journal.pone.0012236

167. Ting Liu et al. "FAN1 Acts with FANCI-FANCD2 to Promote DNA Cross-Link Repair." *Science*, 7/2010

168. M. E. Baker. "Evolution of 11β-hydroxysteroid dehydrogenase-type 1 and 11β-hydroxysteroid dehydrogenase-type 3." *FEBS Letters*, 2010; 584 (11)

169. Bonnedahl et al. "Dissemination of Escherichia coli with CTX-M Type ESBL between Humans and Yellow-Legged Gulls in the South of France." *PLoS ONE*, 2009 .

170. J. P. Boettcher et al. "Tyrosine-Phosphorylated Caveolin-1 Blocks Bacterial Uptake by Inducing Vav2-RhoA-Mediated Cytoskeletal Rearrangements." *PLoS Biology*, 2010; 8 (8)

171. J.-H. Hehemann, G. Correc, T. Barbeyron, W. Helbert, M. Czjzek, G. Michel. "Transfer of Carbohydrate-active Enzymes from Marine Bacteria to Japanese Gut Microbiota." *nature* 464, 908-912 (8 April 2010)

172. K.A. DeMali, A.L.Jue, K. Burridge. "IpaA Targets 1 Integrins and Rho to Promote Actin Cytoskeleton Rearrangements Necessary for *Shigella* Entry." *The Journal of Biological Chemistry, 281, 39534-39541.*

173. Deutsches Krebsforschungszentrum. "Cancer and Infections: Are There Common Mechanisms?" *ScienceDaily.*

174. S.Y. Huang, C. Jeng C, S.C. Kao, J.J Yu, D.Z. Liu. "Improved Haemorrheological Properties by Ginkgo Biloba Extract (Egb 761) in Type 2 Diabetes Mellitus Complicated with Retinopa- thy." *Clin Nutr.* 2004 Aug;23(4):615-21.

175. D. Y. Vargas. "Mechanism of mRNA Transport in Nucleus." *PNAS*, November 22, 2005 vol. 102, no. 47, 17008

176. L. Hernandez et al. "Functional Coupling between the Extracellular Matrix and Nuclear Lamina by Wnt Signaling in Progeria." *Developmental Cell*, 2010; 19 (3): 413-425 DOI:10.1016/j.devcel.2010.08.01

177. American Society for Cell Biology. "Zeroing in on Progeria: How Mutant Laminas Ageing." *ScienceDaily*, 14 Dec.2005.

178. Children's Hospital Los Angeles. "Obesity Accelerates Leukemia, Study Shows." *ScienceDaily*, 10 September 2010

179. http://www.primeorigins.co.za/news/231966.htm

180. D. Heinrich, E. Sackmann. Active Mechanical Stabilization of the Viscoplastic Intracellular Space of Dictyostelia Cells by Microtubule-actin Crosstalk Acta Biomater. 2006 Nov;2(6):619-31. Epub 2006 Aug 30.

181. Genetics Society of America. "Why Fad Diets Work Well for Some, Not Others."ScienceDaily 28/7/ 2010 .

182. Friedrich-Schiller-Universität Jena. Nomadic People's Good Health Baffle Scientists. *ScienceDaily*. Web.

183. Centre National De La Recherche Scientifique. "When Evolution Tends To Maximize The Diversity And Functioning Of Ecosystems." ScienceDaily 2 April 2008.

184. University of Minnesota. "Foods, Not Specific Nutrients,Key To Good Health." ScienceDaily 7/11/ 2007.

185. BMJ-British Medical Journal. "Calcium Supplements Linked to Increased Risk of Heart Attack, Study Finds." ScienceDaily 30 July 2010. Web.

186. A. M. Patel, S. Goldfarb. Got Calcium? Welcome to the Calcium-Alkali Syndrome. *Journal of the American Society of Nephrology*, 2010; DOI: 10.1681/ASN.2010030255

187. USDA/Agricultural Research Service. "Concentration, Timing and Interactions Are Key When It Comes to Dietary Compounds." ScienceDaily 18 July 2010. Web.

188. W. Zukerman Losing Weight May Pollute the Blood. http://www.newscientist.com/article/dn19406-losing-weight-may-pollute-the-blood.html .

189. E. Young. Dieting Could Lead to a Positive Test for Cannabis. http://www.newscientist.com/article/mg20327205.100-dieting-could-lead-to-a-positive-test-for-cannabis.html

190. E. Callaway. Seafood Gave Us the Edge on the Neanderthals. http://www.newscientist.com/article/dn17595-seafood-gave-us-the-edge-on-the-neanderthals.html

191. M. Tandon, R. A. Siddique, S. N. Rai Effect of Nutrients on the Gene Expression. www.pitt.edu/~super7/30011-31001/30951.ppt

192. Are organic foods really beneficial? Nutrition Research Newsletter,May, 2010 http://findarticles.com/p/articles/mi_m0887/is_5_29/ai_n54035519/

193. J. Koening et al. Succession of microbial consortia in the developing infant gut microbiome http://www.pnas.org/content/early/2010/07/27/1000081107

194. C. De Filippoa et al. Impact of Diet in Shaping Gut Microbiota Revealed by a Comparative Study in Children from Europe and Rural Africa. http://www.pnas.org/content/107/33/14691

195. G. A. Koh et al. Exposure to Non-core Foods and Beverages in the First Year of Life: Results from a Cohort Study." *Nutrition & Dietetics*, DOI: 10.1111/j.1747-0080.2010.01445.x

196. Elliott et al. Sweet and salty: nutritional content and analysis of baby and toddler foods. *Journal of Public Health*, 2010; DOI: 10.1093/pubmed/fdq037

197. Elliott et al. Assessing 'fun foods': nutritional content and analysis of supermarket foods targeted at children. *Obesity Reviews*, 2008; 9 (4): 368 DOI:10.1111/j.1467-789X.2007.00418.x

198. University of Arizona. Food Industry Faulted for Pushing High-Calorie, Low-Nutrient Products. ScienceDaily14 December 2009. Web.

199. University of Chicago Medical Center. "Cells Defend Themselves from Viruses, Bacteria With Armor of Protein Errors." ScienceDaily 25 November 2009.

200. Louisiana State University Health Sciences Center. "Early Predictors of Metabolic Syndrome in Healthy 7-9 Year-Olds Identified." ScienceDaily, Web

201. BioMed Central. "Low-Carb Diet Better Than Low-Fat Diet." ScienceDaily 16/11/2005, Web.

202. T. L. Horvath. Synaptic input organization of the melanocortin system predicts diet-induced hypothalamic reactive gliosis and obesity. PNAS, 2010; 107 (33): 14875 DOI:10.1073/pnas.1004282107

203. S.-A. M. Burrell, C. Exley. There is (still) too much aluminium in infant formulas. BMC Pediatrics, 2010; 10 (1): 63 DOI:10.1186/1471-2431-10-63

204. C. Lawson et al. Gene Expression in the Fetal Mouse Ovary Is Altered by Exposure to Low Doses of Bisphenol A. Biology of Reproduction, 2010; DOI:10.1095/biolreprod.110.084814

205. J. A. Taylor, F. S. vom Saal, W. V. Welshons, B. Drury, G. Rottinghaus, P. A. Hunt, C. A. VandeVoort. Similarity of Bisphenol A Pharmacokinetics in Rhesus Monkeys and Mice: Relevance for Human Exposure. Environmental Health Perspectives, 2010; DOI: 10.1289/ehp.1002514

206. H. B. Adewale et al. Neonatal bisphenol-A exposure alters rat reproductive development and ovarian morphology without impairing activation of gonadotropin hormone neurons. Biology of Reproduction, June, 2009

207. Elsevier. "Six Environmental Research Studies Reveal Critical Health Risks From Plastic.ScienceDaily3 October 2008. Web

208. American Thoracic Society. Some Vitamin Supplements Don't Protect Against Lung Cancer. ScienceDaily 21 May 2007. http://www.sciencedaily.com/releases/2007/05/070521113628.htm

209. American Thoracic Society. "Certain Vitamin Supplements May Increase Lung Cancer Risk, Especially In Smokers. ScienceDaily 3 March 2008.

210. University of Helsinki. "Vitamin E May Increase Tuberculosis Risk". ScienceDaily 21 February 2008.

211. Cornell University. "Genetically Engineered Salmon Safe to Eat, but a Threat to Wild Stocks, Expert Says." ScienceDaily 25 September 2010. http://www.sciencedaily.com/releases/2010/09/100925105209.htm

212 D. Grünwald, R. H. Singer. In vivo imageing of labelled endogenous β-actin mRNA during nucleocytoplasmic transport. nature, 2010; DOI:10.1038/nature09438

213. University of Copenhagen. "Stress Can Control Our Genes, Researchers Find." ScienceDaily24/09/ 2010.

214 M. L. Pelchat et al. Excretion and Perception of a Characteristic Odor in Urine after Asparagus Ingestion: a Psychophysical and Genetic Study. *Chemical Senses*, 2010; DOI:10.1093/chemse/bjq081.

215. A. Swarbrick et al. miR-380-5p represses p53 to control cellular survival and is associated with poor outcome in MYCN-amplified neuroblastoma. *nature Medicine*, 2010; DOI:10.1038/nm.2227

216. C. J. Marsit, K. Eddy, K. T. Kelsey MicroRNA Responses to Cellular Stress. Cancer Res. November 15, 2006 66;10843

217. A. D. Maurer et al. Consumption of diets high in prebiotic fiber or protein during growth influences the response to a high fat and sucrose diet in adulthood in rats. *Nutrition & Metabolism*, 2010

218. Nutraceutical http://en.wikipedia.org/

219. M. L. Nelson, A. Dinardo, J. Hochberg, G. J. Armelagos. Brief communication: Mass spectroscopic characterization of tetracycline in the skeletal remains of an ancient population from Sudanese Nubia 350-550 CE. *American Journal of Physical Anthropology*, 2010; DOI: 10.1002/ajpa.21340

220. **P. Ak, A. J. Levine.** p53 and NF- B: different strategies for responding to stress lead to a functional antagonism. *The FASEB Journal*, 2010; DOI:10.1096/fj.10-160549

221. **E. Stice, S. Yokum, K. Blum, C. Bohon.** Weight Gain Is Associated with Reduced Striatal Response to Palatable Food. *Journal of Neuroscience*, 2010; 30 (39): 13105 DOI: 10.1523/JNEUROSCI.2105-10.2010

222. **G. E. Mikhailovsky.** Отрицательная энтропия и диссипативные структуры, порожденные предельными циклами. Журнал физической химии. 1981. Т.55, №7. С.1877-1879.

223. **G. E. Mikhailovsky** Биологическое время, его организация, иерархия и представление с помощью комплексных величин. http://www.chronos.msu.ru/RREPORTS/mikhailovsky1.pdf

224. **J. Holst et al.** Substrate elasticity provides mechanical signals for the expansion of hemopoietic stem and progenitor cells. *nature Biotechnology*, 03 October 2010 DOI: 10.1038/nbt.1687

225. **American Chemical Society.** "Elasticity Of Tissue Environment Plays Role In Determining Stem Cell Growth." ScienceDaily 20 September 2006. http://www.sciencedaily.com/releases/2006/09/060918201126.htm

226. **R. J. O'Sullivan et al.** Reduced histone biosynthesis and chromatin changes arising from a damage signal at telomeres. *nature Structural and Molecular Biology*, 03 October 2010 DOI: 10.1038/nsmb.1897

227. **A. Ceriello, M. A. Ihnat, J. E. Thorpe** The "Metabolic Memory": Is More than Just Tight Glucose Control Necessary to Prevent Diabetic Complications? Journal of Clinical Endocrinology & Metabolism , December 9, 2008 doi:10.1210/jc.2008-1824

228. **Su-Chen Ho, Tzung-Hsun Tsai, Po-Jung Tsai, Chih-Cheng Lin.** Protective capacities of certain spices against peroxynitrite-mediated biomolecular damage. Food and Chemical Toxicology 46 (2008) 920–928.

229. **Michigan State University.** "Fish Oil Linked to Increased Risk of Colon Cancer in Mice."ScienceDaily 6 October 2010.Web.

230. **T. M. Winzenberg, S. Powell, K. A. Shaw, G. Jones.** Vitamin D supplementation for improving bone mineral density in children. *Cochrane Database of Systematic Reviews*, 2010 (10): CD006944 DOI:10.1002/14651858. CD006944.pub2

231. **G. E. Arteel, P. Schroeder, H. Sies** Reactions of Peroxynitrite with Cocoa Procyanidin Oligomers, *Journal of Nutrition.* 2000;130:2100S-2104S

232. **M. A. Mendez et al.** Prenatal Organochlorine Compound Exposure, Rapid Weight Gain and Overweight in Infancy. *Environmental Health Perspectives*, 2010; DOI: 10.1289/ehp.1002169

233. **G. Bistrai.** ТЕРМОДИНАМИКА НЕРАВНОВЕСНЫХ ПРОЦЕССОВ В ОТКРЫТЫХ НЕЛИНЕЙНЫХ СИСТЕМАХ С ДЕТЕРМИНИРОВАННЫМ ХАОСОМ Диссертации на соискание ученой степени доктора физико-математических наук. Екатеринбург, Russia, 2009.

234. **D. A. Case, M. Karplus.** Dynamics of ligand binding to heme proteins Journal of Molecular Biology Volume 132, Issue 3, 15 August 1979, Pages 343-368.

235. **A. Blumenfeld.** Биофизика. 1993. N 1. С. 129. Москва, Россия.

236. **D. Arad, K. Moss, Y. Elias G. Anbar.** Structure-Function Properties of Water Clusters in Proteins, in: C. Taddei-Feretti, P. Marotta High Dilution Effects on Cells and Integrated Systems, Proceedings of the International School of Biophysics Casamicciola, Napoly, 23-28 October (1995a) 313-325.

237. **Max-Planck-Gesellschaft** (2006, October 25). Why Biological Loads Do Not Get Caught Up When Being Transported Through Cells. *ScienceDaily.* http://www.sciencedaily.com/releases/2006/10/061024093427.htm

238. **I. Yu. Galaev,** "'Smart' polymers in biotechnology and medicine", RUSS CHEM REV, 1995, 64 (5), 471–489. DOI: 10.1070/RC1995v064n05ABEH000161

239. **R. Clarke et al.** Effects of Lowering Homocysteine Levels With B Vitamins on Cardiovascular Disease, Cancer, and Cause-Specific Mortality: Meta-analysis of 8 Randomized Trials Involving 37 485 Individuals. *Archives of Internal Medicine*, 2010; 170 (18): 1622 DOI: 10.1001/archinternmed.2010.348

240. **BMJ-British Medical Journal.** "People Wasting Billions Of Dollars On 'Quack' Health Food And Weight Loss Products, Expert Says." ScienceDaily 1 December 2008.Web.

241. **D. Mikitenko.** Коррекция эпигенетических нарушений ДНК. Сибирский медицинский журнал, 2008, № 1, Russia.

242. **Lund University.** "Right Food Effectively Protects Against Risk for Diabetes, Cardiovascular Disease and Cognitive Decline." ScienceDaily 17 October 2010. Web.

243. **J. M. Van Raamsdonk, S. Hekimi.** Deletion of the Mitochondrial Superoxide Dismutase sod-2 Extends Lifespan in Caenorhabditis elegans. *PLoS Genetics*, 2009; 5 (2): e1000361 DOI:10.1371/journal.pgen.1000361

244 **R. Doonan et al.** Against the oxidative damage theory of ageing: superoxide dismutases protect against oxidative stress but have little or no effect on life span in Caenorhabditis elegans. *Genes and Development*, 2008

245. **T. T. Perls.** Anti-ageing medicine: what should we tell our patients? *April 2010, Vol. 6, No. 2, Pages 149-154*, Ageing Health DOI: 10.2217/ahe.10.11

246. **J. M. Lee, M. M. Davis, A. Gebremariam, C. Kim.** Age and Sex Differences in Hospitalizations Associated with Diabetes. *Journal of Women's Health*, 2010; 101012040727072 DOI: 10.1089/jwh.2010.2029

247. **The Peninsula College of Medicine and Dentistry.** "Like Father, Like Son: Childhood Obesity Link To Parents." ScienceDaily 13 July 2009.Web.
248. **Gniuli et al.** Effects of high-fat diet exposure during fetal life on type 2 diabetes development in the progeny. *The Journal of Lipid Research*, 2008; 49 (9): 1936 DOI:10.1194/jlr.M800033-JLR200
249. **K.-M. Schmitz et al.** Interaction of noncoding RNA with the rDNA promoter mediates recruitment of DNMT3b and silencing of rRNA genes. *Genes & Development*, 2010; 24 (20): 2264 DOI: 10.1101/gad.590910
250. **Sh. A. Ross et al.** Introduction: diet, epigenetic events and cancer prevention. Nutr Rev. 2008 August; 66(Suppl 1): S1–S6. doi: 10.1111/j.1753-4887.2008.00055.x.
251. **Sh. A. Ross, J. A. Milner.** Epigenetic modulation and cancer: effect of metabolic syndrome? American Journal of Clinical Nutrition, Vol. 86, No. 3, 872S-877S, September 2007
252. **Ch. A. Rivera et al.** Western diet enhances hepatic inflammation in mice exposed to cecal ligation and puncture .*BMC Physiology,* October 2010.
253. **B. E. Clurman et al.** Loss of the p53/p63 Regulated Desmosomal Protein Perp Promotes Tumorigenesis. *PLoS Genetics*, 2010; 6 (10): e1001168 DOI: 10.1371/journal.pgen.1001168
254. **I. Sack et al.** The impact of ageing and gender on brain viscoelasticity. NeuroImage, 46 (2009), 652–657.
255. **T.S. Cheryl et al.** Two-component protein-engineered physical hydrogels for cell encapsulation. PNAS December 29, 2009,vol. 106, no. 52, 22067-22072.
256. http://en.wikipedia.org/wiki/Phased_array
257. **V. Anichenko et al.** Стохастический резонанс как индуцированный шумом эффект увеличения степени порядка Успехи Физ.Наук,1999. т. 169. № 1.- С. 7 - 38.
258. **A. Karnaukhov.** Диссипативный резонанс и его роль в механизмах действия электромагнитного излучения на биологические и физико-химические системы.- Биофизика, 1997, Т. 42, вып 4, 971 – 979.
259. **I. Shuryak, R. K. Sachs, D. J. Brenner.** Cancer Risks After Radiation Exposure in Middle Age. *Journal of the National Cancer Institute*, 2010; DOI:10.1093/jnci/djq346
260. **J. D. Boice, Jr.** Models, Models Everywhere—Is There a Fit for Lifetime Risks? *Journal of the National Cancer Institute*, 2010; DOI: 10.1093/jnci/djq412
261. **A.C. Steinemann et al.** Fragranced Consumer Products: Chemicals Emitted, Ingredients Unlisted. *Environmental Impact Assessment Review*, 2010; DOI: 10.1016/j.eiar.2010.08.002
262. **Yu. Shen et al.** Blood Peptidome-Degradome Profile of Breast Cancer. *PLoS ONE*, 2010; 5 (10): e13133 DOI: 10.1371/journal.pone.0013133.
263. **P. J. Campbell et al.** The patterns and dynamics of genomic instability in metastatic pancreatic cancer. *nature*, 2010; 467 (7319): 1109 DOI: 10.1038/nature09460
264. **H. Tanizawa et al.** Mapping of long-range associations throughout the fission yeast genome reveals global genome organization linked to transcriptional regulation. *Nucleic Acids Research*, 2010; DOI: 10.1093/nar/gkq955 http://www.rosbalt.ru/2010/10/29/785267.html
265. **J. Hänisch et al.** Molecular dissection of Salmonella-induced membrane ruffling versus invasion. *Cellular Microbiology*, 2010; 12 (1): 84 DOI:10.1111/j.1462-5822.2009.01380.x
266. **E. Lozovskaya.** Жизнь с гравитацией и без нее http://earth.taba.ru/blog/30212_Zhizn_s_gravitaciey_i_bez_nee.html
267. **University of Texas Medical Branch at Galveston.** Shuttle Mice to Boost Disease Research: Experiment on Last Flight of Discovery Will Probe Spaceflight-Induced Immune-System Impairment. ScienceDaily 30/10/2010.
268. **Brown University.** Size of Protein Aggregates, Not Abundance, Drives Spread of Prion-Based Disease. ScienceDaily 1 November 2010, Web.
269. **Biotechnology and Biological Sciences Research Council.** "Human Immune System Assassin's Tricks Visualized for the First Time." ScienceDaily 1 November 2010. Web.
270. **J. C. Choy et al.** Granzyme B Induces Smooth Muscle Cell Apoptosis in the Absence of Perforin. Involvement of Extracellular Matrix Degradation. Arteriosclerosis, Thrombosis, and Vascular Biology. 2004;24:2245-2250
271. **Schieke et al.** Mitochondrial Metabolism Modulates Differentiation and Teratoma Formation Capacity in Mouse Embryonic Stem Cells. *Journal of Biological Chemistry*, 2008; 283 (42): 28506, 10.1074/jbc.M802763200
272. **Basan et al.** Homeostatic competition drives tumor growth and metastasis nucleation. *Advanced Online Publication Articles for HFSP Journal*, 2009; 1 (1): 98 DOI: 10.2976/1.3086732
273. **C. W. Harland et al.** Phospholipid bilayers are viscoelastic. *PNAS, Oct.25, 2010,10.1073/pnas.1010700107*
274. **Rockefeller University.** "Phospholipids In The Cell Membrane Help Regulate Ion Channels." ScienceDaily 28 February 2007. Web.
275. **S. Busch, C. Smuda, L. Carlos Pardo, T. Unruh** . Molecular Mechanism of Long-Range Diffusion in Phospholipid Membranes Studied by Quasielastic Neutron Scattering. J. Am. Chem. Soc., 2010, 132 (10), pp 3232–3233, DOI: 10.1021/ja907581s
276. **O. Y. Borbulevych et al.** T Cell Receptor Cross-reactivity Directed by Antigen-Dependent Tuning of Peptide-MHC Molecular Flexibility. *Immunity*, 2009; 31 (6): 885 DOI: 10.1016/j.immuni.2009.11.003

277. V. Mieulet et al. TPL-2-Mediated Activation of MAPK Downstream of TLR4 Signaling Is Coupled to Arginine Availability. *Science Signaling*, 2010; 3 (135): ra61 DOI: 10.1126/scisignal.2000934

278. X. Peng et al. Unique Signatures of Long Noncoding RNA Expression in Response to Virus Infection and Altered Innate Immune Signaling. *mBio*, 2010; 1 (5): e00206-10 DOI: 10.1128/mBio.00206-10

279. K. B. Petrov, O.C. Kalinina. Соотношение хаоса и порядка в биологических системах. *http://www.medvopros.com/view_story/5080/62*

280. Mikael Knip et al. Dietary Intervention in Infancy and Later Signs of Beta-Cell Autoimmunity. *New England Journal of Medicine*, 2010; 363: 1900-1908 DOI: 10.1056/NEJMoa1004809

281. J. Rittle and M. T. Green. Cytochrome P450 Compound I: Capture, Characterization, and C-H Bond Activation Kinetics. Science, 12 November 2010 330: 933-937 DOI: 10.1126/science.1193478

282. A. Al-Kilani et al. During vertebrate development, arteries exert a morphological control over the venous pattern through physical factors. Physical Review E (May 2008)

283. L. Fontana, S. Klein. Ageing, Adiposity, and Calorie Restriction. *JAMA*. 2007;297:986-994.

284. P. CAÑADAS et al. A Cellular Tensegrity Model to Analyse the Structural Viscoelasticity of the Cytoskeleton. *Journal of Theoretical Biology*, Vol. 218, Issue 2, 21 September 2002, Pages 155-173.

285. A. Bershadsky et al. Involvement of microtubules in the control of adhesion-dependent signal transduction. *Current Biology*, Vol. 6, Issue 10, October 1996, 1279-1289.

286. J. T. Parsons. Integrin-mediated signaling: regulation by protein tyrosine kinases and small GTP-binding proteins. *Current Opinion in Cell Biology*. Vol. 8, Issue 2, April 1996, Pages 146-152.

287. C. E. Elks et al. Thirty new loci for age at menarche identified by a meta-analysis of genome-wide association studies. *nature Genetics*, 21 November 2010 DOI: 10.1038/ng.714

288. T. W. Ridky, J. M. Chow, D. J. Wong, P. A. Khavari. Invasive three-dimensional organotypic neoplasia from multiple normal human epithelia. *nature Medicine*, 21 November 2010 DOI: 10.1038/nm.2265

289. Stanford University Medical Center. "Normal cells transformed into 3-D cancers in tissue culture dishes." ScienceDaily 22 November 2010. Web.

290. R. G. LoCascio, P. Desai, D. A. Sela, B. Weimer, D. A. Mills. Broad Conservation of Milk Utilization Genes in Bifidobacterium longum subsp. infantis as Revealed by Comparative Genomic Hybridization. *Applied and Environmental Microbiology*, 2010; 76 (22): 7373 DOI: 10.1128/AEM.00675-10

291. F. Keesing, L. K. Belden, P. Daszak, A. Dobson, C. Drew Harvell, R. D. Holt, P. Hudson, A. Jolles, K. E. Jones, C. E. Mitchell, S. S. Myers, T. Bogich, R. S. Ostfeld. Impacts of biodiversity on the emergence and transmission of infectious diseases. *nature*, 2010; 468 (7324): 647 DOI: 10.1038/nature09575

292. Cary Institute of Ecosystem Studies. "Loss of species large and small threatens human health, study finds." ScienceDaily 1 December 2010.Web.

293. Gerke J, Lorenz K, Ramnarine S, Cohen B. Gene-environment interactions at nucleotide resolution. *PLoS Genetics*, Sept. 2010 DOI: 10.1371/journal.pgen.1001144

294. J. Todrank, G. Heth, D. Restrepo. Effects of in utero odorant exposure on neuroanatomical development of the olfactory bulb and odour preferences. *Proc. R. Soc. B*, December 1, 2010 DOI: 10.1098/rspb.2010.2314

295. A. A. Fryer et al. Quantitative, high-resolution epigenetic profiling of CpG loci identifies associations with cord blood plasma homocysteine and birth weight in humans. *Epigenetics*, 2011; 6 (1): 86-94.

296. Keele University. "Link between folic acid supplementation in pregnancy, DNA methylation and birth weight in newborn babies." ScienceDaily 6 December 2010. Web.

297. S. Bandyopadhyay et al. Rewiring of Genetic Networks in Response to DNA Damage. *Science*, 2010; 330 (6009): 1385 DOI: 10.1126/science.1195618

298. University of California - San Diego. "Scientists map changes in genetic networks caused by DNA damage." ScienceDaily 7 December 2010. Web.

299. Brandt et al. SCAI acts as a suppressor of cancer cell invasion through the transcriptional control of 1-integrin. *nature Cell Biology*, 2009; DOI: 10.1038/ncb1862

300. http://www.nzhealth.net.nz/poisons/milk_1_2.shtml

301. C. L. Raison, C. A. Lowry, G. A. W. Rook. Inflammation, Sanitation, and Consternation: Loss of Contact With Coevolved, Tolerogenic Microorganisms and the Pathophysiology and Treatment of Major Depression. *Arch Gen Psychiatry*, 2010;67(12):1211-1224 DOI: 10.1001/archgenpsychiatry.2010.161

302. S. P. Mahal, S. Browning, J. Li, I. Suponitsky-Kroyter, C. Weissmann. Transfer of a prion strain to different hosts leads to emergence of strain variants. *PNAS*, 2010; DOI: 10.1073/pnas.1013014108

303. N. Wang, J. D. Tytell, D. E. Ingber. Mechanically coupling the extracellular matrix with the nucleus. nature Reviews Molecular Cell Biology 10, 75-82 (January 2009). doi:10.1038/nrm2594.

304. Volkmer Ward SM, Weins A, Pollak MR, Weitz DA. Dynamic viscoelasticity of actin cross-linked with wild-type and disease-causing mutant alpha-actinin-4. Biophys J. 2008. Nov 15;95(10):4915-23.

305. Komatsu et al. Organelle-specific, rapid induction of molecular activities and membrane tethering. nature Methods, 2010; 7 (3): 206 DOI: 10.1038/nmeth.1428

306. A. Fritsch et al. Are biomechanical changes necessary for tumour progression? nature Physics 6, 730–732 (2010) doi:10.1038/nphys1800

307. J. Sarfati. Origin of life: the polymerization problem. http://creation.com/origin-of-life-the-polymerization-problem

308. R. Vegners et al. Use of a gel-forming dipeptide derivative as a carrier for antigen presentation. Journal of Peptide Science. Volume 1, Issue 6, pages 371–378,November/December 1995

309. D A. Fedosov et al. Quantifying the biophysical characteristics of Plasmodium-falciparum-parasitized red blood cells in microcirculation. PNAS, December 20, 2010 DOI: 10.1073/pnas.1009492108

310. Tracey J. Woodruff et al. Environmental Chemicals in Pregnant Women in the US: NHANES 2003-2004. Environmental Health Perspectives, 2011; DOI: 10.1289/ehp.1002727

311. Ian J. Rickard et al. Food availability at birth limited reproductive success in historical humans. Ecology, 2010; 91 (12): 3515 DOI: 10.1890/10-0019.1

312. B. Crujeiras et al. Weight Regain after a Diet-Induced Loss Is Predicted by Higher Baseline Leptin and Lower Ghrelin Plasma Levels. Journal of Clinical Endocrinology & Metabolism, 2010; 95 (11): 5037 DOI: 10.1210/jc.2009-2566

313. M. Carini Restoring Leptin and Ghrelin Balance for Effective Weight Loss http://www.suite101.com/content/restoring-leptin-and-ghrelin-balance-for-effective-weight-loss-a322870

314. Jae-Hyung Jeon et al. In Vivo Anomalous Diffusion and Weak Ergodicity Breaking of Lipid Granules. Physical Review Letters, 2011; 106 (4) DOI: 10.1103/PhysRevLett.106.048103

315. University of Copenhagen. "Dynamic systems in living cells break the rules." ScienceDaily 25/01/2011.

316. R. D. Gupta et al. Directed evolution of hydrolases for prevention of G-type nerve agent intoxication. nature Chemical Biology, 2011; DOI: 10.1038/nchembio.510

317. Lund University. "Unfolding pathogenesis in Parkinson's: Breakthrough suggests damaged proteins travel between cells." ScienceDaily 26 January 2011. Web.

318. Harvard School of Public Health. "20-year Study Finds No Association Between Low-carb Diets And Risk Of Coronary Heart Disease."ScienceDaily 9 November 2006. Web

319. K. Shkolnik, A. Tadmor, S. Ben-Dor, N. Nevo, D. Galiani, N. Dekel. Reactive oxygen species are indispensable in ovulation. P.N.A.S., 2011; DOI: 10.1073/pnas.1017213108

320. Janel E. Le Belle et al. Proliferative Neural Stem Cells Have High Endogenous ROS Levels that Regulate Self-Renewal and Neurogenesis. Cell Stem Cell, 2011; 8 (1): 59-71 DOI: 10.1016/j.stem.2010.11.028

321. E.J. Okello, G.J. McDougall, S. Kumar, C.J. Seal. In vitro protective effects of colon-available extract of Camellia sinensis (tea) against hydrogen peroxide and beta-amyloid (A (1–42)) induced cytotoxicity in differentiated PC12 cells. Phytomedicine, 2010; DOI: 10.1016/j.phymed.2010.11.004

322. M. Eisenstein. Diversity: Of beans and genes. nature. V.468, P. S13–S15, DOI: doi:10.1038/468S13a

323. Cell Press. "You are what your father ate, too: Paternal diet affects lipid metabolizing genes in offspring, research suggests." ScienceDaily 24 December 2010. http://www.sciencedaily.com/releases/2010/12/101223130149.htm .

324. B. R. Carone et al. Paternally Induced Transgenerational Environmental Reprogramming of Metabolic Gene Expression in Mammals. Cell, 2010; 143 (7): 1084-1096 DOI: 10.1016/j.cell.2010.12.008

325. S. Anava et al. The Regulative Role of Neurite Mechanical Tension in Network Development. Biophysical Journal, 2009; 96 (4): 1661 DOI: 10.1016/j.bpj.2008.10.058

326. R. Sorkin et al. Process entanglement as a neuronal anchorage mechanism to rough surfaces. Nanotechnology, 2009; 20 (1): 015101 DOI: 10.1088/0957-4484/20/1/015101

327. W. Lee, H. Amini, H. A. Stone, D. Di Carlo. Dynamic self-assembly and control of microfluidic particle crystals. PNAS, December 13, 2010 DOI: 10.1073/pnas.1010297107

328. J.A. Pojman. (1999) Self Organization in Synthetic Polymeric Systems. Annals of the New York Academy of Sciences, V. 879: 194–214. doi: 10.1111/j.1749-6632.1999.tb10420.x.

329. D. D. Leipe, L. Aravind, E. Koonin. Did DNA replication evolve twice independently? Nucl. Acids Res. (1999) 27 (17): 3389-3401. doi: 10.1093/nar/27.17.3389

330. M. D. Spencer et al. Association Between Composition of the Human Gastrointestinal Microbiome and Development of Fatty Liver With Choline Deficiency. Gastroenterology, 2010; DOI: 10.1053/j.gastro.2010.11.049

331. M. J. Eisenberg et al. Cancer risk related to low-dose ionizing radiation from cardiac imageing in patients after acute myocardial infarction. Canadian Medical Association Journal, 2011; DOI: 10.1503/cmaj.100463

332. M. Mercuri, T. Sheth, M. K. Natarajan. Radiation exposure from medical imageing: A silent harm? Canadian Medical Association Journal, 2011 DOI: 10.1503/cmaj.101885

333. Fazel et al. Exposure to Low-Dose Ionizing Radiation from Medical Imageing Procedures. New England Journal of Medicine, 2009; 361 (9): 849 DOI: 10.1056/NEJMoa0901249

334. Lauer et al. Elements of Danger -- The Case of Medical Imageing. New England Journal of Medicine, 2009; 361 (9): 841 DOI: 10.1056/NEJMp0904735

335. **National Council on Radiation Protection & Measurements**. "Medical Radiation Exposure Of The U.S. Population Greatly Increased Since The Early 1980s." ScienceDaily 5 March 2009. Web.

336. **Journal of the National Cancer Institute**. "Radiation Exposure In Utero And In Young Children Increases Adult Cancer Risk." ScienceDaily 13 March 2008. Web.

337. **Northwestern University**. "Gonorrhea acquires a piece of human DNA: First evidence of gene transfer from human host to bacterial pathogen." ScienceDaily 13 February 2011, Web.

338. **The Research Council of Norway**. "Folic acid may increase the risk of asthma."ScienceDaily 10/02/ 2011.

339. **F. J. Larsen et al**. Dietary Inorganic Nitrate Improves Mitochondrial Efficiency in Humans. *Cell Metabolism*, 2011; 13 (2): 149-159 DOI: 10.1016/j.cmet.2011.01.004

340. **O. Rocks et al**. The Palmitoylation Machinery Is a Spatially Organizing System for Peripheral Membrane Proteins. *Cell*, 2010; 141 (3): 458 DOI: 10.1016/j.cell.2010.04.007

341. **A. Deem et al**. *Break-Induced Replication Is Highly Inaccurate. PLoS Biology, 2011.*

342. **H. Hemila, J. Kaprio**. Subgroup analysis of large trials can guide further research: a case study of vitamin E and pneumonia. *Clinical Epidemiology*, 2011; 51 DOI: 10.2147/CLEP.S16114

343. **American Academy of Neurology**. "Eating berries may lower risk of Parkinson's." ScienceDaily 17 February 2011. Web.

344. **American Friends of Tel Aviv University**. "Good diets fight bad Alzheimer's genes: Diets high in fish oil have a beneficial effect in patients at risk, researcher says." ScienceDaily 16 February 2011. Web.

345. **Duo Li**. Chemistry behind Vegetarianism. *J.A.F.C.*, 2011; DOI: 10.1021/jf103846u

346. **American Chemical Society**. "Don't blame the pill for estrogen in." ScienceDaily 15 February 2011.Web.

347. **A. Wise, K. O'Brien, T. Woodruff**. Are Oral Contraceptives a Significant Contributor to the Estrogenicity of Drinking Water?. *Environmental Science & Technology*, 2010; 101026133329091 DOI: 10.1021/es1014482

348. **D. Walgraef**. Spatio-Temporal Pattern Formation. Springer, Berlin, 1997.

349. **A. Loskutov, A. Mikhailov**. http://chaos.phys.msu.ru/loskutov/PDF/Loskutov.pdf

350. **C. Krembsa, H. Eickenb, J. W. Deminge**. Exopolymer alteration of physical properties of sea ice and implications for ice habitability and biogeochemistry in a warmer Arctic. *PNAS March 1, 2011 vol. 108 no. 9 3653.*

351. **H. Parsa, R. Upadhyay, S. K. Sia**. Uncovering the behaviors of individual cells within a multicellular microvascular community. *PNAS*, March 7, 2011 DOI: 10.1073/pnas.1007508108

352. **Federation of American Societies for Experimental Biology**. "Obesity and diabetes are a downside of human evolution, research suggests." ScienceDaily 25 February 2011. Web.

353. **J. G. Sutcliffe et al**. Peripheral reduction of β-amyloid is sufficient to reduce brain β-amyloid: Implications for Alzheimer›s disease. *Journal of Neuroscience Research*, March 3, 2011 DOI: 10.1002/jnr.2260

354. **Yu. L. Klimontovich**. Введение в физику открытых систем. Москва. Янус-К. 2002.

355. **Karolinska Institutet**. "Free radicals may be good for you." ScienceDaily 1 March 2011.Web.

356. **D. C. Andersson et al**. Mitochondrial production of reactive oxygen species contributes to the beta-adrenergic stimulation of mouse cardiomycytes. *The Journal of Physiology*, 28 February 2011 DOI: 10.1113/jphysiol.2010.202838

357. **Y. Gambin et al**. Visualizing a one-way protein encounter complex by ultrafast single-molecule mixing. *nature Methods*, 2011; DOI: 10.1038/nmeth.1568

358. **I. Sandovici et al**. Maternal diet and ageing alter the epigenetic control of a promoter–enhancer interaction at the Hnf4a gene in rat pancreatic islets. *P.N.A.S.*, 2011; DOI: 10.1073/pnas.1019007108

359. **M. Desai, T. Li, M. G. Ross**. Hypothalamic neurosphere progenitor cells in low birth-weight rat newborns: Neurotrophic effects of leptin and insulin. *Brain Research*, 2011; 1378: 29 DOI: 10.1016/j.brainres.2010.12.080

360. **University of Rochester Medical Center**. "Air pollution's effects on the heart." ScienceDaily 11 March 2011. Web.

361. **Public Library of Science**. "You Are What You Eat: Some Differences Between Humans And Chimpanzees Traced To Diet." ScienceDaily 3 February 2008. Web.

362. **P. Carpena, J. L. Oliver, M. Hackenberg, A. V. Coronado, G. Barturen, P. Bernaola-Galván**. High Level Organization of Isochores into Gigantic Superstructures in the Human Genome. **Phys. Rev. E 83, 031908**, March, 2011

363. **S. P. Claus et al**. Colonization-Induced Host-Gut Microbial Metabolic Interaction. *mBio*, 2011; 2 (2): 00271-10 DOI: 10.1128/mBio.00271-10

364. **M. T. Bailey et al**. Exposure to a social stressor alters the structure of the intestinal microbiota: Implications for stressor-induced immunomodulation? *Brain, Behavior, and Immunity*, 2011; 25 (3): 397-407. DOI: 10.1016/j.bbi.2010.10.023

365. **L. Balakrishnan, R. A. Bambara**. Eukaryotic Lagging Strand DNA Replication Employs a Multi-pathway Mechanism That Protects Genome Integrity *J. Biol. Chem. 2011 286: 6865-6870.* DOI:10.1074/jbc.R110.209502

366. M. F. Laguna, S. Bohn, E. A. Jagla The Role of Elastic Stresses on Leaf Venation Morphogenesis. PLoS ComputBiol.4(4), 2008:e1000055.doi:10.1371/journal.pcbi.1000055

367. American College of Radiology / American Roentgen Ray Society. "Elastography helps identify patients who need biopsy." ScienceDaily 12 January 2011.Web.

368. American Society for Cell Biology. "Cells 'feel' the difference between stiff or soft and thick or thin matrix." ScienceDaily 14 December 2010. Web.

369. K. M. Neufeld, N. Kang, J. Bienenstock, J. A. Foster. Reduced anxiety-like behavior and central neurochemical change in germ-free mice.*Neurogastroenterology & Motility*, 2011; 23 (3): 255 DOI:10.1111/j.1365-2982.2010.01620.x

370. K. Doheny. Chemicals Linked to Early Menopause. http://www.webmd.com/menopause/news/20110325/chemicals-linked-to-early-menopause

371. N. Jorgensen et al. Recent adverse trends in semen quality and testis cancer incidence among Finnish men. International Journal of Andrology. 2 MAR 2011. DOI: 10.1111/j.1365-2605.2010.01133.x

372. P. Mehta, R. Smith-Bindman. Airport Full-Body Screening: What Is the Risk? Archives of Internal Medicine, 2011; DOI:10.1001/archinternmed.2011.105

373. J. R. Martin, S. B. Lieber, J. McGrath, M. Shanabrough, T. L. Horvath, H. S. Taylor. Maternal Ghrelin Deficiency Compromises Reproduction in Female Progeny through Altered Uterine Developmental Programming. *Endocrinology*, 2011; DOI: 10.1210/en.2010-1485

374 Garvan Institute Of Medical Research. "Cancer Cells Suppress Large Regions Of DNA By A Reversible Process That Can Be Tackled." *ScienceDaily*, 24 Apr. 2006. Web.

375. American Chemical Society. "First report on bioaccumulation and processing of antibacterial ingredient TCC in fish." ScienceDaily 1 April 2011. Web.

376. T. Ichinohe et al. Proc. Natl Acad. Sci. USA doi:10.1073/pnas.1019378108 (2011).

377. P. Schopfer. Biomechanics of plant growth. American Journal of Botany. 2006;93:1415-1425.

378. A.Yakovlev. http://www.vechnayamolodost.ru/pages/vashezdorovye/okimipridi0a.html?print=print

379. Lerner Research Institute. "Common dietary fat and intestinal microbes linked to heart disease." *ScienceDaily*, 6 Apr. 2011. http://www.sciencedaily.com/releases/2011/04/110406131814.htm

380. http://www.sciencenewsline.com/medicine/2010101112000015.html

381. J. Javier Bravo-Cordero et al. A Novel Spatiotemporal RhoC Activation Pathway Locally Regulates Cofilin Activity at Invadopodia. Current Biology, 07 April 2011 DOI:10.1016/j.cub.2011.03.039

382. University of Veterinary Medicine - Vienna. "Changes in 'good' fatty acid concentration of inner organs might be largely independent of diet." *ScienceDaily*, 15 Apr. 2011. Web.

383. M. Stuiver et al. CNNM2, Encoding a Basolateral Protein Required for Renal Mg2 Handling, Is Mutated in Dominant Hypomagnesemia. The American Journal of Human Genetics, 2011; 88 (3): 10.1016/j.ajhg.2011.02.005

384. Barnosky et al. Has the Earth's sixth mass extinction already arrived? *nature* 471, 51–57, (03 March 2011), doi:10.1038/nature09678 .

385. T. J. Key. Fruit and vegetables and cancer risk. *British Journal of Cancer*, 2010; 10.1038/sj.bjc.6606032

386. Public Library of Science. "You Are What You Eat: Some Differences Between Humans And Chimpanzees Traced To Diet." *ScienceDaily*, 3 Feb. 2008. http://www.sciencedaily.com/releases/2008/01/080130092139.htm

387. A.Pross. *The Driving Force for Life's Emergence: Kinetic and Thermodynamic Considerations.* J Theor Biol. 2003 Feb 7;220(3):393-406.

388. D. Browning et al. Short-term weight loss and hepatic triglyceride reduction: evidence of a metabolic advantage with dietary carbohydrate restriction. American Journal of Clinical Nutrition, 2011; DOI: 10.3945/ajcn.110.007674

389. Sh. Menon, K. A. Beningo. Cancer Cell Invasion Is Enhanced by Applied Mechanical Stimulation. *PLoS ONE*, 2011; 6 (2): e17277 DOI: 10.1371/journal.pone.0017277

390. J. Diamond. Guns, Germs and Steel. Vintage, London, 2005

391. M. J. Bolland et al. Calcium supplements with or without vitamin D and risk of cardiovascular events: reanalysis of the Women's Health Initiative limited access dataset and meta-analysis. *BMJ*, 2011; 342 (apr19 1): d2040 DOI: 10.1136/bmj.d2040

392. K. M. Godfrey et al. Epigenetic Gene Promoter Methylation at Birth Is Associated With Child's Later Adiposity. *Diabetes*, 2011; DOI: 10.2337/db10-0979

393. M. Arumugam et al. Enterotypes of the human gut microbiome. *nature*, 2011; DOI:10.1038/nature09944

394. J. L. Round et al. The Toll-Like Receptor 2 Pathway Establishes Colonization by a Commensal of the Human Microbiota. *Science*, 21 April 2011 DOI: 10.1126/science.1206095

395. L. C. M. Antunes et al. Effect of Antibiotic Treatmenton the Intestinal Metabolome. Antimicrobial Agents and Chemotherapy,2011; 55 (4): 1494 DOI: 10.1128/AAC.01664-10

396. K. Richardson et al. The PLIN4 Variant rs8887 Modulates Obesity Related Phenotypes in Humans through Creation of a Novel miR-522 Seed Site. *PLoS ONE*, 2011; 6 (4): e17944 DOI: 10.1371/journal.pone.0017944

397. T. L. Grove et al. A Radically Different Mechanism for S-Adenosylmethionine-Dependent Methyltransferases. *Science*, 2011; DOI: 10.1126/science.1200877

398. American Academy of Pediatrics. "Growth, hormonal profiles differ between breastfed, formula-fed infants: Early nutrition has a long-term metabolic impact." ScienceDaily, 2 May 2011. http://www.sciencedaily.com/releases/2011/05/110502084440.htm .

399. A. C. Bester et al. Nucleotide Deficiency Promotes Genomic Instability in Early Stages of Cancer Development. *Cell*, 2011; 145 (3): 435 DOI: 10.1016/j.cell.2011.03.044

400. K. Koide et al. The Use of 3,5,4'-Tri-O-acetylresveratrol as a Potential Prodrug for Resveratrol Protects Mice from γ-Irradiation. *ACS Medicinal Chemistry Letters*, 2011; 2 (4): 270 DOI: 10.1021/ml100159p

401. Chen-Yu Liao et al. Fat Maintenance Is a Predictor of the Murine Lifespan Response to Dietary Restriction. *Ageing Cell*, 2011; DOI: 10.1111/j.1474-9726.2011.00702.x

402. University of Texas Health Science Center at San Antonio. "Mouse study turns fat-loss/longevity link on its head." *ScienceDaily*, 4 May 2011. http://www.sciencedaily.com/releases/2011/05/110503161409.htm

403. O.J. Rando and K.J. Verstrepen . "Timescales of Genetic and Epigenetic Inheritance". Cell 128 (4): 655-668.doi:10.1016/j.cell.2007.01.023. PMID 17320504

404. D. R. Schrider, J. N. Hourmozdi, M. W. Hahn. Pervasive Multinucleotide Mutational Events in Eukaryotes. Current Biology, June 2, 2011 DOI:10.1016/j.cub.2011.05.013

405. UT Southwestern Medical Center. "More genetic mutations lead to colon cancer."*ScienceDaily*, 18 Jul. 2011. Web

406. U. Eskiocak et al. Functional Parsing of Driver Mutations in the Colorectal Cancer Genome Reveals Numerous Suppressors of Anchorage-Independent Growth.*Cancer Research*, 2011; 71 (13): 4359 DOI: 10.1158/0008-5472. CAN-11-0794

407. K. D. Hansen et al. Increased methylation variation in epigenetic domains across cancer types. nature Genetics, 2011(26June)

408. Johns Hopkins Medical Institutions. "Scientists expose cancer cells' universal 'dark matter'; Findings reveal chaos in biochemical alterations of cancer cells." ScienceDaily, 26 Jun. 2011. Web.

409. Ki-Hyeon Seong et al. Inheritance of Stress-Induced, ATF-2-Dependent Epigenetic Change. Cell, Volume 145, Issue 7, 1049-1061, 24 June 2011 DOI:10.1016/j.cell.2011.05.029

410. Thomas Jefferson University. "Simple sugar, lactate, is like 'candy for cancer cells': Cancer cells accelerate ageing and inflammation in the body to drive tumor growth." 26 May 2011. ScienceDaily. Web

411. U. E. Martinez-Outschoorn et al. Cytokine production and inflammation drive autophagy in the tumor microenvironment. Cell Cycle, 2011; 10 (11):1784-1793.

412. A. K. Witkiewicz et al. Molecular profiling of a lethal tumor microenvironment, as defined by stromal caveolin-1 status in breast cancers. Cell Cycle, 2011; 10 (11): 1794-1809.

413. R. Youshida et al. Design of novel biomimetic polymer gels with self-oscillating function. Science and Technology of Advanced Materials. 3 (2002), 95-102.

414. O. Kuksenok et al. Exploiting gradients in cross-link density to control the bending and self-propelled motion of active gels J. Mater. Chem., 2011, 21, 8360-8371. DOI: 10.1039/C0JM03426F

415. http://www.youtube.com/watch?v=_EQY2x7avWo&feature=related

416. M. Andes-Koback, C. D. Keating. Complete Budding and Asymmetric Division of Primitive Model Cells. Journal of the American Chemical Society, 2011; 110518124742024 DOI:10.1021/ja202406v .

417. H. Scherb, K. Voigt. The human sex odds at birth after the atmospheric atomic bomb tests, after Chernobyl, and in the vicinity of nuclear facilities.E.S.P. R., 2011; 697 DOI: 10.1007/s11356-011-0462-z

418. Springer Science+Business Media. "Radiation affects sex of babies "ScienceDaily, 27 May 2011. Web.

419. B. Meier et al. Chemotactic cell trapping in controlled alternating gradient fields. Proceedings of the National Academy of Sciences, 2011; DOI: 10.1073/pnas.1014853108.

420. Baylor College of Medicine. "Protein interaction network controls gene regulation."ScienceDaily, 26 May 2011. Web.

421. A. Malovannaya et al. Analysis of the Human Endogenous Coregulator Complexome. Cell, Volume 145, Issue 5, 787-799, 27 May 2011 DOI: 10.1016/j.cell.2011.05.006 .

422. A. I. Caplan, D. Correa. The MSC: An Injury Drugstore. *Cell Stem Cell*, Volume 9, Issue 1, 11-15, 8 July 2011 DOI: 10.1016/j.stem.2011.06.008

423. Case Western Reserve University. "A drugstore within." *ScienceDaily*, 7 Jul. 2011. Web.

424. Y. K. Luu. Biomechanical promotion of mesenchymal stem cell proliferation as a countermeasure to the development of obesity and osteoporosis. Doctoral Thesis . State University of New York at Stony Brook, 2007.

425. C. Jousse et al. Perinatal Undernutrition Affects the Methylation and Expression of the Leptin Gene in Adults: Implication for the Understanding of Metabolic Syndrome. FASEB Journal, June 13, 2011.

426. CNRS "Maternal nutrition: gene expression?." ScienceDaily, 4 Jul. 2011. Web.

427. H. Wang et al. Pregnane X receptor activation induces FGF19-dependent tumor aggressiveness in humans and mice. *Journal of Clinical Investigation*, 2011; DOI: 10.1172/JCI41514

428. M. G. Farb et al. Reduced Adipose Tissue Inflammation Represents an Intermediate Cardiometabolic Phenotype in Obesity. J. of the American College of Cardiology, 2011; 58: 232-237 10.1016/j.jacc.2011.01.051 .

429. A. Swanson et al. Enteric commensal bacteria potentiate epithelial restitution via reactive oxygen species-mediated inactivation of focal adhesion kinase phosphatases. *PNAS*, 2011; DOI: 10.1073/pnas.1010042108

430. R. S. Strakovsky et al. Gestational High FatDiet Programs Hepatic Phosphoenolpyruvate Carboxykinase (Pck) Expression and Histone Modification in Neonatal Offspring Rats. The Journal of Physiology, 2011; DOI: 10.1113/jphysiol.2010.203950

431. V. Novikov. http://www.nbuv.gov.ua/portal/natural/nn/2002_2009/statti/vup24/12.pdf

432. J. T. Fox et al. High-throughput genotoxicity assay identifies antioxidants as inducers of DNA damage response and cell death. *PNAS*, 2012; DOI: 10.1073/pnas.1114278109

433. Ch. Chen et al. Mechanical Resuscitation of Chemical Oscillations in Belousov-Zhabotinsky Gels. *Advanced Functional Materials*, 2012; DOI: 10.1002/adfm.201103036

434. J. Choi et al. Emergence of insulin resistance in juvenile baboon offspring of mothers exposed to moderate maternal nutrient reduction. *American Journal of Physiology*, 2011

435. American Physiological Society. "Undernourishment in pregnant, lactating females found key to next generation's disease." *ScienceDaily*, 14 Jun. 2011. Web.

436. T. Imasaki et al. Architecture of the Mediator head module. nature, July 3, 2011.

437. Indiana University School of Medicine. "Protein structure of keymolecule in DNA transcription system deciphered." ScienceDaily, 3 Jul.2011. Web.

438. A. R. Richardson et al. Multiple Targets of Nitric Oxide in the Tricarboxylic Acid Cycle of Salmonella enterica Serovar Typhimurium. *Cell Host & Microbe*, 21 July 2011; 10(1) pp. 33 - 43 DOI: 10.1016/j.chom.2011.06.004

439. University of Washington. "New target found for nitric oxide's attack on salmonella."*ScienceDaily*, 22 Jul. 2011. Web.

440. M. S. Lee et al. Effects of shear stress on nitric oxide and matrix protein gene expression in human osteoarthritic chondrocytes in vitro. J Orthop Res. 2002 May;20(3):556-61.

441. S. Raj et al. Clone history shapes *Populus* drought responses. *Proceedings of the National Academy of Sciences*, July 11, 2011 DOI: 10.1073/pnas.1103341108

442. University of Toronto Scarborough. "Forest trees remember their roots." *ScienceDaily*, 11 Jul. 2011. Web.

443. California Institute of Technology. "First artificial neural network created out of DNA: Molecular soup exhibits brainlike behavior." *ScienceDaily*, 20 Jul. 2011. Web.

444. Public Library of Science. "Differences in precursors of 2 diabetes apparent at an early ." *ScienceDaily*, 26 Apr. 2010. Web.

445. H.-A. Park et al. Natural Vitamin E  -Tocotrienol Protects Against Ischemic Stroke by Induction of Multidrug Resistance-Associated Protein 1. Stroke, 2011; DOI: 10.1161/STROKEAHA.110.608547

446. Ohio State University. "'Gifted' natural vitamin E tocotrienol protects brain against stroke in three ways." ScienceDaily, 5 Jul. 2011. Web.

447. J. M. Seddon et al. Smoking, Dietary Betaine, Methionine, and Vitamin D in Monozygotic Twins with Discordant Macular Degeneration: Epigenetic Implications. Ophthalmology, 2011; DOI:10.1016/j.ophtha.2010.12.020

448. Tufts Medical Center. "Twin study shows lifestyle, diet can significantly influence course of macular degeneration." ScienceDaily, 5 Jul. 2011. Web.

449. S. Cortellino et al. Thymine DNA Glycosylase Is Essential for Active DNA Demethylation by Linked Deamination-Base Excision Repair. Cell, 30 June 2011 DOI: 10.1016/j.cell.2011.06.020

450. C. Allegrucci et al. Epigenetic reprogramming of breast cancer cells with oocyte extracts. Molecular Cancer, 2011; 10: 7 DOI:10.1186/1476-4598-10-7

451. The Peninsula College of Medicine and Dentistry (2009, July 13).Like Father, Like Son: Childhood Obesity Link To Parents. ScienceDaily. http://www.sciencedaily.com¬

452. M. Ezzati et al. National, regional, and global trends in fasting plasma glucose and diabetes prevalence since 1980: systematic analysis of health examination surveys and epidemiological studies with 370 country-years and 2.7 million participants. The Lancet, 25 June 2011 DOI: 10.1016/S0140-6736(11)60679-X

453. Imperial College London. "350 million adults have diabetes: Study reveals the scale of global epidemic." ScienceDaily, 25 Jun. 2011.Web.

454. J. M. Smoliga, J. A. Baur, H. A. Hausenblas. Resveratrol and health - A comprehensive review of human clinical trials. Molecular Nutrition & Food Research, 2011; DOI: 10.1002/mnfr.20110014

455. J. J. Wiens, R. A. Pyron, D. S. Moen. Phylogenetic origins oflocal-scale diversity patterns and the causes of Amazonianmegadiversity. Ecology Letters, 2011; DOI:10.1111/j.1461-0248.2011.01625.x

456. A. Mummert et al. Stature and robusticity during the agriculturaltransition: Evidence from the bioarchaeological record. Economics &Human Biology, Volume 9, Issue 3, July 2011, Pages 284-301

457. H. Ochman et al. Evolutionary Relationships of Wild HominidsRecapitulated by Gut Microbial Communities. PLoS Biol, 8(11): e1000546 DOI: 10.1371/journal.pbio.1000546

458. C. Violle, D. R Nemergut, Z. Pu,L. Jiang.Phylogenetic limiting similarity and competitive exclusion. *Ecology Letters*, 2011; DOI: 10.1111/j.1461-0248.2011.01644.x

459. X. Shen, P. Arratia. Undulatory Swimming in Viscoelastic Fluids. Physical Review Letters, 2011; 106 (20) DOI:10.1103/PhysRevLett.106.208101

460. S. Banwart. Save our soils. nature, 2011; 474 (7350): 151 DOI:10.1038/474151a

461. University of Sheffield. "Planet's soils are under threat, expert warns." ScienceDaily, 8 Jun. 2011. Web.

462. M. Cowperthwaite et al. The Ascent of the Abundant: How MutationalNetworks Constrain Evolution. PLoS Computational Biology, 2008; 4 (7):e1000110 DOI:10.1371/journal.pcbi.1000110

463. Public Library of Science. "Natural Selection May Not Produce The Best Organisms."ScienceDaily, 21 Jul. 2008. Web.

464. S. L. Young, P. W. Sherman, J. B. Lucks,G. H. Pelto. Why on Earth?:Evaluating Hypotheses about the Physiological Functions of Human Geophagy. The Quarterly Review of Biology, 2011; 86: 2

465. University of Chicago - Press Journals (2011, June 2). Eating dirt canbe good for the belly, researchers find. ScienceDaily. Web.

466. http://www.newscientist.com/article/mg20927962.600-faecal-transplant-eases-symptoms-of-parkinsons.html

467. R. E. Beardmore, I. Gudelj, D. A. Lipson, L. D. Hurst. Metabolic trade-offs and the maintenance of the fittest and the flattest. *nature*, 2011; DOI: 10.1038/nature09905

468. University of Florida. "Driving Force Of Evolution? Evolution Of Proteins Linked To Species' Metabolic Rate." *ScienceDaily*, 8 Oct. 2007. Web.

469. T. L. Tollner et al. A Common Mutation in the Defensin DEFB126 Causes Impaired Sperm Function and Subfertility. *Science Translational Medicine*, 2011; 3 (92): 92ra65 DOI: 10.1126/scitranslmed.3002289

470. University of California - Davis. "Sperm coat protein may be key to male infertility."*ScienceDaily*, 21 Jul. 2011. Web.

471. University of Southern California. "Epigenetics May Be The Underlying Cause For Male Infertility." *ScienceDaily*, 13 Dec. 2007. Web

472. L. Carroll. Alice's Adventures in Wonderland. http://www.gutenberg.org/ebooks/11 . Web.

473. R. Acin-Perez et al. *Control of oxidative phosphorylation by vitamin A illuminates a fundamental role in mitochondrial energy homoeostasis. The FASEB Journal*, 2009; DOI:10.1096/fj.09-142281

474. W. B. Liedtke et al. Relation of addiction genes to hypothalamic gene changes subserving genesis and gratification of a classic instinct, sodium appetite.*Proceedings of the National Academy of Sciences*, 2011; DOI: 10.1073/pnas.1109199108

475. T. O. Kilpeläinen et al. Genetic variation near IRS1 associates with reduced adiposity and an impaired metabolic profile. nature Genetics, 2011; DOI: 10.1038/ng.866

476. P. POULSEN, M. ESTELLER, A. VAAG, M. F. FRAGA. The Epigenetic Basis of Twin Discordance in Age-Related Diseases. Pediatric Research: May 2007 - Volume 61 - Issue 5, Part 2 - pp 38R-42R doi: 10.1203/pdr.0b013e31803c7b98

477. J. L. Bayer-Carter et al. Diet Intervention and Cerebrospinal Fluid Biomarkers in Amnesic Mild Cognitive Impairment. Archives of Neurology, 2011; 68 (6): 743-752 DOI:10.1001/archneurol.2011.125

478. D. Mozaffarian et al. Changes in Diet and Lifestyle and Long-Term Weight Gain in Women and Men. New England Journal of Medicine, 2011; 364 (25): 2392 DOI:10.1056/NEJMoa1014296

479. Johns Hopkins University Bloomberg School of Public Health. "Alzheimer's Disease To Quadruple Worldwide By 2050." ScienceDaily, 11 Jun. 2007. Web.

480. BMJ-British Medical Journal. *"Advice to drink eight glasses of water a day 'nonsense." ScienceDaily, 12 Jul. 2011. Web.*

481. http://lifestyle.aol.co.uk/2011/07/25/study-proves-dieters-nearly-always-put-weight-back-on/

482. Queensland University of Technology. "We Are Programmed To Resist Weight Loss." *ScienceDaily*, 31 Aug. 2006. Web.

483. V. Kumaran and G. H. Fredrickson. *(1996) Early stage spinodal decomposition in viscoelastic fluids. In: The Journal of Chemical Physics, 105 (18). pp. 8304-8313.*

484. http://www.polyphys.mat.ethz.ch/education/Semesterarbeiten/project_DRGgels

485. University of Illinois at Urbana-Champaign. "Minority microbes in the colon mapped." *ScienceDaily*, 2 Aug. 2011. Web.

486. G. M. Nava et al. Abundance and diversity of mucosa-associated hydrogenotrophic microbes in the healthy human colon. *The ISME Journal*, 2011; DOI: 10.1038/ismej.2011.90

487. **Georgetown University Medical Center.** "Colon cleansing has no benefit but many side effects including vomiting and death, doctors say." *ScienceDaily*, 2 Aug. 2011. Web.

488. **R. Mishori, A. Otubu, A. A. Jones.** Colon cleansing—a dangerous practice. *The Journal of Family Practice*, August 2011

489. **S.Kaushik et al.** Autophagyin Hypothalamic AgRP Neurons Regulates Food Intake and Energy Balance. Cell Metabolism, Volume 14, Issue 2, 173-183, 3August 2011 DOI:10.1016/j.cmet.2011.06.008

490. **CellPress.** "Why diets don't work: Starved brain cells eatthemselves, study finds."ScienceDaily, 2 Aug. 2011. Web.

491. **C. M. Rico et al.** Interaction of Nanoparticles with Edible Plants and Their Possible Implications in the Food Chain. *Journal of Agricultural and Food Chemistry*, 2011; 59 (8): 3485 DOI:10.1021/jf104517j

492. **American Chemical Society.** "Safety of nanoparticles in food crops is still unclear."*ScienceDaily*, 1 Jun. 2011. Web.

493. **S. J. Pamp, M. Gjermansen, T. Tolker-Nielsen.** The Biofilm Matrix: A Sticky Framework. http://www.open-access-biology.com/biofilms/biofilmsch4.pdf

494. **S. N. Rajpathak et al.** Lifestyle Factors of People with Exceptional Longevity. *Journal of the American Geriatrics Society*, August 3, 2011 DOI: 10.1111/j.1532-5415.2011.03498.x

495. **S. Kohler, V. Schaller, A. R. Bausch.** Collective dynamics of active cytoskeletal Networks. Technische Universitat Munchen, Germany. May 2011, http://arxiv.org/PS_cache/arxiv/pdf/1105/1105.4475v1.pdf

496. **S. Almagro et al.** Individual chromosomes as viscoelastic copolymers. *Europhys. Lett.*, 63 (6), pp. 908–914 (Sept. 2003)

497. **C. J. Delebecque et al.** Organization intracellular reactions with designed RNA . Science. 2011 Jul 22;333(6041):470-4.

498. **Y. Gui et al.** Frequent mutations of chromatin remodeling genes in transitional cell carcinoma of the bladder. *nature Genetics*, 2011; DOI: 10.1038/ng.907

499. **New Drug Application Filed for Recombinant Human Deoxyribonuclease I (rhDNase)** "Pulmozyme" http://www.4-traders.com/CHUGAI-PHARMA-6491165/news/CHUGAI-PHARM-New-Drug-Application-Filed-for-Recombinant-Human-Deoxyribonuclease-I-rhDNase-Pulmozyme-13709452/

500. **N. Levis et al.** Soya Isoflavones in the Prevention of Menopausal Bone Loss and Menopausal Symptoms: A Randomized, Double-blind Trial. *Archives of Internal Medicine*, 2011; 171 (15): 1363 DOI: 10.1001/archinternmed.2011.330

501. **JAMA and Archives Journals.** "Soya tablets not associated with reduction in bone loss or menopausal symptoms in women, study finds." *ScienceDaily*, 8 Aug. 2011. Web.

502. **Bin Xu et al.** Exome sequencing supports a de novo mutational paradigm. *nature Genetics*, 2011; DOI: 10.1038/ng.902

503. **S. L. Girard et al.** Increased exonic de novo mutation rate in individuals. *nature Genetics*, 2011; DOI: 10.1038/ng.886

504. **S. Maeda et al.** Active Polymer Gel Actuators. Int. J. Mol. Sci. 2010, 11, 52-66; doi:10.3390/ijms11010052

505. **S. Maeda et al.** Design of Autonomous Gel Actuators Polymers 2011, 3, 299-313; doi:10.3390/polym3010299

506. **R. M. Turk, N. V. Chumachenko, M. Yarus.** Multiple translational products from a five-nucleotide ribozyme. *Proceedings of the National Academy of Sciences*, Published online February 22, 2010 DOI: 10.1073/pnas.0912895107

507. **Karolinska Institutet.** "Key Protein In Cellular Respiration Discovered." *ScienceDaily*, 14 Apr. 2009. Web.

508. **Editor's Summary to article: T. Savin et al.** On the growth and form of the gutnature. Volume: 476, Pages:57–62. doi:10.1038/nature10277

509. **Loma Linda University Adventist Health Sciences Center.** "Contrary to earlier findings, excess body fat in elderly decreases life expectancy." *ScienceDaily*, 11 Aug. 2011. Web.

510. **M. P. Hitchins et al.** Dominantly Inherited Constitutional Epigenetic Silencing of MLH1 in a Cancer-Affected Family Is Linked to a Single Nucleotide Variant within the 5 UTR. *Cancer Cell*, 2011; 20 (2): 200 DOI: 10.1016/j.ccr.2011.07.003

511. **University of New South Wales.** "'Methyl magnet' genes predispose to increased cancer." *ScienceDaily*, 17 Aug. 2011. Web.

512. **S. E. Torigoe et al.** Identification of a Rapidly Formed Nonnucleosomal Histone-DNA Intermediate that Is Converted into Chromatin by ACF. *Molecular Cell*, 2011; 43 (4): 638-648 DOI: 10.1016/j.molcel.2011.07.017

513. **University of California - San Diego.** "Biologists' discovery may force revision of biology textbooks: Novel chromatin particle halfway between DNA and a nucleosome."*ScienceDaily*, 18 Aug. 2011. Web.

514. **P. B. Gupta et al.** Stochastic State Transitions Give Rise to Phenotypic Equilibrium in Populations of Cancer Cells. *Cell*, 2011; 146 (4): 633-644 DOI: 10.1016/j.cell.2011.07.026

515. **Broad Institute of MIT and Harvard**. "Cancer stem cells made, not born: Experiments and modeling reveal how tumors maintain cellular balance." *ScienceDaily*, 18 Aug. 2011. Web.

516. **R. Fedriga et al**. Relation between food habits and p53 mutational spectrum in gastric cancer patients. Int J Oncol. 2000 Jul;17(1):127-33.

517. **Makoto R. Hara et al**. A stress response pathway regulates DNA damage through β2-adrenoreceptors and β-arrestin-1. *nature*, 2011; DOI: 10.1038/nature10368

518. **J. R. Pietruska et al**. Bioavailability, intracellular mobilization of nickel, and HIF-1α activation in human lung epithelial cells exposed to metallic nickel and nickel oxide nanoparticles.*Toxicological Sciences*, 2011; DOI: 10.1093/toxsci/kfr206

519. **The Research Council of Norway**. "Health risks with nanotechnology? Nanoparticles can hinder intracellular transport." *ScienceDaily*, 24 Aug. 2011. Web.

520. **A. A. Bremer et al**. Adipose Tissue Dysregulation in Patients with Metabolic Syndrome. *The Journal of Clinical Endocrinology & Metabolism*, August 24, 2011 jc.2011-1577 DOI: 10.1210/jc.2011-1577

521. **G. L. Russo et al**. Cellular adaptive response to chronic radiation exposure in interventional cardiologists. *European Heart Journal*, 2011; DOI: **10.1093/eurheartj/ehr263**

522. **T. Gori, T. Munzel**. Biological effects of low-dose radiation: of harm and hormesis. *European Heart Journal*, 2011; DOI: **10.1093/eurheartj/ehr2**

523. **S. Kavaler et al**. Pancreatic β-cell failure in obese mice with human-like CMP-Neu5Ac hydroxylase deficiency. FASEB J., fj.10-175281 DOI: 10.1096/fj.10-175281

524. **J. S.Kooner et al**. Genome-wide association study in individuals of South Asianancestry identifies six new type 2 diabetes susceptibility loci. natureGenetics, 2011; DOI: 10.1038/ng.921

525. **P. M.Abadir et al**. Identification and characterization of a functionalmitochondrial angiotensin system. Proceedings of the National Academy ofSciences, 2011; DOI: 10.1073/pnas.1101507108

526. **L. Actis-Goretta,J.I. Ottaviani, C. G. Fraga**. Inhibition of angiotensin converting enzyme activityby flavanol-rich foods. http://greentea.researchtoday.net/archive/3/1/344.htm

527. **E. Neafsey, M. Collins**. Moderate alcohol consumption and cognitive risk. NeuropsychiatricDisease and Treatment, 2011; 7 (1): 465-484 DOI: 10.2147/NDT.S23159

528. **J. K. Ngo et al**. Impairment of Lon-Induced Protection Against the Accumulation of Oxidized Proteins in Senescent Wi-38 Fibroblasts. The Journals of Gerontology Series A: Biological Sciences and Medical Sciences, 2011; DOI: 10.1093/gerona/glr145

529. **S. Minot etal**. The human gut virome: Inter-individual variation and dynamic response todiet. Genome Res, August 31, 2011 DOI: 10.1101/gr.122705.111

530. **Columbia University's Mailman School of Public Health**. "Obesity to rise: 65 million more obese adults in the US and 11 million more in the UK expected by 2030." ScienceDaily, 29 Aug. 2011. Web.

531. **S. Diano et al**. Peroxisome proliferation–associated control of reactive oxygen speciessets melanocortin tone and feeding in diet-induced obesity. natureMedicine, 2011; DOI: 10.1038/nm.2421

532. **American Chemical Society**. "Super-fruits: Tropical blueberries extremely high inhealthful antioxidants, study suggests." ScienceDaily, 29 Apr. 2011. Web.

533. **American Chemical Society**. "Putting the squeeze on fruit with 'pascalization'boosts healthful antioxidant levels." ScienceDaily, 29 Aug.2011. Web.

534. **E. Sanchez et al**. Influence of Environmental and Genetic Factors Linked to Celiac Disease Risk on Infant Gut Colonization by Bacteroides Species. *Applied and Environmental Microbiology*, 2011; 77 (15): 5316 DOI: 10.1128/AEM.00365-11

535. **D. J. A. Jenkins et al**. Effect of a Dietary Portfolio of Cholesterol-Lowering FoodsGiven at 2 Levels of Intensity of Dietary Advice on Serum Lipids inHyperlipidemia: A Randomized Controlled Trial. JAMA: The Journalof the American Medical Association, 2011; 306 (8): 831 DOI: 10.1001/jama.2011.1202

536. **Boston University Medical Center**. "Consuming vegetables linked to decreased breast cancer risk in African-American women." *ScienceDaily*, 13 Oct. 2010. Web.

537. **Wiley-Blackwell**."Can Diet Alone Control Type 2 Diabetes? ScienceDaily, 16 Jul.2008. Web.

538. **S. A. Wickström et al**. Integrin-linked kinase controls microtubule dynamics required for plasma membrane targeting of caveolae. *Developmental Cell*, 2010; 19 (4): 574-588 DOI: 10.1016/j.devcel.2010.09.007

539. **Society for General Microbiology**. "Gut Bacteria Can Manufacture Defenses Against Cancer And Inflammatory Bowel Disease." ScienceDaily, 9 Feb. 2009. Web.

540. **Marshall University Research Corporation**. "Breast cancer risk drops when diet includes walnuts, researchers find." *ScienceDaily*, 1 Sep. 2011. Web.

541. **A. Jha**. Pulsations reveal which embryos have the best chance of success in IVF. http://www.guardian.co.uk/science/2011/aug/09/pulsations-embryos-success-ivf

542. **A. Ajduk et al**. Rhythmic actomyosin-driven contractions induced by sperm entry predict mammalian embryo viability. *nature Communications* 2, article number: 417. doi:10.1038/ncomms1424

543. R. S. McIsaac, K. C. Huang, A. Sengupta, N. S. Wingreen. Does the Potential for Chaos Constrain the Embryonic Cell-Cycle Oscillator? *PLoS Computational Biology*, 2011; 7 (7): e1002109 DOI:10.1371/journal.pcbi.1002109

544. W. Liu et al. Differential Effects of Daily-Moderate Versus Weekend-Binge Alcohol Consumption on Atherosclerotic Plaque Development in Mice. *Atherosclerosis*, 2011.

545. Q. Sun et al.Alcohol Consumption at Midlife and Successful Ageing in Women: A Prospective Cohort Analysis in the Nurses' Health Study. *PLoS Medicine*, 2011; 8 (9): e1001090 DOI: 10.1371/journal.pmed.1001090

546. Max-Planck-Gesellschaft. "Cellular metabolism self-adapts to protect against free radicals."*ScienceDaily*, 7 Sep. 2011.

547. Nana-Maria Grüning et al. Pyruvate kinase triggers a metabolic feedback loop that controls redox metabolism in respiring cells.*Cell Metabolism*, 2011; 14 (3): 415-427 DOI: 10.1016/j.cmet.2011.06.017

548. J. R. Tollervey et al. Analysis of alternative splicing associated with ageing and neurodegeneration in the human brain. Genome Res, 2011 DOI:10.1101/gr.122226.111

549. Helmholtz Association of German Research Centres. "Plant compound reduces breast cancer mortality, study suggests."*ScienceDaily*, 13 Sep. 2011. Web

550. Cancer rates could be cut by 2.8m with healthier diets. http://www.ingredientsnetwork.com/story/full/cancer-rates-could-be-cut-by-2-8m-with-healthier-diets

551. S. M. Dever et al. Mutations in the BRCT binding site  result in hyper-recombination. *Ageing*, 2011; 3 (5): 515-532.

552. J. Salzman et al. ESRRA-C11orf20 Is a Recurrent Gene Fusion in Serous Ovarian Carcinoma. *PLoS Biology*, 2011; 9 (9): e1001156 DOI:10.1371/journal.pbio.1001156

553. DOE/Lawrence Berkeley National Laboratory. "How genes make healthy skin." *ScienceDaily*, 20 Sep. 2011. Web.

554. C. Becker et al.Spontaneous epigenetic variation in the Arabidopsis thaliana methylome. *nature*, 2011; DOI:10.1038/nature10555

555. J. Wang et al. Inhibition of activated pericentromeric SINE/Alu repeat transcription in senescent human adult stem cells reinstates self-renewal. *Cell Cycle*, Volume 10, Issue 17, Pages 3016 - 3030; September 1, 2011

556. University of Gothenburg. "Human body rids itself of damage when it really matters."*ScienceDaily*, 21 Sep. 2011. Web.

557. G. Kumar et al. The determination of stem cell fate by 3D scaffold structures through the control of cell shape. *Biomaterials*, 2011; DOI:10.1016/j.biomaterials.2011.08.05

558. Donella H. Meadows, Dennis L. Meadows, Jorgen Randers, and William W. Behrens III.(1972).The Limits to Growth. New York: universe Books. ISBN 0-87663-165-0 - ISBN 978-1931498586, ASIN 1 931 498-58-X, 2004 Limits to Growth: The 30-Year Update.

559. University of Loughborough. "Supplements could make athletes unwitting drugs cheats." *ScienceDaily*, 25 Sep. 2011. Web.

560. E. C. Theil et al. Absorption of Iron from Ferritin Is Independent of Heme Iron and Ferrous Salts in Women and Rat Intestinal Segments. *Journal of Nutrition*, 2012; DOI: 10.3945/jn.111.145854

561. M. J. Barry et al. Effect of Increasing Doses of Saw Palmetto Extract on Lower Urinary Tract Symptoms: A Randomized Trial. *Journal of the American Medical Association*, 2011; 306 (12): 1344-1351 DOI:10.1001/jama.2011.1364

562. A. J. Miller, B. L. Gross. From forest to field: Perennial fruit crop domestication. *American Journal of Botany*, 2011; 98 (9): 1389 DOI: 10.3732/ajb.1000522

563. J. U. Guo et al. Hydroxylation of 5-Methylcytosine by TET1 Promotes Active DNA Demethylation in the Adult Brain. Cell, 14 April 2011 DOI: 10.1016/j.cell.2011.03.022

564. Vanderbilt University Medical Center. "New Molecular Mechanism Associated With Arrhythmias Discovered: Possible Novel Target For Treating Arrhythmias." *ScienceDaily*, 22 Jan. 2009. Web.

565. American Institute Of Physics. "Spiral Waves Break Hearts: Importance Of Communication Between Cardiac Cells Is Demonstrated." *ScienceDaily*, 11 Feb. 2002. Web.

566. H. Karcher et al. A Three-Dimensional Viscoelastic Model for Cell Deformation with Experimental Verification. Biophysical Journal Volume 85 November 2003 3336–3349

567. Georgetown University Medical Center. "For The First Time, Patterns Of Excitation Waves Found In Brain's Visual Processing Center." *ScienceDaily*, 4 Aug. 2007. Web.

568. A. Todd et al. Premenopausal hysterectomy is associated with increased brain ferritin iron.*Neurobiology of Ageing*, 2011; DOI:10.1016/j.neurobiolageing.2011.08.002

569. P. Sumazin et al. An Extensive MicroRNA-Mediated Network of RNA-RNA Interactions Regulates Established Oncogenic Pathways in Glioblastoma. *Cell*, 2011; 147 (2): 370-381 DOI: 10.1016/j.cell.2011.09.041

570. M. Cesana et al. A Long Noncoding RNA Controls Muscle Differentiation by Functioning as a Competing Endogenous RNA. *Cell*, 2011; 147 (2): 358-369 DOI:10.1016/j.cell.2011.09.028

571. **Columbia University Medical Center.** "Vast hidden network regulates gene expression in cancer: Study illuminates the 'dark matter' of the genome." *ScienceDaily*, 14 Oct. 2011. Web

572. **Zhang et al.** Exogenous plant MIR168a specifically targets mammalian LDLRAP1: evidence of cross-kingdom regulation by microRNA. *Cell Research*, (20 September 2011) | doi:10.1038/cr.2011.158

573. **N. L. Wicks et al.** UVA Phototransduction Drives Early Melanin Synthesis in Human Melanocytes. *Current Biology*, 2011 DOI: 10.1016/j.cub.2011.09.047

574. **X. Wang et al.** PINK1 and Parkin Target Miro for Phosphorylation and Degradation to Arrest Mitochondrial Motility. *Cell*, 2011; 147 (4): 893-906 DOI: 10.1016/j.cell.2011.10.018

575. **NASA Ames Research Center.** "NASA Find Clues That Life Began In Deep Space." *ScienceDaily*, 31 Jan. 2001. Web.

576. **University of California - Santa Cruz.** "Synthetic Biology Yields Clues to Evolution and Origin of Life." *ScienceDaily*, 15 Feb. 2009. Web.

577. **E. T. Chan et al.** Conservation of core gene expression in vertebrate tissues. *Journal of Biology*, 2009; DOI: 10.1186/jbiol130

578. **C. Colantuoni et al.** *Temporal dynamics and genetic control of transcription in the human prefrontal cortex. nature*, 2011; 478 (7370): 519 DOI:10.1038/nature10524

579. **Hamara Health.** http://hamarahealth.com/news/novel-reprogramming-mechanism-for-tumour-cells-discovered.html

580. **American Association for Cancer Research.** "Intermittent, low-carbohydrate diets more successful than standard dieting, study finds."*ScienceDaily*, 8 Dec. 2011. Web.

581. **R. A. Juergens et al.** Combination Epigenetic Therapy Has Efficacy in Patients with Refractory Advanced Non-Small Cell Lung Cancer. *Cancer Discovery*, 2011; DOI:10.1158/2159-8290.CD-11-0214

582. **M. J. Mendiburo et al.** Drosophila CENH3 Is Sufficient for Centromere Formation. *Science*, 2011; 334 (6056): 686 DOI:10.1126/science.1206880

583. **Suzanne J.P.L. van den Berg et al.** Levels of Genotoxic and Carcinogenic Compounds in Plant Food Supplements and Associated Risk Assessment. *Food and Nutrition Sciences*, Vol. 2, No. 9, 2011, pp. 989-1010.

584. **American Chemical Society.** "High Doses Of Phytochemicals, Including Flavanoids, In Teas And Supplements Could Be Unhealthy."*ScienceDaily*, 30 Apr. 2007. Web.

585. **University of Illinois at Urbana-Champaign.** "Dietary Supplement Genistein Can Undermine Breast Cancer Treatment." *ScienceDaily*, 23 Sep. 2008. Web.

586. **Neumaier et al.** Evidence for formation of DNA repair centers and dose-response nonlinearity in human cells. PNAS December 19, 2011. Doi:10.1073/pnas.1117849108

587. **T. R. Abrahamsson et al.** Low diversity of the gut microbiota in infants with atopic eczema. *Journal of Allergy and Clinical Immunology*, 2011; DOI: 10.1016/j.jaci.2011.10.025

588. **L. Cordain et al.** Origins and evolution of the Western diet: health implications for the 21st century. Am J Clin Nutr February 2005 vol. 81 no. 2 341-354.

589. **E. A. Martens, O. Hallatschek.** Interfering Waves of Adaptation Promote Spatial Mixing. *Genetics*, 2011; 189 (3): 1045 DOI: 10.1534/genetics.111.130112

590. **D. J. Baker et al.** Clearance of p16Ink4a-positive senescent cells delays ageing-associated disorders. *nature*, 2011; DOI:10.1038/nature10600

591. **K. Berer et al.** Commensal microbiota and myelin autoantigen cooperate to trigger autoimmune demyelination. *nature*, 2011; DOI:10.1038/nature10554

592. **I. Keklikoglou et al.** MicroRNA-520/373 family functions as a tumor suppressor in estrogen receptor negative breast cancer by targeting NF-κB and TGF-β signaling pathways. *Oncogene*, 2011; DOI: 10.1038/onc.2011.571

593. **M. S. Stein et al.** A randomized trial of high-dose vitamin D2 in relapsing-remitting multiple sclerosis. *Neurology*, 2011; 77 (17): 1611-1618 DOI:10.1212/WNL.0b013e3182343274

594. **M. Gooyit et al.** Selective Water-Soluble Gelatinase Inhibitor Prodrugs. *Journal of Medicinal Chemistry*, 2011; 54 (19): 6676 DOI: 10.1021/jm200566e

595. **J. D. Clarke et al.** Comparison of Isothiocyanate Levels and Histone Deacetylase Activity Subjects Consuming Broccoli Sprouts or Broccoli Supplement. *Journal of Agricultural and Food Chemistry*, 2011; 110930085450000 DOI: 10.1021/jf202887c

596. **E. A. Klein et al.** Vitamin E and the Risk of Prostate Cancer: Results of The Selenium and Vitamin E Cancer Prevention Trial (SELECT). *JAMA*, 2011; 306 (14): 1549-1556 DOI:10.1001/jama.2011.1437

597. **J. Yoshino et al.** Nicotinamide mononucleotide, a key NAD intermediate, treats the pathophysiology of diet- and age-induced diabetes in mice. *Cell Metabolism*, 5 October 2011; 14(4) pp. 528 - 536 DOI:10.1016/j.cmet.2011.08.014

598. **T. W. Rice et al** Enteral Omega-3 Fatty Acid, -Linolenic Acid, and Antioxidant Supplementation in Acute Lung Injury. *JAMA*, 2011; DOI: 10.1001/jama.2011.1435

599. D. J. Cook, D. K. Heyland. Pharmaconutrition in Acute Lung Injury. *JAMA*, 2011; DOI:10.1001/jama.2011.1470

600. S. Bonn et al. Tissue-specific analysis of chromatin state identifies temporal signatures of enhancer activity during embryonic development. *nature Genetics*, 2012; DOI: 10.1038/ng.1064

601. N. Krishnan et al. Loss of circadian clock accelerates aging in neurodegeneration-prone mutants. *Neurobiology of Disease*, 2011; DOI: 10.1016/j.nbd.2011.12.034

602. American Society for Microbiology. "Bacteria in the gut of autistic children different from non-autistic children." *ScienceDaily*, 9 Jan. 2012. Web.

603. S. Greenblum, P. J. Turnbaugh, E. Borenstein. Metagenomic systems biology of the human gut micro-biome reveals topological shifts associated with obesity and inflammatory bowel disease. *Proceedings of the National Academy of Sciences*, 2011; DOI:10.1073/pnas.1116053109

604. V. Lam et al. Intestinal microbiota severity of myocardial infarction in rats.*FASEB J.*, 2012 DOI: 10.1096/fj.11-197921

605. X. Wang et al. Cryptic prophages help bacteria cope with adverse environments. *nature Communications* 1, Article number: 147, December 2010, doi:10.1038/ncomms1146

606. P. V. Castro et al. Caenorhabditis elegans Battling Starvation Stress: Low Levels of Ethanol Prolong Lifespan in L1 Larvae. *PLoS ONE*, 2012; 7 (1): e29984 DOI:10.1371/journal.pone.0029984

607. R. Nigmatulin. Fractional integral and its physical interpretation. Theoretical and Mathematical Physics. V. 90, 3, 1992, 242-251, Moscow.

608. W. S. Childers et al. Peptides Organized as Bilayer Membranes. *Angewandte Chemie International Edition*, 2010; DOI: 10.1002/anie.201000212

609. E. C. Theil et al. Absorption of Iron from Ferritin Is Independent of Heme Iron and Ferrous Salts in Women and Rat Intestinal Segments. *Journal of Nutrition*, 2012; DOI: 10.3945/jn.111.145854

610. Children's Hospital & Research Center Oakland. "Novel iron source: Newly identified iron absorption mechanism suggests that legumes could provide key to treating iron deficiency worldwide." *ScienceDaily*, 20 Jan. 2012. Web

611. J. N. Savas et al. Extremely Long-Lived Nuclear Pore Proteins in the Rat Brain. *Science*, 2012; DOI:10.1126/science.1217421

612. NYU Langone Medical Center / New York University School of Medicine. "Paradigm changing mecha-nism is revealed for the control of gene expression in bacteria." *ScienceDaily*, 13 Jan. 2010. Web.

613. G. Bange et al. Structural basis for the molecular evolution of SRP-GTPase activation by protein. *nature Structural & Molecular Biology*, 2011; 18 (12): 1376 DOI:10.1038/nsmb.2141

614. Virginia Tech. "Explores Biochemical Landscape To Find Memory Switches." *ScienceDaily*, 19 Jun. 2008. Web.

615. University of Massachusetts Amherst. "Newly Engineered Genetic Switches Enhance Production Of Proteins, Pharmaceuticals."*ScienceDaily*, 26 Jan. 2008. Web.

616. University of California - Santa Barbara. "Chemists Explain The Switchboards In Cells." *ScienceDaily*, 3 Aug. 2009. Web.

617. C. Matito et al. Protective Effect of Structurally Diverse Grape Procyanidin Fractions against UV-Induced Cell Damage and Death. *Journal of Agricultural and Food Chemistry*, 2011; 59 (9): 4489 DOI:10.1021/jf103692a

618. T. Kiyomitsu, I. M. Cheeseman. Chromosome- and spindle-pole-derived signals generate an intrinsic code for spindle position and orientation. *nature Cell Biology*, 2012; DOI: 10.1038/ncb2440

619. G. Shetty et al. Fetal Radiation Exposure Induces Testicular Cancer in Genetically Susceptible Mice. *PLoS ONE*, 2012; 7 (2): e32064 DOI:10.1371/journal.pone.0032064

620. X. Wei et al. Fatty acid synthase modulates intestinal barrier function through palmitoylation of mucin2. *Cell Host & Microbe*, Feb. 16, 2012

621. J. Gretchen et al. Oral exposure to polystyrene nanoparticles affects iron absorption. *nature Nanotechnology*, 2012; DOI: 10.1038/nnano.2012.3

622. Michigan State University. "Studying the evolution of life's building blocks." *ScienceDaily*, 20 Feb. 2012. Web.

623. L. Burroughs et al. Asymmetric organocatalytic formation of protected and unprotected tetroses under po-tentially prebiotic conditions. *Organic & Biomolecular Chemistry*, 2012; DOI:10.1039/C1OB06798B

624. Northern Arizona University. "Evolution of staph 'superbug' traced between humans." *ScienceDaily*, 21 Feb. 2012. Web

625. University of Copenhagen. "Researchers unravel the molecular dance of DNA repair."*ScienceDaily*, 15 Mar. 2012. Web.

626. J. Lopez-Contreras et al. An extra allele of Chk1 limits oncogene-induced replicative stress and promotes transformation. *Journal of Experimental Medicine*, 2012; 209 (3): 455 DOI: 10.1084/jem.20112147

627. **Methodist Hospital, Houston.** "More than 500 genes that may cause pancreatic cancer." *ScienceDaily*, 12 Mar. 2012. Web.

628. **American Academy of Neurology.** "Diet patterns may keep brain from shrinking."*ScienceDaily*, 29 Dec. 2011. Web.

629. **D. Gilbert-Diamond et al.** Rice consumption contributes to arsenic exposure in US women. *Proceedings of the National Academy of Sciences*, 2011; DOI:10.1073/pnas.1109127108

630. **E. A. Hu, A. Pan, V. Malik, Q. Sun.** White rice consumption and risk of type 2 diabetes: meta-analysis and systematic review. *BMJ*, 2012; 344 (mar15 3): e1454 DOI: 10.1136/bmj.e1454

631. http://www.lifeaftertheoilcrash.net/

632. **G. Begum et al.** Epigenetic changes in fetal hypothalamic energy regulating pathways are associated with maternal undernutrition and twinning.*The FASEB Journal*, 2012; 26 (4): 1694 DOI: 10.1096/fj.11-198762

633. **Federation of American Societies for Experimental Biology.** "Epigenetic changes in twins of dieting mothers increases risk of obesity and diabetes." *ScienceDaily*, 1 Apr. 2012. Web.

634. **N. J. Roberts et al.** The Predictive Capacity of Personal Genome Sequencing. *Sci Transl Med*, 2 April 2012 DOI:10.1126/scitranslmed.3003380

635. **Mayo Clinic.** "Dramatic rise in skin cancer in young adults." *ScienceDaily*, 2 Apr. 2012. Web.

636. **S. P. Shah et al.** The clonal and mutational evolution spectrum of primary triple-negative breast cancers. *nature*, 2012; DOI: 10.1038/nature10933

637. **B. D. Pfeiffer et al.** Using translational enhancers to increase transgene expression Drosophila. *PNAS*, 201210.1073/pnas.1204520109

638. **NewYork-Presbyterian Hospital/Weill Cornell Medical Center.** "Two genetic deletions in human genome linked to the development of aggressive prostate cancer." *ScienceDaily*, 9 Apr. 2012. Web.

639. **University of Colorado at Boulder.** "Babies' First Bacteria Depend on Birthing Method." ScienceDaily 24 June 2010.

640. **N. P. Sharp, A. F. Agrawal.** Evidence for elevated mutation rates in low-quality genotypes. *PNAS*, 2012; DOI:10.1073/pnas.1118918109

641. **University of Toronto.** "Low-quality gene cause mutational meltdown." *ScienceDaily*, 16 Apr. 2012.

642. **S. Bandyopadhyay et al.** Rewiring of Genetic Networks in Response to DNA Damage. *Science*, 2010; 330 (6009): 1385 DOI:10.1126/science.1195618

643. **D. H. Atha et al.** Copper Oxide Nanoparticle Mediated DNA Damage in Terrestrial Plant Models. *Environmental Science & Technology*, 2012; 46 (3): 1819 DOI: 10.1021/es202660k

644. **I. Martincorena et al.** Evidence of non-random mutation rates suggests an evolutionary risk management strategy. *nature*, 2012; DOI: 10.1038/nature10995.

645. **Hebrew University of Jerusalem.** "Two distinguishable gene groups identified: One 'normal' and the other problematic."*ScienceDaily*, 23 Apr. 2012. Web.

www.ingramcontent.com/pod-product-compliance
Lightning Source LLC
Chambersburg PA
CBHW051209170526
45166CB00005B/1815